T0350145

# Doping Engineering for Front-End Processing

MATERIALS RESEARCH SOCIETY
SYMPOSIUM PROCEEDINGS VOLUME 1070

# Doping Engineering for Front-End Processing

Symposium held March 25–27, 2008, San Francisco, California, U.S.A.

## EDITORS:

**B.J. Pawlak**
NXP Semiconductors
Leuven, Belgium

**M.L. Pelaz**
University of Valladolid
Valladolid, Spain

**M. Law**
University of Florida
Gainesville, Florida, U.S.A.

**K. Suguro**
Toshiba Corporation
Yokohama, Japan

Materials Research Society
Warrendale, Pennsylvania

# CAMBRIDGE
## UNIVERSITY PRESS

University Printing House, Cambridge CB2 8BS, United Kingdom

One Liberty Plaza, 20th Floor, New York, NY 10006, USA

477 Williamstown Road, Port Melbourne, VIC 3207, Australia

314-321, 3rd Floor, Plot 3, Splendor Forum, Jasola District Centre, New Delhi - 110025, India

79 Anson Road, #06-04/06, Singapore 079906

Cambridge University Press is part of the University of Cambridge.

It furthers the University's mission by disseminating knowledge in the pursuit of education, learning and research at the highest international levels of excellence.

www.cambridge.org
Information on this title: www.cambridge.org/9781605110400

Materials Research Society
506 Keystone Drive, Warrendale, PA 15086
http://www.mrs.org

First published 2008
First paperback edition 2012

Single article reprints from this publication are available through University Microfilms Inc., 300 North Zeeb Road, Ann Arbor, MI 48106

CODEN: MRSPDH

*A catalogue record for this publication is available from the British Library*

ISBN 978-1-605-11040-0 Hardback
ISBN 978-1-107-40854-8 Paperback

# CONTENTS

## *ULTRA SHALLOW JUNCTIONS I*

*Invited Paper

## SHALLOW JUNCTION CONTACTING

## POSTER SESSION

\*Invited Paper

*ULTRA SHALLOW JUNCTIONS II*

*Invited Paper

\*Invited Paper

\*Invited Paper

*Invited Paper

x

# PREFACE

This volume contains papers presented at Symposium E, "Doping Engineering for Front-End Processing," held March 25–27 at the 2008 MRS Spring Meeting in San Francisco, California. The scope of the symposium was to bring together researchers from the field of materials science and technology to review the state-of-the-art in doping engineering and activation methods, metal-semiconductors contacting for integrated circuits, to discuss the current achievements, remaining challenges, and to identify future research directions for fundamental investigation and technology development. These proceedings document the recent developments in the areas of experiments, modeling and metrology related to planar and FIN transistor source and drain regions.

It has been a pleasure and privilege to have the opportunity to organize the symposium and edit this volume. This would not be possible without the support of the speakers and authors, all of whom are gratefully acknowledged.

Many exciting research achievements have been presented at this symposium. We hope that the readers will enjoy reading this proceedings volume, and find its contents both informative and interesting.

B.J. Pawlak
M.L. Pelaz
M. Law
K. Suguro

July 2008

# MATERIALS RESEARCH SOCIETY SYMPOSIUM PROCEEDINGS

# MATERIALS RESEARCH SOCIETY SYMPOSIUM PROCEEDINGS

**Prior Materials Research Society Symposium Proceedings available by contacting Materials Research Society**

# Ultra Shallow Junctions I

Mater. Res. Soc. Symp. Proc. Vol. 1070 © 2008 Materials Research Society 1070-E01-02

# Strengths and Limitations of the Vacancy Engineering Approach for the Control of Dopant Diffusion and Activation in Silicon

Alain Claverie[1], Fuccio Cristiano[2], Mathieu Gavelle[2], Fabrice Sévérac[2], Frédéric Cayrel[3], Daniel Alquier[3], Wilfried Lerch[4], Silke Paul[4], Leonard Rubin[5], Vito Raineri[6], Filippo Giannazzo[6], Hervé Jaouen[7], Ardechir Pakfar[7], Aomar Halimaoui[7], Claude Armand[8], Nikolay Cherkashim[1], and Olivier Marcelot[1]

[1]nMat Group, CEMES-CNRS, 29, rue J. Marvig, BP4347, Toulouse, 31055, France
[2]LAAS / CNRS, 7 av. du Col. Roche, toulouse, 31077, France
[3]LMP, université de Tours, 16 rue Pierre et Marie Curie, BP 7155, Tours, 37071, France
[4]Mattson, Mattson Thermal Products GmbH, Daimlerstr. 10, Dornstadt, D-89160, Germany
[5]Axcelis, Axcelis Technologies, 108 Cherry Hill Drive, Beverly, MA, MA 01915
[6]CNR / IMM, CRN / IMM, Stradale Primosole 50, Catania, 95121, Italy
[7]STMicroelectronics, 850 rue Jean Monnet, Crolles, 38926, France
[8]Genie Physique, INSA, 135, Avenue de Rangueil, toulouse, 31077, France

## ABSTRACT

The fabrication of highly doped and ultra-shallow junctions in silicon is a very challenging problem for the materials scientist. The activation levels which are targeted are well beyond the solubility limit of current dopants in Si and, ideally, they should not diffuse during the activation annealing. In practice, the situation is even worse and when boron is implanted into silicon excess Si interstitial atoms are generated which enhance boron diffusion and favor the formation of Boron-Silicon Interstitials Clusters (BICs). An elegant approach to overcome these difficulties is to enrich the Si layers where boron will be implanted with vacancies before or during the activation annealing. Spectacular results have been recently brought to the community showing both a significant control over dopant diffusion and an increased activation of boron in such layers. In general, the enrichment of the Si layers with vacancies is obtained by Si$^+$ implantation at high energy. We have recently developed an alternative approach in which the vacancies are injected from populations of empty voids undergoing Ostwald ripening during annealing. While different, the effects are also spectacular. The goal of this work is to establish a fair evaluation of these different approaches under technologically relevant conditions. The application domains of both techniques are discussed and future directions for their development/improvement are indicated.

## INTRODUCTION

Dopant diffusion and activation phenomena in silicon both involve point defects. For boron, diffusion occurs by pairing a B atom with a silicon interstitial atom (Is) and consequently, the diffusivity of boron is directly proportional to the concentration of Is in the region. Transient Enhanced Diffusion (TED) of boron, a technologically undesirable effect, is often observed when annealing of B implanted silicon [1] [2]. This behavior is due to the large supersaturations of Is which evolve in time and space during the Ostwald ripening of Is clusters, {113} defects and then dislocation loops [3]. The limited activation of boron often observed after such annealing, at concentration values well below its solid solubility limit, is also due to the interaction of boron atoms with Is in large supersaturations which favour the formation of immobile, electrically inactive, small $B_nSi_m$ clusters (BIC's) [4]. Recent strategies for attaining the required high activation levels are aimed at either "breaking" these BIC's while limiting boron diffusion for example using very high temperature annealing for a very short time, or at forming highly doped but metastable layers by Solid Phase Epitaxy (SPE). However both approaches have their limits. TED of boron greatly increases the junction depths even during "spike" or "flash" annealing ([5], Fig. 1) while boron deactivation occurs during high temperature annealing of SPE regrown layers [6].

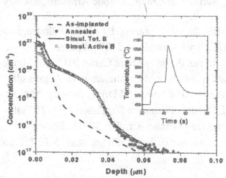

*Fig; 1 : Comparison between experiments and simulations for boron implanted at 0.5 keV with a dose of $1.10^{15}$ /cm$^2$ and spike annealed at 1050°C. Enhanced boron diffusion occurs even for such short annealing times. Insert is the temperature profile used for annealing (after ref. 1).*

An elegant approach to overcome these difficulties is to enrich the Si layers where boron will be implanted with vacancies before or during the activation annealing. Spectacular results have been recently brought to the community showing both a significant control over dopant diffusion and an increased activation of boron in such layers. In general, the enrichment of the Si layers with vacancies is obtained by Si$^+$ implantation at high energy. We have recently developed an alternative approach in which the vacancies are injected from populations of empty voids undergoing Ostwald ripening during annealing. While different, the effects are also spectacular.

It is the goal of this paper to review the results found in the literature and as well as our own results and to provide a fair evaluation of different vacancy engineering methods in view of fabricating improved ultra-shallow p+/n junctions under technologically relevant conditions in bulk and SOI wafers.

## VACANCY ENGINEERING USING Si$^+$ IRRADIATION

### A bit of History

The concept of "point defect imbalance" due to ion implantation has been introduced decades ago based on single collision arguments. Indeed, when the nuclear energy transfer between a moving atom or ion and a target atom is larger than a certain threshold a Frenkel pair is created. The vacancy (V) stays where the collision took place while the recoiling atom moves further as an interstitial (I). When the energy transfer is small (typically < 1 keV), the recoiling atom gains a small momentum and thus their "exchange angle" is large and they may move in almost every direction. This argument was used by Brinkman in the 50's to predict the structure of low energy sub-cascades i.e., the formation of zones consisting of a vacancy-rich core surrounded by an interstitial-rich shell (7). Alternatively, when the energy transfer is much larger, the "exchange angle" is small and recoiling atoms predominantly moves along the direction of the impinging particle. It has been suspected for long that, in this case, the concentration of vacancies and interstitials in the target should not be constant along the depth but instead should show enrichment of vacancies close to the surface and of interstitials deeper in the target. However, this finding could not be demonstrated until Monte Carlo simulations of the slowing down process of ions into materials and that the effect of the subcascades created by the recoiling target atoms was included. This happened when TRIM was rendered available to the community in 1984 (8). It took another couple of years for the computers to become powerful enough so that information about a possible point defect imbalance could be extracted from the simulation of the slowing down of several tens of thousand of incident ions. This was done for the first time by Holland et al. in 1991 (9) and, although the profiles were still very noisy, the concept of vacancy engineering was introduced as well as the possible benefits for the fabrication of shallow junctions. Two years later, Laanab et al. used the same type of calculation to explain the origin of the interstitials defects found after annealing of amorphous layers created by ion implantation and known as "End of Range" (EOR) defects (10).

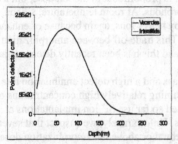

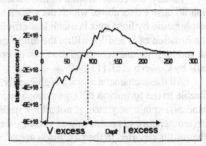

*Fig. 2a: Depth distributions of vacancies and interstitials after a 150 keV Ge⁺ implantation into silicon (from TRIM)*

*Fig. 2b: Depth distribution of Vs and Is "in excess" after the same implant.*

We have plotted in Fig 2, the result of the simulation of a Ge⁺ implantation at 150 keV in silicon. On the left, we have plotted both the vacancy and the interstitial profiles obtained by averaging the results over 500 000 ions. These profiles cannot be distinguished one from the other by eye. However, when subtracting one from the other, a profile showing an excess of vacancies close to the surface and an excess of interstitials close the end of range can be evidenced. In general, the vacancy-rich profile exhibits two components, a very steep high concentration region at the surface and a plateau extending further down to a depth approximately equal to that corresponding to the maximum of both the vacancy and the interstitial profiles. The first one is due to the fact that there is simply no silicon before the surface and that the corresponding Is are "missing". Only vacancies mostly originating from primary knock out collisions are left in this region. The plateau found deeper is due to generation of high energy recoiling atoms in this region which preferentially propagate in directions close to that defined by the incident ions i.e., towards greater depths, leaving behind their corresponding vacancies. Finally, the width of this plateau increases with the incident energy of the ions. For very high energies, the V-rich region can be displaced towards greater depth leaving an almost defect-free surface region because only "electronic" collisions take place at the beginning of the slowing down of the ions.

In all cases, assuming that, after ion implantation at room temperature in silicon, a "vacancy-rich region exists close to the surface" deserves discussion. It must be kept in mind that, in most practical cases, the concentrations of point defect "in excess" one can calculate weights for only $10^{-4}$ to $10^{-2}$ of the total concentrations of point defects created by the implantation. Thus, manipulating such numbers requires a good understanding of the physical meaning and final impact of the parameters injected into the simulation. Moreover, the regions showing an excess of vacancies or an excess of interstitials can "transform" into vacancy-rich or interstitial-rich regions only if the total recombination of Vs and Is takes place at every depth where they are created during implantation or annealing. At room temperature, this can occur within "dilute" cascades where Vs and Is are almost randomly distributed. Single Vs and Is being not stable at RT in Si, they diffuse until they recombine either to form Vn and In (with n>2) or annihilate, the latter phenomenon being favored within dilute cascades. Such cascades result from the interaction of "light" ions with silicon, light meaning low mass and high energy as in the LSS theory.

Finally, the concentration of this vacancy excess is proportional to the ion dose, but a threshold dose exists above which the silicon turns amorphous. At room temperature, silicon amorphization by light ions is much less efficient than by heavy ions, again because of enhanced recombination of Is and Vs within the dilute cascades. This trade-off between amorphization dose and high vacancy concentration can be optimized and this has been recently discussed in details by Cowern et al. [11].

All these characteristics, a high energy, a low mass and a high defect annihilation rate, favorable to the formation of large enough regions containing relatively high concentrations of vacancies, explain why most if not all the studies reported so far use $Si^+$ ion implantations at room temperature to produce these vacancy-rich regions. Unfortunately, we suspect that several of the results reported in the past made use of Si doses equal or above the amorphization dose.

<u>Literature review</u>

One of the first conclusive reports of the effect of $Si^+$ irradiation onto boron diffusion has been published by Raineri in 1991 [12]. In this paper, it is shown that a minimum dose of $5.10^{13}$ $Si^+/cm^2$ at 1 MeV is required to almost suppress the diffusion of boron implanted at 10 keV and at low dose and during annealing at 900°C for 10s. It is to be noticed that the maximum concentration of boron involved in this experiment was quite small ($<10^{18}/cm^3$). In 1999, Venezia et al. [13] have shown that the "famous" TED seen from the evolution of lightly doped B-marker layers ($<2.10^{18}/cm^3$) and related to the dissolution of {113} defects could be suppressed when the structure was previously irradiated with 1 MeV. From these experiments it was concluded that the Is emitted by the defects were recombining with the vacancies in excess in the region. However, the dose used in this experiment was dangerously high ($10^{16}$ $Si^+/cm^2$) and we doubt that part of the silicon matrix was not amorphized by this Si implant. Interestingly, in 2003, Neijim et al. [14] have shown that not only TED but also "regular", equilibrium, diffusion could be suppressed by pre-irradiating silicon with Si ions at 1 MeV. From this experiment, it can be concluded that the enrichment of the region with Vs results in a lowering of the concentration of Is in the same region, below the equilibrium concentration $Ci^*$. In 2002, Shao et al. [15] have shown that the surface region is less damaged (measured by RBS) after a sequential implantation with 500 keV Si+ and 2 keV B+ than after a 2 keV B+ implantation alone. From this results, it was inferred that BIC's already form at room temperature and that immediately after Si+ irradiation, Vs are available and prevent the formation of BICs when boron is implanted later.

More recently, Cowern and co-workers have studied the activation of boron in such V-rich layers [11, 16] after isochronal anneals. They have shown for the first time that for the same annealing conditions, the activation rate of boron was an increasing function of the irradiation dose. Another important result from their work is that for sufficiently high Si irradiation doses, the activation of boron can be as good after a 600°C than after a 1000°C annealing. Unfortunately, the benefit of this irradiation reduces for increasing temperature and after annealing at 1000°C, the effect of Si irradiation is marginal.

Finally, it is interesting to know whether this approach is suitable to increase the sheet resistivities attainable in SOI layers. Indeed, carrier mobilities are often reduced in SOI layers and this could be compensated by increasing the doping levels. Actually, an answer to this question cannot be simply found in the literature. For the materials scientist, the idea of "isolating" the vacancy-rich region from the Is-rich region found immediately below it is tempting as it could prevent possible recombinations between the two antinomic species. Again, controversial results can be found in the literature [17, 18] which focus only on the behavior of boron diffusion in these SOI layers. As we will show it later in this paper, diffusion in SOI layers is mostly governed by trapping at the $Si/SiO_2$ interfaces and a clear demonstration of the usefulness of this "barrier" was still lacking.

## Experimental details

In our experiments, bulk and SOI (70/140 nm) wafers were implanted at RT with 250 keV $Si^+$ at doses up to $1.10^{15}/cm^2$ i.e., below the amorphization dose. Fig. 3 shows the depth distribution of excess point defects created by this irradiation. For a dose of $1.10^{15}/cm^2$, there exists a plateau extending from the surface and towards a depth of about 300 nm within which the concentration of excess Vs is relatively constant at $1.10^{19}/cm^3$. The total number of Vs available in the region is thus about $3.10^{14}/cm^3$ in the bulk Si wafers. After this irradiation step, B was implanted at 3 or 0.5 keV for doses ranging from $1.10^{14}/cm^2$ to $3.10^{15}/cm^2$. Finally, samples from these wafers were annealed by RTA for 10 s or by spike annealing at 800 or 1000°C under nitrogen gas. The samples were analyzed by TEM, SIMS, Hall effect or Scanning Capacitance. We report below a selection of important and representative results.

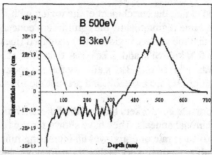

*Fig. 3: Depth distribution of excess Vs and Is resulting from the irradiation of silicon with 250 keV $Si^+$ at a doses of $1.10^{15}/cm^2$*

## Low Boron concentrations, small thermal budgets

Fig. 4a shows the B SIMS profiles obtained after spike annealing at 800°C of B implanted at 3 keV and with a dose of $1.10^{15}/cm^2$.

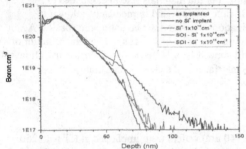

*Fig. 4a: SIMS profiles showing the effect of Si irradiation onto diffusion of B implanted at 3 keV $1.10^{15}/cm^2$ and spike annealed at 800°C.*

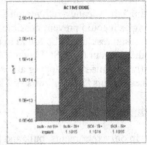

*Fig. 4b: Corresponding active doses measured by Hall effect.*

Under equilibrium conditions, B should not diffuse for such a small thermal budget. The large diffusion of B evidenced after annealing is due to TED. However, it is clear that Si irradiation with a dose of $1.10^{15}/cm^2$ totally suppresses this diffusion. In the case of the SOI samples, B diffusion is mostly governed by trapping at the SOI interface. After Si irradiation, this trapping is almost suppressed what is an indirect proof that boron diffusivity is extremely small in this SOI layer. Fig. 4b shows the boron active doses we have extracted from Hall measurements on these layers. After such an annealing, only about 4 % of the total B dose implanted in bulk Si is activated. After Si irradiation this activation rate reaches more than 20 % i.e., the injection of vacancies results in a 5 time increase of the concentration of boron atoms sitting on substitutional sites. In SOI, although this effect is somehow reduced, the activation rate can be improved by a factor of 4 when using Si irradiation.

To summarize our results, in the case of boron implanted at relatively low doses and annealed at low temperatures and/or for short times, the effect of Si irradiation is spectacular both in bulk and SOI wafers. Boron TED can be suppressed and its activation rate largely increased.

## High Boron concentrations, large thermal budgets

Fig. 5a shows the B SIMS profiles obtained after RTA annealing at 1000°C for 10 s of B implanted at 0.5 keV and with a dose of $3.10^{15}/cm^2$. Under equilibrium conditions, B largely diffuses for such a high thermal budget, giving rise to the observed shoulder in the annealed profile. The tail extending towards greater depths is due to B TED. After Si irradiation with a dose of $1.10^{15}/cm^2$, we only note a small reduction of TED. In the case of the SOI samples, B diffusion seems to be reduced but these profiles can be explained by the very large B trapping at the SOI interfaces. After Si irradiation, we do not note any significant reduction of B trapping at the interfaces, an indirect proof that B diffusivity has not been sensibly reduced by the irradiation in these layers.

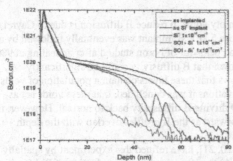

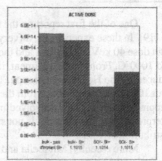

*Fig. 5a: SIMS profiles showing the effect of Si irradiation onto diffusion of B implanted at 0.5 keV and with a dose of $3.10^{15}/cm^2$ and during 1000°C, 10 s RTA annealing.*

*Fig. 5b: Corresponding active doses measured by Hall effect.*

Fig. 5b shows the boron active doses we have extracted from Hall measurements on these layers. After such an annealing, about 10 % of the total B dose implanted in bulk Si is activated. After Si irradiation this activation rate is unchanged. In SOI, we only note a slight improvement of the activation rate of boron.

To summarize our results, in the case of boron implanted at high doses and annealed at high temperatures and for long times, the effect of Si irradiation is only marginal both in bulk and SOI wafers.

## Conclusions

We confirm here what could be guessed from the literature i.e., that spectacular effects due to Si irradiation are always evidenced when the boron doses are relatively small and/or that the annealing times and temperature are also small. In other words, Vs engineering by Si irradiation is an efficient mean of reducing B diffusion and increasing B activation only if the total number of Is injected during annealing from the B profile is equal or smaller than the total number of Vs available in the region. In our experiments (Si$^+$, 250 keV, $1.10^{15}$/cm$^2$), the B dose under which these beneficial effects are maximum is limited to $3.10^{14}$/cm$^2$, significantly lower than the dose targeted by the industry. In SOI, Si irradiation can however be used to compensate the degraded mobility in these layers and drive their sheet resistance back to the value normally obtained in non-irradiated bulk Si wafers.

## VACANCY ENGINEERING USING EMPTY VOIDS

Voids i.e., precipitates of vacancies, have been largely used in the past to getter metallic impurities. They are however recent reports suggesting they could also be used for vacancy engineering. Voids are, in general, fabricated by using He implantation followed by some thermal annealing aimed at precipitating He then exodiffusing it from the implanted layer, leaving a population of empty voids behind.

### Literature review

One of the first report on the use of empty voids to reduce B diffusion is due to Cayrel et al. [19]. In these experiments, a (very) low dose 5 keV B implant was eventually followed by a high dose 40 keV He implant and the diffusion behavior of boron studied after annealing at 900 and 1000°C. From this experiment, it was shown that B diffusivity could be significantly reduced by the He co-implantation. TEM shows that these layers contain a population of voids of bout 10-20 nm in diameter. From these observations it was concluded, that these voids act as trapping centers for the Is involved in B TED driving B diffusivity back to normal. However, the interpretation of these results is not straightforward as the B profiles overlap with the depth distribution of voids.

More recently, Mirabella and Bruno [20, 21], have refined this experiment by spatially separating the voids from the B profiles. Again, they found a dramatic reduction of B diffusion when He was implanted in the layers. They ascribed this reduced diffusion to the presence of "nano-voids" located between the surface of the wafer and the region where larger voids exists, again acting as "dead sinks" and trapping the Is normally involved in B TED.

However, these experiments are difficult to understand and explain mostly because B and He were implanted together before annealing, He precipitation, the growth of these precipitates, the exodiffusion of He towards the surface take place at the same time as B diffusion and activation. To solve this problem and study the influence of vacancies only on boron diffusion and activation, we have designed a set of experiments in which a controlled supersaturation of vacancies can be maintained constant during annealing of a boron implant. For this purpose, we first fabricate empty voids by helium implantation at 40keV followed by an anneal which is sufficiently energetic to totally desorb helium from the layers. B is then implanted at energies and doses of interest for the fabrication of ultra-shallow junctions into these "void-containing" wafers and its diffusion and activation behaviour are studied in detail [22, 23]. We recall here the main results of these studies..

### Characteristics of the populations of voids

Fig. 6 shows the two populations of (empty) voids we have fabricated by annealing the He implanted wafers at either 800°C or 1000°C. The two populations are depth-distributed and centred on the projected range of the He ions, i.e. at about 400 nm below the surface. They extend over a width of about 200 nm. After annealing at 800°C, few nanovoids are found between the surface and the layer where most of the voids are found. These nanovoids have a diameter of typically 2-4 nm. In contrast, after 1000°C annealing, no nanovoids survive [23].

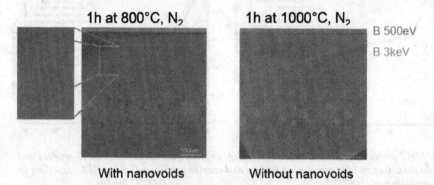

Fig. 6 : Set showing the structure of the He implanted then annealed layers and the distribution of implanted boron in these layers. TEM images taken in off-Bragg, underfocused conditions.

## Boron diffusion and activation in presence of voids

Fig. 7a is a set of SIMS results showing the influence of the voids on the diffusion of boron implanted at 0.5 keV, $3 \times 10^{15} \text{B.cm}^{-2}$ and annealed by RTA at 1000°C for 10s. Boron diffusion occurs for all concentrations lower than $1 \times 10^{20} \text{at/cm}^{-3}$, close to the theoretical value of the solid solubility of boron in silicon at this temperature. Also shown on the same figure is the result of the simulation of boron diffusion under equilibrium conditions. TED is evidenced in the samples not containing voids. Clearly, we note a strong reduction but not the suppression of boron TED in the layers containing voids. This reduction is larger in the sample containing the population of small voids but is still significant in the sample containing only large voids very far from the B profile. This clearly shows that direct trapping of the Is emitted from the B implanted region by the voids cannot be the principal mechanism leading to the observed reduction of TED. Indeed, when the voids are located much deeper than B, the Is can be trapped only after having traveled all along towards the voids, a phenomenon during which B diffusion should have been mostly unaffected. On the contrary, our observations suggest that the voids are able to reduce significantly the concentration of Is available in the B implanted region and far from them. This observation is the key to understand the effect of voids on dopant diffusion as it will be discussed later.

Fig. 7b shows the comparison between the chemical boron profile (SIMS) and the carrier concentration profile (SCM) obtained on these samples. In the sample with no voids, the carrier concentration shows a plateau extending almost all along the chemical profile. This carrier concentration is much smaller than the actual concentration of boron. The active dose is of about $1 \times 10^{14} \text{B.cm}^{-2}$, i.e. only 3% of the implanted dose. The voids differently affect boron activation depending on their size. The activation of boron is is much better in the sample containing small voids. The boron active dose is increased by about 80% compared to the reference. The carrier profile follows the SIMS profile from the tail towards the middle of the region where boron is free to diffuse and then decreases again towards the surface.

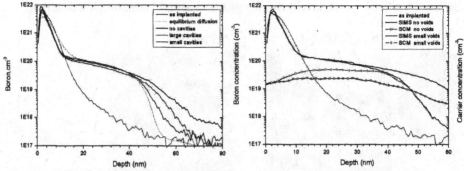

Fig. 7a: SIMS profiles showing the effect of voids onto the diffusion of boron during RTA annealing at 1000°C for 10s.

Fig. 7b: Comparison between SIMS profiles and and carrier profiles following RTA annealing at 1000°C for 10s.

Comparing the relative shapes of the SIMS and SCM profiles, we observe that boron activation is "easier" on the deep side where the cavity stands but gets more difficult as the boron atoms are getting closer to the region where the BICs stand. This observation suggests that boron activation proceeds from the deep side of the boron profiles, i.e. from the voids region, and towards the BICs region. However, it is demonstrated that voids can be used to increase boron activation in ion implanted silicon.

The results presented here and elsewhere [22, 23] cannot be understood only by assuming that voids are "dead sinks" only being able to passively trap Is which diffuse to them. Alternatively, we believe that they can be explained by assuming that during annealing the population of voids emits vacancies which diffuse towards the surface and recombine with the Is involved in B diffusion and/or limiting its activation. This emission rate is however a function of the void size (through the Gibbs-Thomson equation) in accordance with the results presented here. Implications and details are discussed somewhere else (Marcelot to be published).

## COMPARISONS AND POSSIBLE IMPROVEMENTS

Putting together results from the literature with our own, a general picture emerges in which the characteristics of both vacancy engineering approaches can be compared. This is summarized in Fig. 8.

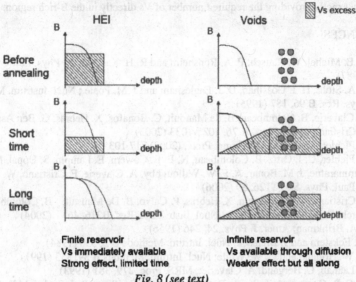

Vs concentration evolution versus annealing time

Fig. 8 (see text)

In the Si irradiation approach, there exists, prior to dopant introduction, a limited number of Vs but they are immediately "available" during activation annealing. At the beginning of this annealing, these Vs recombine with the Is which would have been involved in boron diffusion and the formation of BICs. For this reason, the effects are spectacular and virtually no diffusion occurs during this period of time. However, in this process, these Vs are consumed and after a while there are no more Vs available and B diffusion again occurs; TED can still exist if the implanted boron dose was larger than the number of Vs available.

In the case of voids, before annealing, all the Vs are stored within the voids. During annealing, the population is in equilibrium with a supersaturation of Vs only function of their sizes and of the temperature. There exists a vacancy concentration gradient extending from the voids and towards the surface which drives the Vs in the region where boron stands. While the Vs are apparently not in high enough concentrations to suppress the large TED occurring at the very beginning of the anneal, this concentration stays almost constant all along the anneal. Indeed, the voids provide a quasi infinite reservoir of Vs and do not evolve in size during typical activation annealing.

Finally, both methods could be improved or even combined. For the Si irradiation approach, the need is to increase the number of Vs available before B implantation. Since amorphization limits the maximum dose that can be used with Si irradiation, one could use conditions for which the thermal recombinations of Is and Vs are favored. This could be done by using ions with a lower mass than Si and/or by increasing the wafer temperature during implantation. Although the last proposal might not be please too much to the industry, an additional advantage would be to store many more Vs in the form of multi-vacancies. For the approach involving voids, the need is to increase the concentration and the arrival rate of Vs in the B-rich region. This can be done by reducing the distance between the population of voids and by reducing their size. Both approaches could merge and add their benefits by using He at a dose and energy exactly providing the required number of Vs directly in the B-rich region.

# REFERENCES

1. A. E. Michel, W. Rausch, P. A. Ronsheim and R. H. Kastl, Appl. Phys. Lett. 50 (7), 416 (1987)
2. P. A. Stolk, H. J. Gossman, D. J. Eaglesham and J. M. Poate ; Nucl. Instrum. Methods Phys. Res. B 96, 187 (1995)
3. A. Claverie, B. Colombeau, B. de Mauduit, C. Bonafos, X. Hebras, G. Ben Assayag and F. Cristiano, Appl. Phys. A 76, 1025-1033 (2003)
4. P. Pichler, Mat. Res. Soc. Symp. Proc., (2002) 717 103
5. P. Pichler, C. J. Ortiz, B. Colombeau, N. E. B. Cowern, E. Lampin, S. Uppal, M. S. A. Karunaratne, J. M. Bonar, A. F. W. Willoughby, A. Claverie, F. Cristiano, W. Lerch and S. Paul; Phys. Scr. T126, 89 (2006)
6. F. Cristiano, N. Cherkashin, X. Hebras, P. Calvo, B. De Mausuit,   B. Colombeau, W. Lerch, S. Paul and A. Claverie; Nucl. Instr. Phys. Res. B216, 46    (2004)
7. J. A. Brinkman; Amer. J. Phys. 24, 246 (1956)
8. W. Eckstein and J. Biersack; Nucl. Intrum. Methods B 2, 550 (1984)
9. O. W. Holland and C. W. White; Nucl. Intrum. Methods B 59, 353 (1991)
10. L. Laânab, C. Bergaud, A. Claverie:, MRS. Proc. 279, 381 (1993)
11. N. E. B. Cowern, A. J. Smith, B. Colombeau, R. Gwilliam, B. J. Sealy and E. J. H. Collart; Electron Devices Meeting (2005)
12. V. Raineri, R. J. Schreutekamp, F. W. Saris, K. T. F. Janssen and R. E. Kaim; Appl. Phys. Lett. 58, 922 (1991)
13. V. C. Venezia, T. E. Haynes, A. Agarwal, L. Pelaz, H. J. Gossmann, D. C. Jacobson and D. J. Eaglasham ; Appl. Phys. Lett. 74, 1299 (1999)
14. A. Nejim and B. J. Sealy, Semicond. Sci. Technol. 18; 839 (2003)
15. L. Shao, X. Wang, J. Bennet, L. Larsen and W. K. Chu; J. Appl. Phys. 92, 4307    (2002)
16. R. Gwilliam, N. E. B. Cowern, B. Colombeau, B. Sealy and A. J. Smith, Nucl. Instrum. Methods B 261; 600 (2007)
17. E. G. Roth, O. W. Holland, V. C. Venezia and B. Nielsen ; J. Electon. Mater. 26, 1349 (1997)
18. A. J. Smith, B. Colombeau, R. Gwilliam, E. Collart, N. E. B. Cowern and B. J. Sealy ; Mat. Res. Soc. Symp. Proc. 810 (2004)
19. F. Cayrel, D. Alquier, D. Mathiot, L. Ventura, L. Vincent, G. Gaudin and R. Jerisian ; Nucl. Instrum. Methods B 216, 291 (2004)
20. S. Mirabella, E. Bruno, F. Priolo, F. Giannazzo, C. Bongiorno, V. Raineri, E. Napolitani and A. Carnera; Appl. Phys. Lett. 88, 191910 (2006)
21. E. Bruno, S. Mirabella, E. Napolitani, F. Giannazzo, V. Raineri and F. Priolo ; Nucl. Instr. Meth. Phys. Res. B 257, 181 (2007)
22. O. Marcelot, A. Claverie, F. Cristiano, F. Cayrel, D. Alquier, W. Lerch, S. Paul, L. Rubin, H. Jaouen, C. Armand; Nucl. Intr. Phys. Res. B 257, 249 (2007)
23. O. Marcelot, A. Claverie, D. Alquier, F. Cayrel, W. Lerch, S. Paul, L. Rubin, V. Raineri, F. Giannazzo, H. Jaouen; Sol. Stat. Phen. 131, 357 (2008)

Mater. Res. Soc. Symp. Proc. Vol. 1070 © 2008 Materials Research Society     1070-E01-03

# Modeling and Experiments of Dopant Diffusion and Defects for Laser Annealed Junctions and Advanced USJ

Taiji Noda[1], Wilfried Vandervorst[2,3], Susan Felch[4], Vijay Parihar[4], Christa Vrancken[2], and Thomas Y. Hoffmann[2]

[1]Matsushita Electric Industrial Co., Ltd., 19 Nishikujyo-Kasugacho, Minami-ku, Kyoto, 601-8413, Japan

[2]IMEC, Kapledreef 75, Leuven, B-3001, Belgium

[3]K. U. Leuven, Kasteelpark, Arenberg 10, Leuven, B-3001, Belgium

[4]Applied Materials, 974 E. Arques Ave. MIS 81280, Sunnyvale, CA, 94085

## ABSTRACT

Laser annealed junctions and advanced ultra shallow junctions are studied in both atomistic modeling and experiments. SIMS and sheet resistance measurement for spike-RTA + Laser annealing show that additional laser annealing after spike-RTA ("+Laser") improve the dopant activation level without increasing in junction depth. "+Laser" effect become effective in the combination of low spike-RTA temperature and high laser temperature. This effect is significant for As doped layer. Spike-RTA based junction has a limitation in viewpoint of Rs-Xj trade-off. Laser-only annealing is promising candidate to overcome this limitation. Boron diffusion with laser-only annealing is investigated. An atomistic kinetic Monte Carlo modeling show that $B_nI_m$ complexes and End-of-Range (EOR) defects are formed during sub-millisecond annealing time range. Impact of F co-implant on Boron diffusion and EOR defect evolution during sub-millisecond annealing are also investigated.

## INTRODUCTION

Achieving diffusion-less annealing is required for the 32 nm technology node and beyond. For the formation of highly activated and ultra-shallow junctions, millisecond (ms) annealing, such as non-melt laser annealing (NLA), is a promising candidate.[1-13] Co-implant can also suppress the transient enhanced diffusion (TED) of dopants and create shallow junctions. The role of co-implanted species is also an interesting topic.[14-20]

Millisecond annealing can achieve temperatures up to ~ 1350 °C within 1 ms. But the dopant diffusion behavior and defect evolution in such very short annealing time range is not known yet. The understanding of dopant diffusion and defect evolution behavior during the sub-millisecond annealing time range is very important. In this article, our recent studies on modeling and experiments of non-melt laser annealing are shown. An atomistic diffusion model using a kinetic Monte Carlo (KMC) approach [6, 10, 21, 22] is used for the understanding of dopant diffusion and defect evolution.

## MODELING DESCRIPTION

For the accurate modeling of defect evolution and dopant activation during sub-ms annealing with co-implant effects, an atomistic kinetic Monte Carlo (KMC) diffusion model was used. In our KMC model, (1) the depth of the amorphous-Si (a-Si) layer and (2) solid phase epitaxial regrowth (SPER) velocity, which are important for accurate prediction of EOR defect behavior, are modified using experimental results [6, 10]. For the dopant activation/deactivation modeling, dopant-defect complexes, such as $B_n I_m$ complexes for boron, are considered. For F co-implant modeling, $F_n V_m$ complexes are considered.

## RESULTS AND DISCUSSION

### Spike + Laser annealing:

Spike-RTA + Laser annealing is a bridging technology to combine the spike-RTA and laser annealing. The concept of this combination annealing is the improvement of dopant activation level without changing junction depth after spike-RTA using additional "+Laser". This combination annealing between spike-RTA and laser annealing was applied to 45nm technology node device integration work and shows about 5% - 8% nFET improvement and pFET shows almost no improvement (0% - 3% improvement). Figures 1, 2 show the sheet resistance (Rs) as a function of laser temperature with different spike-RTA temperature for F + B implants, and As implants, respectively [6]. In those figures, broken lines show sheet resistance with spike-RTA only. When sheet resistance is lower than this broken line, dopant activation level is improved by additional laser annealing after spike-RTA ("+Laser" effect). With sufficiently high spike-RTA temperature, for instance, spike-RTA at 1050 °C, sheet resistance shows almost no improvement due to "+Laser". As spike-RTA temperature is reduced, sheet resistance decrease as a function of laser peak temperature.

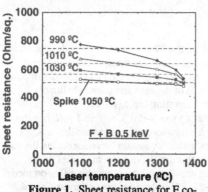

**Figure 1.** Sheet resistance for F co-implant + B 0.5 keV, $7\times10^{14}$/cm$^2$.

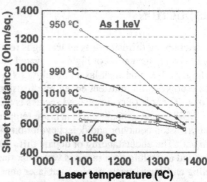

**Figure 2.** Sheet resistance for As, 1 keV, $1\times10^{15}$/cm$^2$.

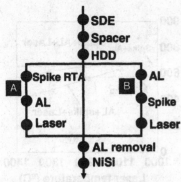

**Figure 3.** Laser annealing sequence comparison in this work. AL denotes "Absorbing layer".

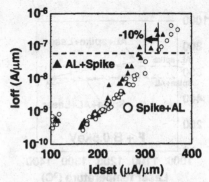

**Figure 4.** PFET Ion-Ioff comparison between AL before spike-RTA and AL after spike-RTA (Vdd=1.0V). [9]

Thus "+Laser" becomes effective in the combination between low spike-RTA temperature and high laser temperature.

The sheet resistance measurement results show that "+Laser" impact on sheet resistance improvement is more significant on As doped layer than that of B doped layer. This is in good agreement with the device experimental observation.

SIMS profiles for boron and arsenic are compared between after spike-RTA and after spike-RTA + Laser. SIMS measurements show that both B and As show no additional diffusion during laser annealing after the spike-RTA. Dopant diffusion and junction depth is determined by spike-RTA. Thus sheet resistance and SIMS measurements reveal that dopant activation level is improved by "+Laser" without increase in junction depth. KMC simulations also show that no additional diffusion during laser annealing after spike-RTA and junction depth is determined by spike-RTA.

**Impact of annealing sequence:**

For the actual device integration flow, the impact of annealing sequence is important. In order to improve the thermal uniformity during high temperature laser annealing, an absorbing layer (AL) is deposited before laser annealing. During absorbing layer deposition, low temperature thermal treatment takes place. The additional thermal budget during AL deposition is possible issue for ultra shallow junction formation, dopant TED and deactivation. We compare the laser annealing sequence between (1) AL deposition after spike-RTA and (2) AL deposition before spike-RTA (Figure 3). In the device work, AL before spike-RTA show about 10% pFET device performance degradation (Figure 4).[9] Extension resistance is higher and Cov is smaller than those of spike+AL. It indicates that extension of pFET is degradaded due to AL before spike-RTA. On the other hand, AL before spike-RTA shows better nFET device performance than the reference spike-RTA process. Figure 5 shows the sheet resistance comparison between AL before spike-RTA and AL after spike-RTA for B doped layer. For B doped layer, AL before spike-RTA shows higher sheet resistance value than that of AL after spike-RTA.

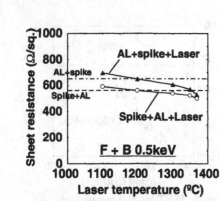

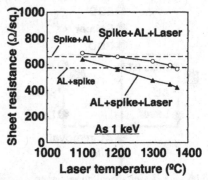

**Figure 5.** Sheet resistance comparison between AL + spike + Laser and spike + AL + Laser for F co-implant + B 0.5 keV, $7\times10^{14}/cm^2$.

**Figure 6.** Sheet resistance comparison between AL + spike + Laser and spike + AL + Laser for Arsenic 1.0 keV, $1\times10^{15}/cm^2$.

It indicates that spike-RTA with capping layer may induce degradation of junction property for B doped layer. Figure 6 shows the sheet resistance comparison between AL before spike-RTA and AL after spike-RTA for As implanted sample. For As doped layer, AL before spike-RTA shows a significant sheet resistance improvement in comparison with AL after spike-RTA. This behavior is opposite to B doped layer. It indicates that optimum annealing sequence is different between pFET and nFET. Boron SIMS profile comparison between spike + AL and AL + spike show that Boron dose loss looks enhanced during spike-RTA with capping layer. It is reported that Boron dose loss during spike-RTA is influenced in the presence of capping layer. This phenomenon is also related with PMOS degradation due to SMT process. For As doped layer, SIMS profiles shows that AL before spike-RTA shows higher As shoulder position than that of spike + AL. It indicates that active As is increased with AL before spike-RTA.

**Limitation of "+Laser" effect:**

Laser annealing after spike-RTA can improve the sheet resistance significantly without increasing in junction depth. Sheet resistance measurement results show that the "+Laser" effect for As doped layer is larger than that of F + B doped layer. In addition to this, the optimum annealing sequence looks different between pFET and nFET.

Figures 7, 8 show the Rs-Xj trend plot for p-type, and n-type with various annealing, respectively. As spike-RTA temperature decreases, improvement of sheet resistance due to "+Laser" effect becomes larger. The combination between low temperature spike-RTA and high laser peak temperature looks interesting.

In those figures, the Rs-Xj trend lines for ideal box profile are also plotted. When data are compared with ideal Rs-Xj trend line, the dopant activation limitation is seen to be around 1E20 (atoms/cm$^3$). To overcome this limitation, Laser-only annealing looks promising annealing technique. But the additional thermal budget during AL deposition is a possible issue for the ultra shallow junction formation.

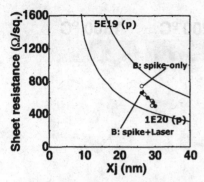

**Figure 7.** Rs-Xj trade off plot for Boron doped layer. Lines are Rs-Xj trend line for ideal box profile.

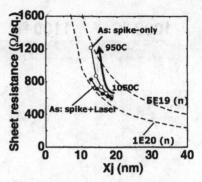

**Figure 8.** Rs-Xj trade-off plot for Arsenic doped layer. Broken lines are Rs-Xj trend line for ideal box profile.

<u>**Laser-only annealing:**</u>

Low temperature thermal treatment, such as AL deposition, is used before laser-only annealing. XTEM observation shows that amorphous-Si layer, which is formed during implantation. is partially regrown during low temperature thermal treatment. Figure 9 shows Boron SIMS profiles during laser annealing process.[6] Boron SIMS profiles show that Boron diffuses during partial SPER at 550 °C and then diffuses again during laser annealing. Thus Boron diffusion in the laser annealing process is 2-step diffusion. Boron diffusivity enhancement in amorphous-Si layer during partial SPER is observed. This observation agrees with the previous report showing Boron diffusivity is a-Si layer is 5 orders of magnitude larger than in crystalline-Si.[17] The sheet resistance as a function of Ge PAI energy shows that deep PAI shows lower sheet resistance than that of shallow PAI. Figure 10 shows KMC simulation results of sub-ms annealing at 1300 °C.

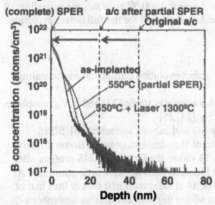

**Figure 9.** Boron SIMS profiles during laser annealing with partial SPER.

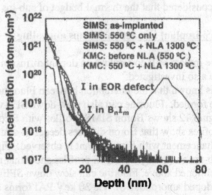

**Figure 10.** Comparison between SIMS and KMC simulations for Laser annealing.

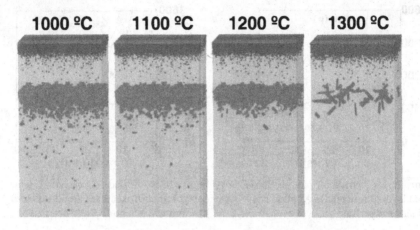

**Figure 11.** KMC atomistic modeling of end-of-range defects during sub-ms annealing.

KMC with various size of $B_nI_m$ can predict well Boron diffusion behavior during sub-ms annealing.[6] KMC simulations show that $B_nI_m$ complexes are already formed at 1300 °C with sub-ms annealing. The control of $B_nI_m$ complex is important for sub-ms annealing. KMC simulations also show that EOR defects are formed during sub-millisecond annealing. Defect formation and evolution behavior during sub-ms annealing is also very important because they influence the junction leakage and junction thermal stability. XTEM shows that EOR defects are clearly formed during sub-ms annealing time. At low laser annealing temperature, the defect size is small and the defect density is very high. Figure 11 shows the KMC atomistic simulations of defects during sub-ms annealing.[6] KMC atomistic model predicts that {311} defects cannot completely evolve into the dislocation loops at 1300 °C with sub-ms annealing. It is considered that the thermal budget of sub-ms annealing too small for the full defect evolution.

### F co-implant impact on sub-ms annealing:

The impact of F co-implant on dopant diffusion and defect evolution during sub-ms annealing are also investigated.
It is known that the interaction between Fluorine and Vacancy is stable [23] and $F_nV_m$ complexes are formed. Fluorine can also slow down SPE velocity.[24, 25]
Figure 12 shows Boron SIMS profiles with F co-implant and sub-ms annealing.[6] SIMS profiles show that Boron diffuses deeper as a function of F co-implant energy. Boron diffusivity enhancement with F co-implant is observed. Figure 13 shows the Fluorine SIMS profiles after sub-ms annealing.[6] SIMS profiles show that most Fluorine atom remains in Si-substrate. As mentioned above, Fluorine can slow down SPER velocity at maximum x10 slower than that of undoped amorphous-Si. Ge 30 keV PAI forms about 46 nm depth of continuous amorphous-Si layer.

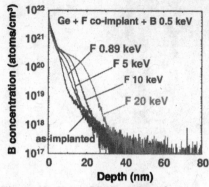

**Figure 12.** Boron SIMS profiles after sub-ms laser annealing with various F co-implant energies.

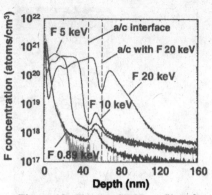

**Figure 13.** Fluorine SIMS profiles after sub-ms laser anneal.

When F co-implant is used, no partial SPER during low temperature treatment is observed. The estimated time to complete SPER at 1300 °C is from 4.4 μs ~ 40 μs (Figure 14). Then it is considered that most Boron diffusion at 1300 °C occurs in crystalline-Si.

KMC simulations with $F_n V_m$ model show that in the presence of Fluorine, vacancy clusters remain in the Si-substrate after ms-annealing and form $F_n V_m$ complexes (Figure 15). This can explain the reason why much Fluorine atom remains in Si after sub-ms annealing.

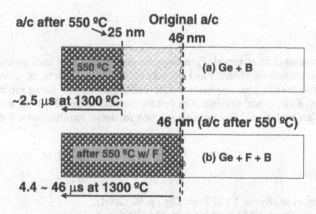

**Figure 14.** Time for complete SPER at 1300 °C for (a) Ge + B, and (b) Ge + F + B, respectively.

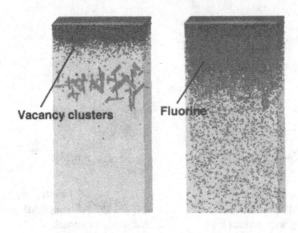

**Figure 15.** KMC atomistic simulation of sub-ms annealing with $F_nV_m$ model.

During SPER annealing stage, small vacancy clusters and $F_nV_m$ complexes are formed. Then $F_nV_m$ complexes can capture the free interstitials and annihilate a couple of point defect (I-V) pair. Finally, a single Fluorine atom is released and can migrate very rapidly toward the Si-surface. In the case of sub-ms annealing time range, $F_nV_m$ complexes are still remaining.

## CONCLUSIONS

Modeling and experiments of laser annealed junctions are shown. Non melt laser annealing can improve the dopant activation significantly and can achieve the shallow junctions. Atomistic KMC model shows that both $B_nI_m$ complexes and EOR defects are formed during sub-ms annealing time range. KMC model also indicates that the thermal budget of sub-ms annealing is too small for full defect evolutions. One possible solution for defect stabilization is F co-implant.

## REFERENCES

1. S. K. H . Fung, *et al.*, Symp. VLSI Tech. Dig., p. 92 (2004).
2. A. Shima, *et al.*, Symp. VLSI Tech. Dig., p. 174 (2004).
3. M. Hane, *et al.*, IEDM2004 (2004).
4. K. Adachi, *et al.*, Symp. VLSI Tech. Dig., p. 142 (2005).
5. A. Pouydebasque *et al.*, IEDM2005, p. 663 (2005).
6. T. Noda, *et al.*, IEDM2006, p. 377 (2006).

7. E. Josse, *et al.*, IEDM2006, p. 693 (2006).
8. S. Severi, *et al.*, IEDM2006, p. 859 (2006).
9. T. Y. Hoffmann, *et al*, IWJT2007, S8-3 (2007).
10. T. Noda, *et al*, IEDM2007, p. 955 (2007).
11. C. Ortolland, *et al.*, VLSI2008, p. 186 (2008).
12. T. Noda, *et al.*, Mat. Res. Soc. Symp. Porc. Vol. 912, 0912-C05-06 (2006).
13. T. Noda, *et al.*, IIT2006, p. 21 (2006).
14. B. Pawlak, *et al.*, Appl. Phys. Lett. **89**, p. 062110 (2006).
15. B. Pawlak, *et al.*, Appl. Phys. Lett. **89**, p. 062101 (2006).
16. B. Colombeau, *et al.*, IEDM2006, p. 381 (2006).
17. J. Jacques, *et al.*, Appl. Phys. Lett., **82**, p. 3469 (2003).
18. R. Duffy, *et al.*, Appl. Phys. Lett. **84**, p. 4283 (2004).
19. T. Noda, J. Appl. Phys. **96**, p. 3721 (2004).
20. N. Cowern, et al., Appl. Phys. Lett. **86**, p. 101905 (2005).
21. M. Jaraiz, *et al.*,Mat.Res.Soc.Symp.Proc. **532**,43 (1998).
22. T. Noda, J. Appl. Phys. **94**, p. 6396 (2003).
23. G. Lopez, *et al.*, Phys. Rev. B **72**, p. 045219 (2005).
24. G. Olson, *et al.*, Mater. Sci. Rep. 3, p. 1 (1988).
25. S. Mirabella, *et al.*, Appl. Phys. Lett. **86**, p. 121905 (2005).

Mater. Res. Soc. Symp. Proc. Vol. 1070 © 2008 Materials Research Society

# Surfaces and Interfaces for Controlled Defect Engineering

Edmund G. Seebauer

Chemical & Biomolecular Engineering, University of Illinois, 600 S Mathews Ave, 114 RAL, Box C3, Urbana, IL, 61801

## ABSTRACT

The behavior of point defects within silicon can be changed significantly by controlling the chemical state at the surface. In ultrashallow junction applications for integrated circuits, such effects can be exploited to reduce transient enhanced diffusion, increase dopant activation, and reduce end-of-range damage.

## INTRODUCTION

Engineering of point defects semiconductors is important for a variety of applications, including ion implantation/annealing technology and crystal growth. In particular, forming increasingly shallow *pn* junctions in silicon-based microelectronic logic devices is critical as device dimensions continue to diminish. More electrically active dopant is also required in the implanted regions. Current technology for junction formation relies mainly on ion implantation followed by rapid thermal annealing to introduce dopants into the substrate. Post-implant annealing technologies have struggled in their ability to simultaneously increase dopant activation and decrease post-implant diffusion, especially for the key dopant boron. In addition, end-of-range defects left over from pre-amorphizing implants need to be reduced.

As junctions move progressively closer to nearby surfaces and interfaces, the possibility arises for using these boundaries themselves for defect engineering. Such engineering could also prove useful in the formation of three-dimensional devices such as FINFETs. However, up until recently the science base required for performing such defect engineering was scanty. However, it has now been shown [1-2] that the behavior of point defects within silicon can be engineered by controlling the chemical state at the surface. Experiments combined with modeling of dopant diffusion/activation have suggested that certain chemical treatments of the surface induce it to act as a large controllable "sink" for defects that removes Si interstitials selectively over impurity interstitials, leading to less diffusion and better dopant activation. In addition, end-of-range damage can be reduced. There are two separate mechanisms for such effects: addition to dangling bonds and electrostatic attraction/repulsion.

## ADDITION TO DANGLING BONDS

The first mechanism for interaction of a surface or interface with bulk defects involves insertion of defects into dangling bonds at the boundary. Various surfaces and interfaces differ markedly in their ability to annihilate defects. For example, an atomically clean surface is chemically active and can annihilate interstitial atoms by simple addition of the interstitials to dangling bonds. However, if the same surface becomes saturated with a strongly bonded adsorbate, annihilation requires the insertion of interstitials into existing bonds. Such insertion should have a higher activation barrier, and the surface becomes less chemically active toward defects. A schematic diagram of this idea appears in Figure 1a.

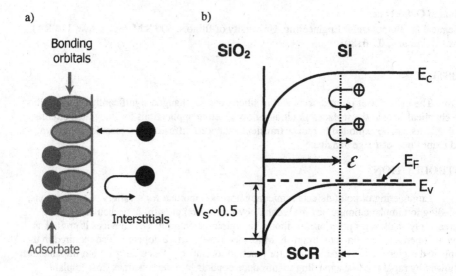

**Figure 1.** (a) Bond insertion mechanism, showing how bulk interstitials should react relatively easily with surface dangling bond sites, but less easily with sites saturated by a strongly bonded adsorbate. (b) Potential energy diagram for electrons in p-type silicon, illustrating the electrostatic mechanism. Defects at the Si/SiO$_2$ interface absorb charge from the underlying bulk, creating a narrow space charge region (SCR) and a corresponding electric field E that pointing into the bulk. The field repels positively charged interstitials.

Behavior of dopants such as boron during annealing is usually determined by the interplay between lone interstitials and interstitial clusters/extended defects that render the lone interstitials immobile. During implantation, numerous lone interstitials are created that diffuse quickly and accrete into clusters and extended defects. Subsequent annealing dissociates many of these aggregates. However, the Si and dopant interstitials are released together. The Si interstitials tend to keep the dopant atoms from entering and remaining in the desired substitutional atomic sites, and also promote the unwanted transient enhanced diffusion (TED) of the dopant atoms. Annealing protocols that avoid these problems also leave most of the dopant locked within clusters, rendering it useless.

A large additional "sink" that removes Si interstitials selectively over dopant interstitials solves this problem. The loss rate of bulk interstitials through incorporation into the surface ranges from very high at a chemically active surface with many dangling bonds to nearly zero at a surface whose dangling bonds are already saturated with a tightly bonded adsorbate. When the surface loss probability is adjusted to be large, Si interstitials diffuse toward the surface about 100 times faster than the dopant because dopant interstitials are impeded by exchange with the bulk lattice atoms in a way that Si interstitials are not. The reason is statistical. A dopant interstitial diffusing toward the surface periodically kicks into the lattice, and becomes immobile and electrically active. The kick-in process almost always releases an interstitial of the majority species in the lattice (Si), and the immobilized dopant atom must wait for another Si interstitial to come along in order to become mobile again. Thus, the lattice serves to impede the motion of dopant interstitials toward the surface. Silicon interstitials also exchange with the lattice. However, at typical doping levels near 1%, a lattice exchange event simply yields another Si interstitial atom. The remaining 1% of the events results in kickout of dopant. Thus, the lattice does not impede the net motion of Si interstitials nearly as much as for dopant. Based on crude statistics assuming equal Si-Si and Si-dopant exchange rates, the surface will extract Si interstitials about 100 times faster than dopant interstitials, even if the surface is equally active toward both species. This preferential loss of Si interstitials keeps electrically active dopants fixed in the lattice by inhibiting the "kick-out" reaction that makes such dopant atoms mobile and inactive. The chemical activity of solid-solid interfaces toward defects can vary as well, though the physical picture for predicting such effects is less clear.

## Experimental demonstration of dangling bond exchange

To examine the effects of bond insertion on the surface loss probability, one set of experiments [13] was performed on <100>-oriented, Czochralski grown, n-type silicon wafers with a resistivity of 6~9 ohm·cm. Ge was implanted at 15 keV to a dose of $3 \times 10^{14}$ atoms/cm$^2$, resulting in the formation of a continuous amorphous layer at a depth approximately 22 nm (measured from Rutherford backscattering (RBS) channeling measurements and transmission electron microscopy (TEM)). The wafers were then implanted with boron at 500 eV to a dose of $1 \times 10^{15}$ atoms/cm$^2$. All implants were performed in single-quad mode at 0° tilt and 0° twist, and with native oxide thickness of 11±2 Å. After implantation, some samples were pre-treated with 49% aqueous HF to remove the native oxide and thus create the atomically clean surfaces.

Annealing was carried out in an ultrahigh vacuum environment using Ta clips for resistive heating. The pressure in the chamber was maintained around $10^{-8}$~$10^{-9}$ torr during annealing to prevent formation of native oxide and contamination of the surface. In contrast, many annealing studies reported in the literature have been performed in an inert N$_2$ ambient at atmospheric pressure. Such environments typically contain low, ill-defined levels of reactive gases such as oxygen or moisture [6]. This is especially important since any form of adsorption may deactivate the dangling bonds at the atomically clean surface. The annealing conditions were in the range of 750°C to 900°C for 60 minutes.

Boron profiles were analyzed ex-situ with secondary ion mass spectroscopy (SIMS) with oxygen source. Sheet resistance ($R_s$) was measured by standard four point probe, accurate to within about 10%, while the active carrier concentration ($N_s$) was obtained from Hall measurement by assuming a unity Hall scattering factor. Cross-sectional TEM (XTEM) was performed to analyze the extent of amorphization and the induced end-of-range (EOR) defects.

Another set of experiments [12] used p-type Si (100) (5 ohm·cm) implanted with arsenic at 2 keV and dose of $2 \times 10^{15}$ ions/cm$^2$. There was no surface treatment of the wafers prior to the implantation, and the native oxide thickness was 11±3 Å. Specimens of approximate dimensions 1.7 cm × 0.7 cm were cut from the wafers and mounted in an ultrahigh vacuum chamber using Ta clips for resistive heating. Temperature was monitored with a chromel-alumel thermocouple junction pressed into a small pit drilled into the silicon. To induce diffusion, all specimens were annealed at 750°C for 1 hr. Specimens were exposed only to atmospheric oxidation, and were not subjected to HF treatment to remove oxide before annealing. In one case, the oxidation was simply the native oxide that built up at room temperature over a period of several weeks after implantation. In another case, native oxide was removed by degreasing and exposure to dilute HF. Thermal oxide was then grown at 630°C for 20 min in moist air. Ellipsometric measurements determined the resulting $SiO_2$ thickness to be 6 nm. Arsenic diffusion profiles were measured *ex situ* with SIMS using a Cameca 4f, with a 700 eV Cs beam incident at 50°.

Figure 2 shows data illustrating the difference in arsenic diffusion based on the type of oxide (native vs grown). The native oxide promotes significantly more profile spreading than the grown oxide. The grown oxide has higher $R_s$ than the native oxide, meaning that both the diffusional spreading and dopant activation decreased for the grown oxide. This dataset clearly shows that the surface chemical bonding state exerts large effects on dopant diffusion in the bulk.

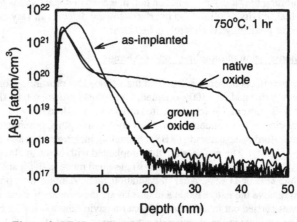

**Figure 2.** SIMS profile of implanted arsenic diffusion in silicon for various types of surface oxide. The grown oxide was 6 nm thick.

Figure 3(a) shows the boron diffusion profiles after annealing at 700°C for 60 minutes. Both the native oxide-covered and atomically clean samples exhibit surface-directed diffusion (boron uphill diffusion) at the high concentration portion of the boron profiles as well as dopant dose loss. This is in agreement with the findings reported by Duffy *et al.* and Wang *et al* [7,8]. The phenomenon is attributed to the interstitial flux induced by the evolution of end-or-range (EOR) defects located around the a/c interface during the annealing cycle. The results show that clean surface sample has a dopant profile at which high B concentration region diffuses more towards the surface, while the tail of profile is diffusing deeper into the substrate for the native oxide sample. In addition, gettering of boron around the EOR defect band can only be seen in the oxide-covered surface.

Figure 3(b) shows the boron SIMS profiles annealed at a higher temperature of 800°C. Boron TED is observed in both samples and is due to the emitted interstitials from EOR bands diffusing back towards the surface during annealing. The emitted Si interstitials cause boron

a)

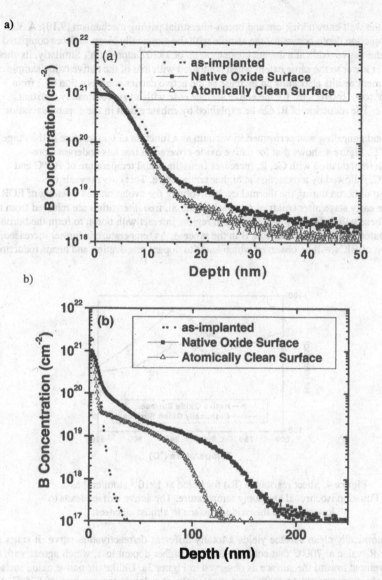

b)

**Figure 3.** SIMS profiles of 500eV B implant with prior 15keV Ge pre-amorphizing implant. Annealing was performed at (a) 700°C and (b) 800°C for 60 minutes. The atomically clean surface leads to reduced TED in both cases.

diffusion via the well known kick-out and boron-interstitial pairing mechanism [9,10]. A 30% reduction in junction depth occurs for the sample with the atomically clean surfaces compared to the one with the native oxide at a dopant concentration of $1 \times 10^{18}$ atoms/cm$^2$. Similarly, its sheet resistance $R_s$ is lower to the same extent when compared with that of the native oxide sample. Hall measurement results also show an increase in the active carrier concentration ($N_s$) from $2.85 \times 10^{14}$ cm$^2$ to $4.16 \times 10^{14}$ cm$^2$ for the active surface case with an approximately constant mobility value. The reduction of $R_s$ can be explained by enhancement in the dopant activation level.

Isochronal annealing was performed in vacuum as a function of temperature in the range of 700°C to 950°C. Figure 4 shows that for native oxide-covered Si that has undergone pre-amorphization implantation with Ge, $R_s$ increases from the initial temperature of 700°C and peaks at 850°C, followed by a reduction at higher temperature. This curve reveals the de/reactivation of boron during the thermal cycle involving the evolution and ripening of EOR defects. In the early stage after epitaxial regrowth of the Si, free interstitials are released from the EOR band. These will diffuse towards the B-rich region, interact with boron to form the boron-interstitial clusters (BICs) and deactivate B in the process. As temperature is further increased, the dissolution of BICs will be observed which leads to dopant reactivation and hence reduction in $R_s$.

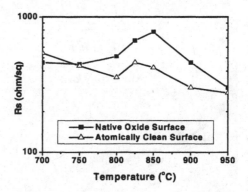

**Figure 4.** Sheet resistance ($R_s$) for boron at $1 \times 10^{15}$ atom/cm$^2$ as a Function isochronal annealing temperature. The active surface leads to lower $R_s$ than the oxided surface in almost all cases.

The atomically clean surface yields a totally different de/reactivation curve. It starts out with a higher $R_s$ value at 700°C that could be due to a higher dopant loss, which agrees with the higher B movement toward the surface as observed in Figure 3a. Unlike the native oxide surface, $R_s$ continuously decreases up to 800°C and then peaks at a lower temperature of 825°C. The activated dangling bonds at the clean surface may contribute to a higher "sink" efficiency in the removal of free silicon interstitials. The lower deactivation level, as seen from overall lower $R_s$ values for the clean surface indicates the possibility of higher EOR defect dissolution rates, resulting in smaller sizes or lesser amounts of extended defects remaining after thermal annealing.

The higher rate of interstitial annihilation at the active surface promotes higher EOR defect dissolution rates, resulting in smaller numbers and sizes of these defects after epitaxial regrowth. Figure 5 shows this effect directly for the native oxide and the atomically clean surface, taken after annealing at 750°C for 60 minutes. A dark band corresponding to the EOR defects shows up clearly for the native oxide-covered case in Figure 5(a). For the active, atomically clean surface in Figure 3(b), no EOR defects are evident.

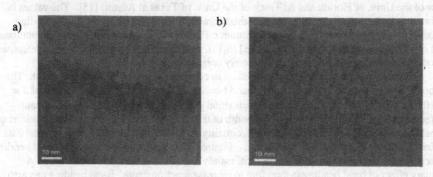

**Figure 5.** XTEM micrographs for (a) native oxide and (b) atomically clean surfaces annealed at 750°C for 60 minutes. EOR defects are essentially eliminated in the presence of the atomically clean surface. Micrographs for as-implanted material (not shown) revealed a 20nm continuous amorphous top layer.

## ELECTROSTATIC MECHANISM

The second mechanism for interaction between the surface and bulk involves the reflection of charged interstitials from the surface due to electrically active surface defects that set up a repulsive electric field. The driving force for reflection is surface band bending that results from dangling bonds at a free Si surface (no oxide) or a damaged Si-oxide interface [3-5]. Changing the number of electrically charged dangling bonds (for example, by adsorption) modifies the degree of band bending. Interstitial atoms of B and Si are positively charged in implanted material. Band bending sets up a near-interface electric field pointing into the bulk. The field repels the interstitials from the surface. (An analogous effect would be observed for negatively charged defects diffusing in n-type material.) The opposing field can transform the interface from a significant sink into a good reflector, which changes the average concentration of interstitials in the underlying bulk, which in turn influences the degree of dopant activation and transient-enhanced diffusion. A schematic diagram of this idea appears in Figure 1b.

For investigations of the electrostatic mechanism, simulation results were compared to experimental data reported previously [11] for Si wafers implanted with B through screen oxide at 0.60 keV with a fluence of $2\times10^{15}$ ions/cm$^2$ at 0°tilt. The heating program was a conventional "spike anneal" described in Ref. [11], with heating rates varying from 75 to 350°C/s. We employed a model for boron diffusion developed previously [14]. The model utilizes continuum equations to describe the reaction and diffusion of boron, self-interstitial atoms, and related defects in silicon. These equations have the general form for species $i$:

$$\frac{\partial C_i}{\partial t} = -\frac{\partial J_i}{\partial x} + G_i \qquad [14]$$

where $C_i$, $J_i$, and $G_i$ denote the concentration, flux, and net generation rate of species $i$, respectively. The model was implemented using the process simulator FLOOPS (by Mark E. Law of the Univ. of Florida and Al Tasch of the Univ. of Texas at Austin) [15]. The values of activation energies were determined by methods drawn from systems engineering, including Maximum Likelihood estimation and Maximum *a Posteriori* estimation based on computational and experimental results from the literature [16,17]. The model has no adjustable parameters, and has demonstrated predictive, rather than merely correlative, capabilities.

Figure 6 shows simulated boron profiles in comparison with a typical experiment. The figure indicates that that incorporation of band bending in concert with a best-fit value of $S = 2\times10^{-5}$ for both $B_i$ and $Si_i$ matches the experimental profile quite well – the only significant difference being a small difference in the width of the profile in the high-concentration region of dopants near the surface. Matches of similar quality were obtained for other experimental data at different peak temperatures and ramp rates. Figure 6 also shows that exclusion of band bending effects greatly degrades the quality of the fit, mainly due to substantially reduced TED. A primary effect of band bending is therefore to increase junction depth. Band bending can also induce pileup of highly active dopants near the surface, as discussed elsewhere [5].

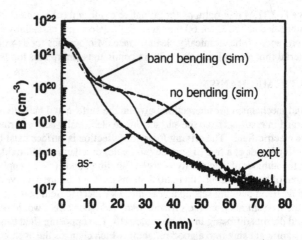

**Figure 6.** Simulation fits to a typical experimental TED profile. Combination of a surface sink with $S = 2\times10^{-5}$ and band bending yields good fit to experiment, mimicking an essentially perfect reflecting boundary condition for interstitials. The sink condition alone results in a much poorer fit with a shallower junction.

These electrostatic effects operate essentially in parallel with dangling bond exchange as described in earlier sections. In fact, specialized experiments were required to disaggregate the two phenomena from each other for independent examination. In ion implantation applications involving screen oxide, the electrostatic effects vary with time because the interfacial bond rupture that leads to band bending will disappear over several minutes at elevated temperatures [3]. Generally speaking, the effects of dangling bond exchange (when the surface or interface is active) tend to dominate those of electrostatic coupling, and offer more possibilities for beneficial effects on dopant diffusion and activation as well as EOR mitigation.

## CONCLUSION

Simulations and experiments suggest that surfaces and interfaces can affect the behavior of bulk defects by more than one mechanism. In particular, a chemically active, atomically clean surface with many dangling bonds preferentially removes silicon interstitials from the bulk. This phenomenon keeps the dopant atoms in an immobilized and electrically active form. The approach offers promise for simultaneous reduction in TED and increase in dopant activation, as well as reducing problems with EOR defects. Obtaining a relatively clean surface in the processing environment does not need ultrahigh vacuum. It can be simply achieved by cleaning the surface by a procedure that leaves the surface H-terminated (such as an HF-last etch). Most adsorbates such as hydrogen then desorb early during annealing, resulting in an atomically clean surface at the peak temperature where most diffusion and activation take place. Strong surface annihilation may help reduce the dopant deactivation in subsequent processing steps by dissociating most clusters. The beneficial effects of this method should be observed for any dopant atom and annealing scheme (e.g., spike, flash or submelt laser anneal). It should also work in the presence of strain (for SiGe) as well as co-implanted species such as C and F. Moreover, the effect of surface sink should be felt strongly in 3-dimensions (i.e., under the gate), though only 1-dimension profiles have been studied in this work.

## ACKNOWLEDGMENTS

This work was partially supported by A*STAR in Singapore through the National University of Singapore – University of Illinois (UIUC) research collaboration program. Characterization support from the Institute of Material Research and Engineering (IMRE), Singapore are gratefully acknowledged. Electron microscopy was performed at the UIUC Center for Microanalysis of Materials, which is partially supported by the U.S. Department of Energy under grant DEFG02-96-ER45439. Experiments with arsenic benefited from collaboration with Majeed Foad and Houda Graoui at Applied Materials. This work was also partially supported by NSF (DMR 07-04354) and the ACS PRF (43651-AC5).

## REFERENCES

1. E. G. Seebauer, K. Dev, M. Y. L. Jung, R. Vaidyanathan, C. T. M. Kwok, J. W. Ager, E. E. Haller, and R. D. Braatz, *Phys. Rev. Lett.*, **97**, 055053 (2006).
2. X. Zhang, M. Yu, C. T. M. Kwok, R. Vaidyanathan, R. D. Braatz, and E. G. Seebauer, *Phys. Rev. B* **74**, 235301 (2006).
3. K. Dev, M. Y. L. Jung, R. Gunawan, R. D. Braatz and E. G. Seebauer, *Phys. Rev. B*, **68**, 195311 (2003).
4. K. Dev and E. G. Seebauer, *Surface Sci.*, **550**, 185 (2004).
5. M. Y. L. Jung, R. Gunawan, R. D. Braatz and E. G. Seebauer, *J. Appl. Phys.*, **95**, 1134 (2004).

6.  H. -J.Gossman, C. S. Rafferty, F. C. Unterwald, T. Boone, T. K. Mogi, M. O. Thompsom, and H. S. Luftman, *Appl. Phys. Lett.* **67**, 1558 (1995).
7.  R. Duffy, V. C. Venezia, A. Heringa, T. W. T. Hu¨sken and M. J. P. Hopstaken, N. E. B. Cowern, P. B. Griffin, and C. C. Wang, *Appl. Phys. Lett.* **82**, 21 (2003).
8.  H. C. H. Wang, C. C. Wang, C. S. Chang, T. Wang, P. B. Griffin, and C. H. Diaz, *IEEE Electron Device Lett.* **22**, 65 (2001).
9.  P. Stolk, H. -J. Gossmann, D. Eaglesham, D. Jacobson, C. Rafferty, G. Gilmer, M. Jaraiz, J. Poate, H. Luftman, and T. Haynes, *J. Appl. Phys.* **81**, 6031 (1997).
10. N. Cowern, K. Janssen, and H. Jos, *J. Appl. Phys.* **68**, 6191 (1990).
11. M. Y. L. Jung, R. Gunawan, R. D. Braatz and E. G. Seebauer, *J. Electrochem. Soc.*, **150**, (2003) G838.
12. R. Vaidyanathan, H. Graoui, M. Foad and E. G. Seebauer, *Appl. Phys. Lett.* **89** (2006) 152114.
13. S. H. Yeong, M. P. Srinivasan, B. Colombeau and Lap Chan, Ramam Akkipeddi, C. T. M. Kwok, R. Vaidyanathan, and E. G. Seebauer, *Appl. Phys. Lett.*, **91** (2007) 102112.
14. R. Gunawan, M. Y. L. Jung, E. G. Seebauer and R. D. Braatz, *AIChE J.* 49, 2114 (2003).
15. See Mark Law, http://www.swamp.tec.ufl.edu/
16. M. Y. L. Jung, R. Gunawan, R. D. Braatz and E. G. Seebauer, *AIChE J.* 50, 3248 (2004).
17. R. Gunawan, M. Y. L. Jung, R. D. Braatz, and E. G. Seebauer, *J. Electrochem. Soc.* 150, G758 (2003).

Mater. Res. Soc. Symp. Proc. Vol. 1070 © 2008 Materials Research Society 1070-E01-08

# Experimental Investigation of the Impact of Implanted Phosphorus Dose and Anneal on Dopant Diffusion and Activation in Germanium

Vincent Mazzocchi, Stéphane Koffel, Cyrille Le Royer, Pascal Scheiblin, Jean-Paul Barnes, and Marco Hopstaken
CEA-LETI MINATEC, 17 rue des Martyrs, Grenoble, 38054, France

## ABSTRACT

In this work, we have investigated the influence of low energy ions implantation, variations of dose implant and low temperature anneal on diffusion, exo-diffusion and activation of phosphorus into germanium. Experimental results show that we achieved a high electrical activation level, around $5 \times 10^{19}$ at.cm$^{-3}$. We tuned both dose implant and annealing temperature in order to limit the exo-diffusion with practically no in-diffusion of the dopant. We also showed that very abrupt profiles can be achieved with appropriate implant and thermal anneal conditions. To limit the leakage current in Ge MOS devices [1], the defects generation during the implantation has to be limited. For annealing temperature below 550°C, we have observed by cross-sectional Transmission Electron Microscopy (TEM) that the defects were totally removed by addition of a pre-annealing step of 1 hour at 400°C.

## INTRODUCTION

Germanium has regained attention in the semiconductor industry for MOSFET application [2] because of its larger carrier mobility values – two times higher for electrons and four times for holes - as compared to silicon [3]. Two of the major issues with germanium are limiting the n-dopant diffusion [4] and increasing the electrical activation level. Because of its high chemical solubility limit, phosphorus seems to be the best choice for n-dopant in germanium [5]. However a large gap between the electrical activation and the equilibrium solid solubility of P in Ge has been observed. High level activation is still to be reached. Recent publications indicate that a high level of activation can not be obtained without the diffusion and exo-diffusion of the P into Ge [4]. In the present study, 2.5 µm germanium epitaxial layers on Si (001) [6] were subjected to several phosphorus implant conditions, and different activation anneals in order to determine the best conditions to obtain a high activation level without dopant diffusion. Furthermore cross-sectional TEM were performed to study conditions so as to obtain a crystalline germanium without any defects.

## EXPERIMENT

In this study we used 2.5µm non-intentionally doped (NID) Ge layer epitaxially grown by Reduced Pressure-Chemical Vapor Deposition (RPCVD) onto 200mm Si (001) wafers [6]. No special surface preparation or cleaning was performed prior to implantation. Wafers were implanted with phosphorus at 7° tilt and 27° twist to avoid channelling effects. Implantations were performed at 0°C with energies from 15keV to 40keV and doses from $8 \times 10^{13}$cm$^{-2}$ up to $1 \times 10^{15}$ cm$^{-2}$. Selected implanted wafers were capped with 5 nm thick SiO$_2$ deposited by Plasma Enhanced Chemical Vapor Deposition (PECVD) at a temperature of 380°C during 30s. Then wafers were annealed in nitrogen atmosphere using Rapid Thermal Processing (RTP), from

515°C to 600°C with holding times between 10s and 60s. Some of them were also annealed in Forming Gas atmosphere using hotplate processing from 280°C to 400°C with holding times from 5min to 1h. Then the $SiO_2$-capped wafers were stripped using (1%Hcl+0.2%HF) solution. Dopant chemical profiles were analysed by Secondary-Ion-Mass Spectroscopy (SIMS). The electrical activation was measured using sheet resistance measurement ($R_S$) with a four-point probe (4PP). We have checked that the 4PP method is efficient to achieve accurate resistivity measurement for our samples. The implantation-induced defects deep in the Ge layers were studied by cross-sectional TEM.

## DISCUSSION

We have investigated the influence of annealing temperature on the diffusion for different implant conditions as shown in SIMS (figure 1). We observe that diffusion, calculated as shown on figure 1(c), depends on implantation conditions.

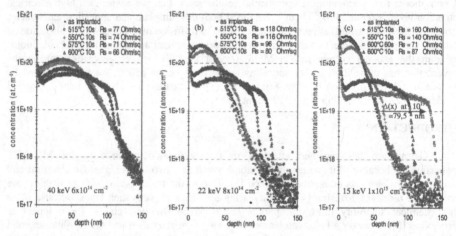

Figure 1. Example of SIMS profiles of implanted phosphorus in Ge for various activation annealing temperatures (a) 40 keV $6\times10^{14}$ cm$^{-2}$ (b) 22 keV $8\times10^{14}$ cm$^{-2}$ (c) 15 keV $1\times10^{15}$ cm$^{-2}$.

The first point to mention is the influence of the thermal budget on the in-diffusion and exo-diffusion of implanted P in Ge. As shown in figure 2, the increase of the thermal budget by increasing the time at constant temperature (a) or by increasing the temperature with the same anneal duration (b) leads to large phosphorus atoms diffusion (up to 50nm for 15keV $1\times10^{15}$ phosphorus implantation). The exo-diffusion, which is more sensitive to temperature (figure 2 c)) than to annealing duration (figure 2 a)), can be dissociated from the in-diffusion, which is driven by both the increase of temperature and annealing duration. In good agreement with previous studies [7], we observe in figure 2 b) that the implant dose has a strong impact on exo and in-diffusion, with an exponential dependence when increasing the dose value. We also observe that reducing the implant dose delays the diffusion for a given annealing temperature.

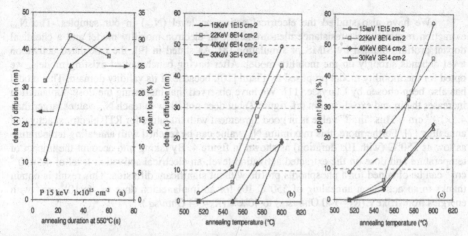

Figure 2. Impact of anneal duration and temperature on phosphorus in-diffusion and dopant loss: (a) In and exo-diffusion versus annealing duration at 550°C (for phosphorus 15 keV $1\times10^{15}$cm$^{-2}$ implantation in Ge); Impact of annealing temperature on in-diffusion (b) and exo-diffusion (c) with 10s annealing duration (for different implant conditions).

One can note in figure 3, that diffusion can be used in order to obtain a more abrupt dopant profile with box-shape profile. In this case, the activation anneal for a given implant dose has to be very precisely controlled to avoid over-diffusion. Indeed, a small variation of the temperature and/or of the annealing duration has an impact on the diffusion, as shown in figure 2(a)&(b). Therefore, adjusting the implant dose is mandatory in order to limit (or to control) the dopant diffusion.

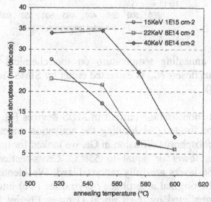

Figure 3. Annealing temperature influence on phosphorus profile abruptness (for various implant conditions). The abruptness extraction is based on the $10^{18}$ to $10^{19}$ cm$^{-3}$ decade slope.

37

We have also studied the electrical activation level ($N_{act}$) in our samples. The $N_{act}$ extraction requires sheet resistance measurements, an electron mobility model and a chemical dopant profile obtained by SIMS or simulation [8]. As stated in [9], the calculated activation level depends strongly on the mobility model. After having benchmarked existing models, we opted for the mobility model proposed by Fistul [10], because of its validity domain. This model has also been chosen by Clarysse [11]. We have observed that increasing the dose implantation increases the in and exo-diffusion (cf. figure 2) but does not allow to reach $N_{act}$ values larger than $4.5 \times 10^{19}$ cm$^{-3}$. This "limit" value is in good agreement with literature for RTP electrical activated annealing [7]. Furthermore, this maximum $N_{act}$ value can be reached with annealing temperature as low as 550°C (with 10s duration) as shown in figure 4. By taking into account the impact of temperature and dose on the extracted activation level, an electrical activation level of $4.5 \times 10^{19}$ cm$^{-3}$ can be obtained for a phosphorus profile without significant diffusion. This result is obtain thanks to an activation annealing of 550°C 10s for an implantation dose of $6 \times 10^{14}$ cm$^{-2}$ with energies from 15keV ($R_S$= 140 Ohm/sq.) to 40keV ($R_S$= 74 Ohm/sq.).

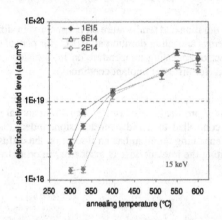

Figure 4. Impact of annealing temperature on the chemical activation level for 15keV implantation and various doses. $N_{act}$ is extracted either by SIMS (solid symbols) or simulated P-profile (empty symbols).

To limit the junction leakage current in MOS devices [1], defects generated during the implantation have to be limited. As limiting the activation annealing temperature could be a solution to reduce the phosphorus diffusion in Ge, we looked for implant defects in samples with annealing temperatures lower or equal to 550°C. Cross-sectional TEM images (figure 5) demonstrate that the Ge layers are fully recrystallized. We identify for some samples implant damages localized at the original amorphous / crystalline (a/c) interface (figure 5 a,b,c), whose depth is successfully predicted by the Critical Damage Energy Density (CDED) model [12] (these damages remain End Of Range (EOR) defects in Silicon). The observation of these defects is not in agreement with previous studies with comparable anneal conditions [7].

We have investigated the influence of several thermal budgets on theses defects and we observed that at 550°C 10s annealing, damages are removed as can be seen in figure 5(e). In contrast, a 515°C 10s annealing (figure 5(b)) requires a pre-annealing at 400°C during 1h to remove defects (figure 5(d)). A pre-annealing at 400°C for only 5min is not sufficient to totally eliminate these defects (figure 5(c)).

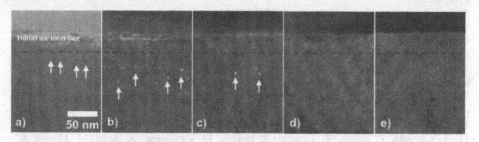

Figure 5. TEM micrographs of phosphorus implanted Ge layers with 15keV 1x10$^{15}$ at.cm$^{-2}$ then annealed at 385°C 30s (same scale for all pictures) : (a) bright field, no additional annealing (b) dark field, 515°C 10s (c) dark field 400°C 5min + 515°C 10s (d) dark field 400°C 1h + 515°C 10s (e) dark field 550°C 10s.

## CONCLUSIONS

In this work, we have investigated the doping of Ge samples with phosphorus using SIMS, sheet resistance and TEM analysis. We have highlighted the effect of implantation dose and activation annealing temperature on the phosphorus diffusion in Ge. We have observed that reducing the implanted dose delays the diffusion for a given annealing temperature. By tuning both the implanted dose and the activation annealing temperature, an electrical activation level of $4.5 \times 10^{19}$ cm$^{-3}$ can be achieved with almost no diffusion. Finally, by using TEM micrograph we have determined that for an activation annealing temperature lower than 550°C, a pre-annealing step at 400°C for 1h in necessary to remove all the defects observed at the initial location of the amorphous / crystalline interface.

## ACKNOWLEDGMENT

This work was supported by the French National Research Agency (ANR) through Carnot institute funding. The authors would like to thank their colleagues L. Hutin, A. Pouydebasque, C. Laviron for fruitful discussions, J.-M. Hartmann for the growth of Ge epitaxial layers, E. Deronzier for hot plate annealing.

## REFERENCES

1. A. Pouydebasque, C. Le Royer, C. Tabone, K. Romanjek, E. Augendre, L. Sanchez, J.-M. Hartmann, H. Grampeix, V. Mazzocchi, S. Soliveres, R. Truche, L. Clavelier, and S. Deleonibus, VLSI TSA 2008 (accepted).
2. H. Shang, MM. Frank, E. P. Gusev, J. O. Chu, S. W. Bedell, K. W. Guarini, M. Leong, IBM J. Res. & Dev. Vol. 50 N°4/5, (2006).
3. S.M. Sze, K.K. Ng, "Physics of Semiconductor devices - Third edition", ed. Wiley Interscience, p.30 (2007).

4. E Simoen, A. Satta, A. D'Amore, T. Janssens, T. Clarysse, K. Martens, B. De Jeager, A. Benedetti, I. Hflijk, B. Brijs, M. Meuris and W. Vandervorst, *MSSP* edited by Elsevier 2006, pp.634-639.

5. F. A. Trumbore, *Electrical Society*, (1959) 205-233.

6. J.M. Hartmann, J-F. Damlencourt, Y. Bogumilowicz, P. Holliger, G. Rolland, T. Billon, J. of Crystal Growth 274, 90-99 (2005).

7. A. Satta, T. Janssens, T. Clarysse, E. Simoen, M. Meuris, A. Benedetti, I. Hoflijk, B. De Jeager, C. Demeurisse and W. Vandervorst, J. Vac Sci Tech B, (2006) 24(1).

8. S. Koffel, P. Scheiblin, V. Mazzocchi, E-MRS spring 2008 (accepted)

9. L. Hutin, S. Koffel, C. Le Royer, P. Scheiblin, V. Mazzocchi, S. Deleonibus, EuroSOI Proc., p111-112 (2008).

10. V. I Fistul, M. I. Iglitsyn, E. M. Omelyanovskii, Soc. Phys. Solid State 4, 4 (1962).

11. T. Clarysse, P. Eyben, T. Janssens, I. Hoflijk, D. Vanhaeren, A. Satta, M. Meuris, W. Vandervorst, J. Bogdanowicz, G. Raskin, J. Vac. Sci. Tech. B. 24 (1), 381 (2006).

12. S. Koffel, A. Claverie, G. BenAssayag, P. Scheiblin, Mat. Sc. Semi. Proc. 9, 664-667 (2006).

Mater. Res. Soc. Symp. Proc. Vol. 1070 © 2008 Materials Research Society                1070-E01-10

## Micro-uniformity during laser anneal : metrology and physics

W. Vandervorst[1,2], E. Rosseel[1], R. Lin[3], D. H. Petersen[3,4], T. Clarysse[1], J. Goossens[1], P. F. Nielsen[3], and K. Churton[5]

[1]IMEC, Kapeldreef 75, Leuven, B3001, Belgium
[2]IKS, K.U.Leuven, Celestijnenlaan 200D, Leuven, B3001, Belgium
[3]Scion-DTU, CAPRES A/S, Building 373, Kgs. Lyngby, DK-2800, Denmark
[4]Dept. of Micro-and Nanotechnology, Technical University of Denmark, DTU Nanotech, Building 345 East, Kgs. Lyngby, DK-2800, Denmark
[5]Applied Materials, 974 East Arques Avenue, Sunnyvale, CA, 94085

## ABSTRACT

Maintaining or improving device performance while scaling semiconductor devices, necessitates the development of extremely shallow (< 20 nm) source/drain extensions with a very high dopant concentration and electrical activation level. Whereas solutions based on RTA with cocktail implants have been proposed in previous generations, sub-45 nm technologies will require even shallower junctions which motivates the research effort on milli-second anneal approaches as these hold the promise of minimal diffusion coupled with high activation levels [1]. Laser annealing is one of these concepts proposed to achieve the junction specifications and is typically described as a msec anneal process. Different from lamp based concepts which illuminate a full wafer simultaneously, the laser has an illuminated area which is much smaller than the wafer size thus necessitating a dedicated scanning pattern. In such a case one is potentially faced with areas subject to multiple overlaps and/or different temperatures and thus issues related to within wafer and within die uniformity need to be addressed.

In this work we use optimized metrology to probe such macro- and micro non-uniformity and determine the origin of the various components contributing to the observed non-uniformity patterns (laser stitching patterns, laser beam uniformity, optical path) and their impact on the local sheet resistance.

## INTRODUCTION

One of the major challenges in sub-45nm technologies is the formation of highly active (low resistivity) source and drain regions combined with a very well controlled overlap between junction and gate. As conventional ion implantation followed by rapid thermal annealing (RTA) results in excessive dopant diffusion and limited electrical activation levels, high temperature millisecond annealing is considered as an alternative approach to reach very high (metastable) dopant activation with minimal dopant diffusion. Whereas issues such as final sheet resistance, junction depth, lateral diffusion, defect evolution and junction leakage are of prime importance when assessing the prospects of laser annealing, one can not ignore more manufacturing related issues such as within wafer and within die uniformity as well. These issues are of concern with laser based annealing as one needs to apply a scanning pattern to anneal the entire wafer and thus overlap regions will be inevitable and may induce non-uniformities. Recent studies have indeed identified the impact of these micro-uniformities on the Vt-distribution of nearby devices [2]. Attempts to optimize throughput and minimize the stitching frequency by enlarging the laser beam spot, can only be successful if the power density distribution within the laser beam itself satisfies stringent requirements as otherwise micro-non uniformities may appear. The importance of the stitching pattern and the laser beam uniformity hinges on the sensitivity of the sheet resistance to the laser power and on its response to multiple laser illuminations. Both of them will be studied in this paper. Faced with the shallow junctions after laser anneal and the requirement for a high localized analysis, probing these patterns with the conventional four-point probe systems (FPP) is inappropriate as probe penetration and probe spacing (typically 0.5 mm) hamper the correct assessment of the (micro)-non uniformities in laser annealed junctions [3, 4]. Hence we have also optimized our methodology for the proper detection of these effects.

## EXPERIMENTS

In the present work, we used 300 mm n-type device wafers which received standard 0.5 or 3 keV B implants ($1 \times 10^{15}$ at/cm$^2$). Laser annealing was done in an Applied Materials DSA chamber [5, 6] which has a spot size of ~ 11 mm x 75 µm. The 808 nm high power laser beam is generated from stacked laser diode arrays and can be scanned across the wafer in the x-direction at a speed varying from 50-300 mm/sec. The standard stitching pattern (in the y-direction) uses a stepping distance of 1/3 of the laser beam (~ 3.65 mm) leading nominally to a triple illumination of any point on the wafer. In the present studies, overlapping stripes with no stepping distance were also made in order to study in more detail the effects of the inhomogeneity of the laser beam and the impact of multiple laser illuminations.

In order to get an accurate evaluation of the sheet resistance with high spatial resolution, conventional four-point probe measurements are inappropriate and one needs to rely on results obtained with the miniaturized four-point probe system (M4PP) manufactured by Capres [7]. This system consists of micromachined probes mounted on a force cantilever providing measurements at extremely low contact forces (~ $10^{-5}$ N, no probe penetration!) whereby the probe pitch can be as small as 1 µm. The impact of the reduced probe pitch has already been discussed in detail in ref [4] and the results illustrate that the apparent uniformity changes from ~ 7-8% (when measured with a small probe pitch) to ~ 1-2 % (when measured with a conventional four point probe system operating at 0.5 mm spacing). Typical results [4] from an anneal of the 0.5 keV B-implant with stitching, are shown in Fig. 1a and indicate three kind of variations ; in the y-direction: 1) relative large Rs changes (~ 7-10 %) with a periodicity of 3.65 mm and  2) finer Rs oscillations (~ 2-3%) in between the major variations (apparent periodicity ~ 750 μm) ; in the x-direction even finer (but clearly visible) oscillations (periodicity to be discussed further). In order to unravel the small variations related to item 2 (y-direction), an M4PP-analysis across the laser scan direction was performed on stripes annealed without any stitching. The results represented in Fig. 1b show these oscillations within regions which have been illuminated 1x-7x.

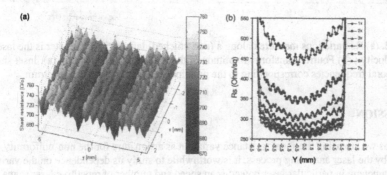

**Figure 1.** (a) 45 x 101 point area scan measured with a 10 μm pitch M4PP. Sample: 0.5 keV B, $1x10^{15}$ at/cm² [4]. (b) Rs variation measured across a stripe annealed without stitching. The data are taken perpendicular to the scan direction. Multiple laser scans (1x-7x) are performed without displacing the laser beam in the y-direction.

Obviously there is a global variation (10-20%) across the laser beam spot with additional finer scale oscillations (~2%, ~ 750 μm periodicity) which remain present regardless of the number of overlapping laser scans. As these measurements are taken on stripes without stitching, the Rs-variations observed must be correlated to inhomogeneities within the laser spot. Similarly we measure Rs variations in the x-direction (i.e. along the laser scan direction) on annealed stripes without stitching (Fig. 2a). The minima/maxima observed in Fig. 1b remain present along the laser stripe but their absolute values seem to fluctuate slightly (~ 1%). By varying the laser scan velocity and performing a Fourier transform of the Rs(x) traces (Fig. 2b), it becomes clear that the spatial distribution of these oscillations is directly linked to the laser velocity and thus to temporal variations in (global) laser power. Multiplying the frequency of the principal peak in the Fourier spectrum with the scan velocity shows that all curves yield a ~ 300 Hz oscillation in laser power, independent of the velocity.

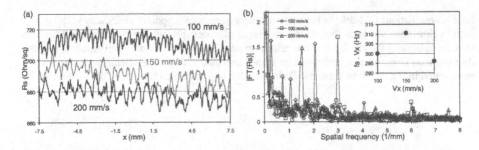

**Figure 2.** (a) Rs variations measured along a (non-stitched) laser strip. Parameter is the laser beam velocity ; (b) Fourier transform amplitude of the Rs(x) profiles shown in (a). Inset shows the temporal frequencies corresponding to the principal peaks in the Fourier spectrum.

## DISCUSSION

As we are using the sheet resistance variation as a signature for the non-uniformity induced by the laser annealing process, it is worthwhile to study its dependence on the various laser parameters, in particular laser power, scan speed and number of repetitive laser scans. The power and scan speed dependence has been measured for a 0.5 and a 3 keV B-implant and is shown in Fig. 3. The observed linear change of the sheet resistance with either the laser power (Fig. 3a, at a fixed scan velocity, 150 mm/sec) or the scan speed (at fixed laser power, 82.3 %, Fig. 3b) indicates that their effect can be correlated back to the same phenomena i.e. the actual maximum temperature experienced by the wafer. A higher laser power will inevitably lead to a higher anneal temperature and thus a higher dopant activation. Similarly it is clear that a change in scan velocity impacts on the time integrated power density seen by each point on the wafer which again translates into a different maximum temperature. The actual wafer temperature for all these conditions can be determined independently by the integrated pyrometer [8] which is calibrated by the onset of Si melt at 1412°C. By doing so, the Rs data for a wide velocity range can be combined into one Rs-Temperature correlation plot (Fig. 3c). Obviously a higher anneal temperature leads directly to a higher activation of the implanted boron and thus a reduced Rs-value. Fig. 3c is very interesting as it indicates that when studying Rs uniformity distributions, one can translate these back into local temperature variations due to non-uniform power densities within the laser beam spot irrespective of the scan velocity. In that respect it is important to note that a 0.5 keV B-implant shows a much stronger dependence on the laser power density than a deeper) 3 keV B-implant. Hence as a tool for probing variations in laser power density or anneal temperature, the shallower B-profile is the most sensitive one and to be preferred. Different from a lamp based anneal where the sample is exposed only once to the high temperature step, the scanning concept of a laser implies that each area is exposed to multiple laser illuminations.

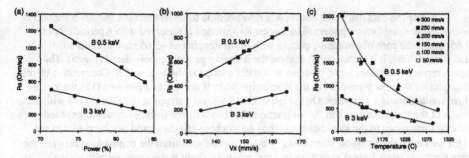

**Figure 3.** (a) Rs versus laser power (Vx = 150 mm/sec) ; (b) Rs versus laser scan velocity (power = 82.3 %) ; (c) Rs versus pyrometer temperature for a wide range of scan speeds.

As the total thermal budget will increase with the number of exposures, it is fair to expect that the dopant activation and diffusion (and thus the sheet resistance) will reflect the increasing thermal budget with increasing number of scans. In order to determine the impact of multiple exposures, Rs was determined from stripes which had been exposed to 1x to 7x non-stitching laser scans. For all stripes, Rs was taken as the average of the two central symmetric minima in the Rs(y) profiles of Fig. 1b. The results in Fig. 4a were taken for two different laser powers

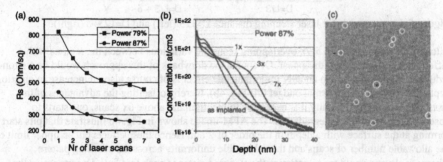

**Figure 4.** (a) Rs variation as a function of overlapping laser scans. Parameter is the laser power (79%, 87%) corresponding to respective maximum temperatures of 1210 °C and 1300 °C ; (b) Selected SIMS profiles for 87% power ; (c) AFM image of 7x annealed (1300 °C) surface.

but basically reflect a very similar behavior. With increasing laser scans, Rs decreases initially quite rapidly, a decline which starts to saturate at ~ 5-7 scans. The decrease of Rs is primarily the result of the dissolution of B-clusters and the subsequent outdiffusion of the B-atoms, as evidenced in the SIMS profiles in Fig. 4b. It is also clear from Fig. 4a that the temperature plays a decisive role in the Rs value as the curve for the lower laser power is always higher than the one for the higher power. In fact even after multiple scans the Rs remains higher than the initial Rs values after one scan at the higher temperature.

Using the data discussed above, it is now possible to understand the various components within the non-uniformity pattern. The largest Rs variation is observed with a periodicity of ~ 3.65 mm in the case of a stitching pattern with beam stepping of ~3.65 mm. In order to understand these, it is necessary to analyse the scanning pattern in more detail (Fig. 5). The boxes represent the laser beam position as it starts a stripe in the x-direction. Consecutive boxes then picture the new y-position of the laser stripe when it steps over a distance D in the y-direction and starts a new scan. On the left in Fig. 5 we use a stepping distance D=L/3 with L the width of the laser beam spot in the y-direction. As shown by the lines, each area is exposed exactly three times to the laser beam and thus no stitching effects should be observed. On the right we sketch the situation when D=L/3+ $\delta$. As can be seen from the comparison between the new positions (dark boxes) versus the original positions (light boxes) regions with varying number of overlaps appear, in this particular case 2 or 3 times respectively. Based on the results of Fig. 4a it is then clear that the Rs values in those regions will be different leading to the observed stitching patterns. In principle one could argue based on Fig. 4a, that with an increased

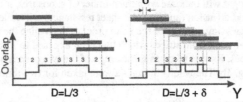

**Figure 5.** Laser overlap for laser stepping distance D=L/3 (left) and D=L/3+$\delta$ (right).

stitching density leading to 5-7x overlapping scans the variations between the regions with different exposures should disappear. Of course the drawback of this approach would be a strong reduction in throughput. An equally important side effect is of course also an increase in junction depth due to the higher thermal budget (cfr Fig. 4b), thereby offsetting the advantages of the laser anneal approach. Finally it is necessary to mention that above 6x scans, one starts to observe a serious surface degradation. The AFM-image shown in Fig. 4c illustrate that pits start forming at the surface with a depth in the order of 2 nm. Hence these effects put an upper limit to the allowable number of scans and the achievable uniformity across the stitching pattern. Alternatively one could try to optimize the stepping distance such that $\delta$≈0. Unfortunately the latter solution would only be valid if the power density distribution within the laser beam would be perfectly homogenous. If not, stitching patterns would still occur. A non-uniformity in power density implies that the local temperature will be different and this will translate directly in different activation levels and outdiffusion, cfr the different Rs curves in Fig. 4a. In fact the distributions reported in Fig. 1b are a direct result of such a power density distribution. If we take the result for one scan and use the correlation between Rs and temperature from Fig. 3c, it is possible to estimate the temperature (and thus the power density variation) variation across the laser spot. The results, shown in Fig. 6, indicate a global temperature variation of 20°C across the full laser beam spot with, in this particular case, additional smaller oscillations in the order of 5°C. Such a variation corresponds to an apparent power density variation of respectively 5 % (20°C) and 1 % (5°C). The global power density variation or "smile" is often observed in laser diode arrays and is related to the fabrication process itself [9]. The smaller oscillations are intimately linked to the details of the optical system and can be removed by improving the laser design. The Rs(y) profile taken from an anneal with an improved laser (laser #2 in Fig. 6b), clearly shows that the small oscillations can be removed although the smile remains. The variations observed in the x-direction (300 Hz) are clearly also a signature of some small

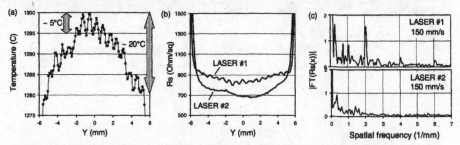

**Figure 6.** (a) Estimated temperature distribution across the laser beam spot ; (b) Rs variation across the laser beam spot for two different laser designs ; (c) Amplitude spectrum of the Rs(x) Fourier Transform for both lasers. The 300 Hz peak (~2/mm) is absent for laser #2.

oscillations in power density as the laser moves across the wafer surface. With the improved laser design, the 300 Hz signature can be removed as well (laser #2 in Fig.6. c) demonstrating how micro-uniformity is closely linked to the details of the laser system and its implementation. It also emphasizes the importance of micron scale resolution for the Rs metrology in both x and y directions.

## CONCLUSIONS

The use of laser annealing may lead to advantages in terms of junction depth and activation level. Using advanced four-point probe concepts it is possible to observe characteristics features in the Rs-maps which can be correlated with the impact of the stitching pattern (lower Rs due to multiple overlaps) as well as details of the laser design and the optical path (lower/higher Rs due to higher/lower power density/temperature). The results of this paper show that a careful optimization of the laser design and the scan pattern is required in order to limit the micro-uniformities to less than a few %. Whereas the laser stability and optical path problems can be improved, it appears improbable that the stitching pattern can be removed completely.

## REFERENCES

[1] P. Timans et al., Mat. Res. Soc. Symp. **912**, (2006) C1.1.
[2] C. Ortolland et al., VLSI'08
[3] T. Clarysse et al., Mat. Res. Soc. Symp. **912**, (2006) 197.
[4] D. H. Petersen et al., JVST **B 26(1)**, (2008) 362.
[5] D. Jennings et al., Proc. IEEE RTP 2004, 47.
[6] A. Hunter et al., Proc. IEEE RTP 2007, 13.
[7] C. L. Petersen et al., Sens. Actuators **A 96**, (2002) 53.
[8] B. Adams et al., Proc. IEEE RTP 2005, 105.
[9] N.U. Wetter, Optics and Laser Technology **33**, (2001) 181.

Mater. Res. Soc. Symp. Proc. Vol. 1070 © 2008 Materials Research Society

# Scanning Spreading Resistance Microscopy for 3D-Carrier Profiling in FinFET-Based Structures

Jay Mody[1,2], Pierre Eyben[2], Wouter Polspoel[1,2], Malgorzata Jurczak[2], and Wilfried Vandervorst[1,2]

[1]Electrical Engineering Department, INSYS, Katholieke Universiteit Leuven, Kasteelpark Arenberg 10, Leuven, B-3001, Belgium

[2]IMEC vzw, Kapeldreef 75, Leuven, B-3001, Belgium

## ABSTRACT

Junction formation in FinFET-based 3D-devices is a challenging problem as one targets a complete conformal doping of the source/drain regions in order to produce equal gate-profile overlaps (and thus transistor behavior) on all sides of the fins. Due to the lack of predictive modeling for several of the doping strategies explored (plasma immersion, cluster implants, vapor phase deposition, etc...) it becomes difficult to correctly predict the performance of the devices and hence, accurate 3D-doping profile determination is desired. Although several dopant/carrier profiling methods exist with excellent one- or two-dimensional resolution and properties, there is an urgent need to extend these towards a quantitative three-dimensional geometry. In this work, we use scanning spreading resistance microscopy (SSRM) with dedicated FinFET test structure to obtain three-dimensional information from successive two-dimensional scanning spreading resistance maps. We also assess the validity of our methodology by comparing various sections along the fins which represent the variability due to the processing and measurement procedure.

## INTRODUCTION

Due to the excellent scaling properties of a FinFET over its planar counterpart, these 3D-devices are being considered as promising candidates for the 32nm technology node and beyond. The performances of FinFET-based devices are determined by the size of the fins and as with any transistor, the distribution of electrically active dopants (carriers) in the source/drain regions and the gate-overlap [1-3]. Given the small size of the fin (typically 15nm x 65nm), the metrology requirements are somewhat different as compared to planar devices. Indeed since the source/drain regions will in general no longer have a junction depth but will be (ideally) homogenously doped, the most relevant information is the lateral dopant diffusion under the 3D-gate material. The processes governing this lateral extension (on all sides of the fin) are inherently linked with the technology used to dope the sidewall (implant vs. plasma doping vs. vapor phase doping, etc...) and the subsequent thermal treatments. Unfortunately, these processes are not necessarily isotropical and thus simple one-dimensional experiments to mimic the doping process in the fin are of very little value. As a consequence, there is an urgent need for a quantitative 3D-carrier concentration mapping technique which can particularly probe in FinFET-based devices with sub-nm resolution.

Presently there is no accepted technique that can quantitatively map 3D-carrier distribution and only the Tomographic Atomprobe [4] contains the potential to probe the 3D-dopant distribution. As all dopants are not necessarily active (in particular when very short annealing approaches are used), being able to probe the 3D-carrier distribution is of utmost importance. Scanning Spreading Resistance Microscopy (SSRM) in the recent past has shown the capability of quantitatively mapping 2D-carrier distributions with sub-nm spatial resolution and less than 2nm/decade gradient resolution with good dopant sensitivity. More recently, it was also shown that when performed in high-vacuum improved reproducibility and repeatability can be obtained [5-8].

In this work, we propose to use the SSRM-methodology for obtaining a 3D-carrier concentration map for FinFET-based devices by using a dedicated test structure concept and the first results presented in this paper successfully demonstrate the extraction of 3D-carrier distribution with SSRM.

## METHODOLOGY

When targeting a 3D-methodology it is instructive to realize that the problem might be simplified if one would be able to obtain a 2D-carrier concentration map on successive planes from one of the three planes (A-A, B-B' and C-C' planes) shown in Figure 1. Indeed collecting for instance the 2D-distributions on successive A-A' planes, as the later moves down the structure, represents in essence the acquisition of the complete 3D-distribution. This can be viewed as collecting a 2D-map on one plane then removing a small amount of material in a plane parallel way and repeating the data collection again. Upon completion of the data collection on all planes, one only needs to combine the information from the various planes to reconstruct the 3D-distribution. Of course the practical problem which arises now is how to have to move such a plane through the small structure, ideally with incremental steps of 1nm. It is clear that due to the limited dimensions (typically 15nm x 65nm) of the FinFET-based structures it becomes extremely difficult to obtain multiple cross-sections with nanometer resolution along the two planes (A-A', B-B' and C-C') of the device either by cleaving or polishing.

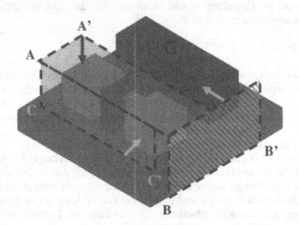

**Figure 1: Planes to obtain carrier concentration maps in FinFET-based structures.**

To overcome these limitations a different methodology was needed to obtain carrier concentration maps in FinFET-based structures. As it is virtually impossible to produce successive cross-sections on the same device, we intend to make cross-sections on multiple devices whereby we intend to place that cross-section at each device a little deeper (along one for the intended direction). By assuming then that the processing and thus 3D-profiles for nearby fins are identical one can combine the results from the different fins into one 3D-profile.

To facilitate this concept we have (as first demonstration attempt) designed a test pattern based on staggered FinFET's. As shown in Figure 2, this is composed of multiple fins whereby the gate in each set of fins is shifted laterally by 50nm. By cleaving through this staggered set of fins we actually produce a series of cross-sections with incremental steps towards the gate region. In this feasibility study (see Figure 2), there are eight sets of 10 fins and with each set of the fins the gate shifts laterally by 50nm to the right. Thus, when cleaved and measured with SSRM on fins in each set we would obtain 2D-carrier concentration maps of fins which are 50nm stepped in lateral depth. Once these 2D-carrier concentration maps are obtained they can be can corrected for drift, aligned and reconstructed to form a 3D-carrier concentration map along the lateral depth of the fin using MATLAB©.

## STRUCTURES & INSTRUMENTATION

In order to truly assess the validity of the methodology and to obtain first results we fabricate the bulk fins without any gate stack. In practice this means that we expect to find an identical profile on each successive cross-section. Their agreement can be viewed as an assessment of the combined variability in processing and measurement procedure.

The bulk fins were fabricated on a *p*-type substrate with 100nm in height and width of 30nm. These fins were doped with PH3 plasma and annealed (RTA) at 1050°C. The lengths of the fins were kept around 9.5mm for ease in cross-sectioning of the devices and manual placing of the back-contact to the fins. The back-contacts were made manually by scribing the structures and filling it with Ga-In eutectic and covering it with silver paint.

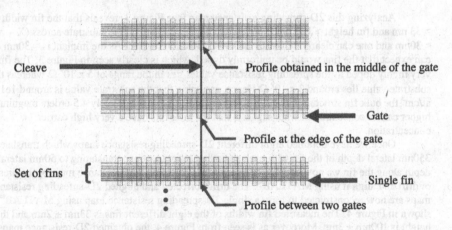

Cleave — Profile obtained in the middle of the gate

Gate

Profile at the edge of the gate

Set of fins { — Single fin

Profile between two gates

**Figure 2: Design of staggered fins.**

For all SSRM measurements, we used a Veeco EnviroScope atomic force microscope (AFM) equipped with an SSRM module. Measurements were realized in vacuum (around $10^{-5}$

torr). The voltage applied between the tip and the back-contact was -50mV in order to obtain a high resolution while maintaining a good signal to noise ratio.

## RESULTS & DISCUSSION

Eight different 2D-spreading resistance maps of the bulk fins doped with PH3 plasma were obtained from 8 different staggered bulk fin sets shown in Figure 2. A detail of one is shown in Figure 3.

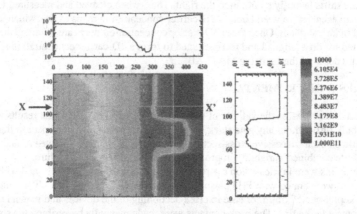

**Figure 3: 2D-spreading resistance map of PH3 plasma doped bulk fin.**

Analyzing this 2D-spreading resistance map (see Figure 3) reveals that the fin width was ≈ 33 nm and fin height ≈ 100 nm. The implant depth analyzed on the substrate across (X – X') is ≈ 30nm and one can clearly delineate the $n++/p$ junction formed. As the implant is ≈ 30nm one can expect that the fins would be uniformly doped, which is clearly seen in Figure 3. The fins are very highly doped as the spreading resistance values lies in the range of 5 x $10^4$ Ω whereas the substrate value lies around 3 x $10^9$ Ω Hence, assuming that the substrate value is around 1e15 at/cm$^3$ the bulk fin structure has a carrier concentration of approximately 4.5 orders magnitude higher carrier concentration what suggests that the bulk fins have a very high carrier concentration.

Once we have obtained eight different 2D-spreading resistance maps which translate to 350nm lateral depth of the fin with each spreading resistance map translating to 50nm lateral depth along the fin we now correct each individual 2D-spreading resistance map for the presence of drift and align it using MATLAB©. The drift-corrected and aligned 2D-spreading resistance maps are now reconstructed to form a single 3D-spreading resistance map using MATLAB© as shown in Figure 4. The measured fin widths of the eight different fins is 33nm ± 2nm and the fin height is 100nm ± 3nm. Moreover as is seen from Figure 4, the obtained 2D-resistance maps are quite reproducible as is seen in Figure 4. Thus, as shown above in Figure 4, we can now translate multiple 2D-spreading resistance maps to a 3D-spreading resistance maps over a volume of a bulk-fin device.

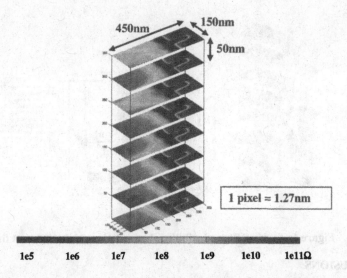

450nm — 150nm

50nm

1 pixel = 1.27nm

1e5    1e6    1e7    1e8    1e9    1e10    1e11Ω

**Figure 4: Reconstructed 3D-spreading resistance map of bulk-fin device.**

The assessment and viability of this methodology can be done by comparing cross-sections along the fins. As shown in Figure 5, we take three sections along a single fin. The standard deviation of spreading resistance values along the fin is approximately 20%. We believe the high value of standard deviation results from the extremely low bias (-50mV) used to maintain good resolution. The average reproducibility of the spreading resistance values were extracted by taking sections along eight different fins used to construct the 3D-spreading resistance map (see Figure 6). The average reproducibility values were obtained around 18%. It has to be noted that all the SSRM measurements were obtained using a single full diamond tip thus, eliminating the tip variability factor.

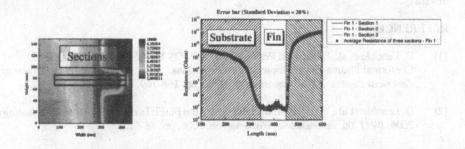

**Figure 5: Standard deviation along single Fin**

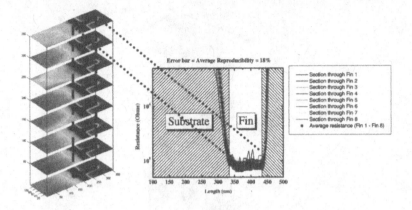

**Figure 6: Average Reproducibility obtained along eight different fins**

## CONCLUSIONS

A methodology based on SSRM and dedicated test pattern is presented which should enable to obtain a full 3D-carrier profile. In this paper we demonstrate the validity of the method by comparing consecutive sections along a single fin where we obtain a standard deviation of approximately 20% and by comparing sections along eight different fins used for 3D-spreading resistance map reconstruction with approximately 18% average reproducibility. Thus, as a proof of concept our methodology to obtain 3D-carrier concentration map is promising for FinFET-based devices.

## ACKNOWLEDGMENTS

The authors would like to thank C. Demeulemeester for the fabrication of full-diamond tips for this study.

## REFERENCES

[1]    D. Lenoble et al., "Enhanced Performance of PMOS MUGFET via Integration of Conformal Plasma-Doped Source/Drain Extensions," *VLSI Technology, 2006. Digest of Technical Papers. 2006 Symposium on*, 2006, pp. 168-169.

[2]    D. Lenoble et al., "The junction challenges in the FinFETs device," *Junction Technology, 2006. IWJT '06. International Workshop on*, 2006, pp. 78-83.

[3]    A. Burenkov and J. Lorenz, "3D simulation of process effects limiting FinFET performance and scalability," *Simulation of Semiconductor Processes and Devices*, München, Germany: 2004, pp. 125-128.

[4]     T.F. Kelly and M.K. Miller, "Invited Review Article: Atom probe tomography," *Review of Scientific Instruments*, vol. 78, Mar. 2007, pp. 031101-20.

[5]     P. Eyben et al., "Probing Semiconductor Technology and Devices with Scanning Spreading Resistance Microscopy," *Scanning Probe Microscopy*, Springer New York, 2007, pp. 31-87.

[6]     P. Eyben et al., "Impact of the environmental conditions on the electrical characteristics of scanning spreading resistance microscopy," *Journal of Vacuum Science & Technology B: Microelectronics and Nanometer Structures*, vol. 26, Jan. 2008, pp. 338-341.

[7]     L. Zhang et al., "High-resolution characterization of ultrashallow junctions by measuring in vacuum with scanning spreading resistance microscopy," *Applied Physics Letters*, vol. 90, May. 2007, pp. 192103-3.

[8]     L. Zhang et al., "Reproducible and High-Resolution Analysis on Ultra-Shallow-Junction CMOSFETs by Scanning Spreading Resistance Microscopy," *Junction Technology, 2006. IWJT '06. International Workshop on*, 2006, pp. 108-111;

Mater. Res. Soc. Symp. Proc. Vol. 1070 © 2008 Materials Research Society    1070-E01-12

## USJ Dopant Bleaching and Device Effects in Advanced Microelectronic Plasma Enhanced Resist Strip Processing

Frank Wirbeleit[1], Volker Grimm[1], Christian Krüger[1], Christoph Streck[1], Roger Sonnemans[2], and Ivan Berry[2]

[1]Technology Department, Advanced Micro Devices (AMD), Wilschdorfer Landstrasse 101, Dresden, 01109, Germany

[2]Axcelis Technologies, Inc., Beverly, MA, 01915

## Abstract

The impact of low temperature plasma resist strip on doped silicon surfaces and microelectronic device performance is investigated using different chemical gas mixtures. In this investigation, different plasma treatments where applied on non-structured and structured silicon on insulator (SOI) wafers, post ultra shallow surface implants. The dopant bleaching and oxide loss effects in conjunction with plasma surface treatments were analyzed by time of flight secondary ion mass spectrometry (TOF-SIMS) and electrical measurements of microelectronic test devices. As the result, a long range plasma radiation induced dopant activation and diffusion is separated from the effects from surface oxide loss and re-oxidation processes. The data suggests the necessity for optimization of plasma resist strip processes for device improvements.

## Introduction

When polymeric photoresist materials are subject to ion bombardment, volatile compounds and gases are released[1,2]. At doses greater than about $1 \times 10^{15}$ ions/cm$^2$, the surface of the resist has been transformed into a carbonaceous material to the depth of about twice the ion range[3,4]. Removal of this carbonaceous, or "crust" region is difficult and is typically requires a plasma process, because of its ability to efficiently remove the crust material, followed by an acid SPM/APM clean to remove residue and particulates[4,5,6,7,8,9,10,11]. Scaling devices down in the sub 100 nm range requires ultra shallow implants, which makes devices very sensitive to surface effects.

It is well known that plasma processes can impact the junction in multiple ways: (1) plasma processes oxidize the surface or etch the surface oxide causing junction loss,[12,13] (2) plasma reactants can bleach out near surface dopants, and (3) plasma can change the chemical bonding of the junction surface possibly leading to vacancy enhanced diffusion[14]. As junctions depths scale below 150nm, effects of the resist strip and wet clean on junction performance can be significant, especially for heavily amorphized junctions where the enhanced chemical reactivity accelerates the effects[11,15,16,17].

Multiple resist strips are applied in microelectronic technology during the integrated circuit manufacturing process, clean and surface treatment processes gain more and more attention therefore. The selected plasma resist processes for this work are commonly used in advanced microelectronic technology. All of these processes are proofed in production for complete resist removal at low defectivity level in combination with a post wet clean. The investigation of surface oxidation and dopant loss effects in advanced resist strip processing is the analysis of a kind of second order effects, which becomes more and more important in leading edge high performance device tuning for upcoming small pitch structures in microelectronic technology.

## .Experiments

For this investigation a down-stream microwave plasma ash tool system is used. In the first experimental run non-structured silicon wafers with a 1200 A PECVD polycrystalline silicon film implanted with either $BF_2$ - 3keV, $5x10^{14}$ cm² or arsenic - 3keV and $5x10^{14}$ cm² and are then treated with a remote plasma at 240°C under either an oxygen rich (90% $O_2$ + 10% forming gas), a forming gas rich (97% forming gas + 3% $O_2$) or an ammonia atmosphere ($NH_3$). Before treatment all samples were cleaned by an SPM/APM followed by an RCA clean. For reference an untreated film and one treated by a wet RCA clean only is used. From these samples TOF-SIMS spectra are taken. Figure 1 shows the resulting spectra for SiO, B, As and C. As can be seen from figure 1a on the reference samples, a native oxide of approx 12Å was found which is about the same as the ammonia treated sample. Depending on the amount of oxygen the samples show 30Å or 40Å surface oxide for forming gas and oxygen rich plasma atmosphere respectively.

In a follow up experiment non-structured silicon wafers are implanted with $BF_2$, 3keV, $5x10^{14}$ cm² and arsenic, 3keV and $5x10^{14}$ cm². These wafers are processed together with full integrated wafers for which one additional plasma resist strip process step is added after the shallow transistor extension implant. The $BF_2$ implanted samples were annealed for 10s at 1025°C, 1050°C and 1075°C in nitrogen, while, the arsenic implanted wafers were annealed for 10s at 1000°C, 1025°C and 1050°C. For each of these implant and anneal conditions, four different, additional, plasma treatments were used: pure forming gas plasma (FG), forming gas rich plasma with 3% oxygen (FG/$O_2$), oxygen rich atmosphere with 10% FG ($O_2$/FG) and oxygen rich atmosphere with addition of 1% $CF_4$ ($O_2$/FG/$CF_4$). The non-structured wafers were measured by four probe point, while the full integrated device wafers were electrically tested after metalization. Figure 2 shows the obtained dependencies of the sheet resistance versus the anneal temperature for the $BF_2$ and arsenic implanted samples.

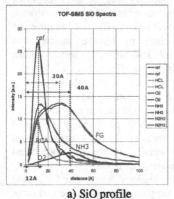

a) SiO profile

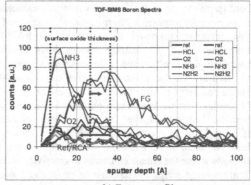

b) Boron profile

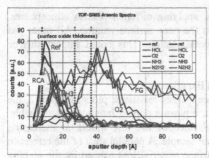
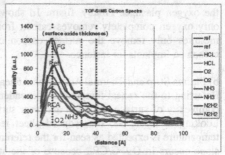

c) Arsenic profile         d) Carbon profile

**Figure 1:** TOF-SIMS spectra for SiO, boron, arsenic and carbon after different plasma surface treatments.    Dotted lines indicate approximate location of surface oxide/silicon interface, for each plasma condition.

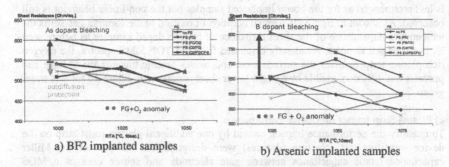

a) BF2 implanted samples       b) Arsenic implanted samples

**Figure 2:** Four probe sheet resistance measurement after plasma surface treatments in different gas atmosphere under different thermal anneal temperatures. The splits are: noPS – no plasma strip, PS(FG) – forming gas only, PS(FG/O2) – 97% forming gas + 3% O2, PS(O2/FG) – 90%O2 + 10% forming gas, PS(O2/FG/CF4) – 90%O2 + 9%forming gas + 1% CF4

## Results
### a) TOF Sims Spectra Implant Investigation
For all samples treated in oxygen rich plasma atmosphere a surface plasma oxide three to four times thicker than the native oxide is found according to figure 1a. This increased oxidation causes poly silicon material loss and it protects the surface against native oxide growth afterwards. Additionally, this plasma oxide can cap the silicon and reduce dopant out diffusion during anneal. For the boron implanted samples, the oxygen rich plasma atmosphere bleaches out boron because of the known segregation effect[18] (see. figure 1b). There appears to be enhanced diffusion of boron in the case of a forming gas plasma (N2H2), visible by a smoothed diffusion profile of boron in TOF-SIMS spectra in figure 1b. This effect is also seen for forming gas the arsenic TOF-SIMS profile given in figure 1c. Applying an oxygen rich plasma results in a shifted arsenic peak towards the volume by a well known pile-up mechanism. No significant plasma enhanced diffusion is seen in

the oxygen plasmas. Finally, figure 1d shows, applying different plasma atmospheres, only oxygen completely removes carbon from the near surface volume. All other gas mixtures only remove the surface carbon contamination. This could impact surface carbon co-implants for instance.

b) Plasma Strip and Thermal Activation Anneal Experiment
Referring to the anneal temperature dependence - sheet resistance plots in figure 2, the complexity of surface oxidation and near surface plasma doping interaction are visible. Looking at figure 2a, the reduced sheet resistance of the test wafers at increasing temperature is expected as seen for the reference sample. A parallel shift of the reference slope occurs in $O_2$/FG/$CF_4$, because of the surface material loss caused by the etching $CF_4$ gas component. However, if forming gas is part of the plasma gas atmosphere the sheet resistance dependence shows a non linear behavior, indicating that at a critical post thermal anneal temperature, the hydrogen component in the surface volume enhances dopant activation and/or dopant diffusion. This could explain, why the sheet resistance is lower for the forming gas related plasma strips as compared to the non-forming gas splits. The impact of forming gas plasma atmosphere on arsenic doped samples sheet resistance is less pronounced as for the boron implanted samples but the non-linear behavior is still observed. As shown in figure 2b, a small addition of oxygen to the forming gas shifts the temperature dependence of the sheet resistance of arsenic doped samples as before seen on boron doped samples. As already discussed from the TOF-SIMS spectra, the oxygen rich plasma strip oxidizes the surface most and moves arsenic towards the volume and prevents out diffusion, visible by a slight lower sheet resistance in figure 2b for this wafer group.

c) Plasma Strip Impact on Device Performance
To quantify the performance impact, caused by one additional plasma resist strip on the device, sensitive test structure (devices) were designed and the plot of the Miller capacitance (static capacitance between gate electrode and source area of a MOS transistor) versus the extension sheet rho of the MOS test transistor is used. Figure 3a shows this dependence for an NMOS transistor. Comparing all split groups in this plot, the impact on device is minor, except for the $O_2$/FG/$CF_4$ related split. All other plasma gas atmospheres causes a slight shift in extension sheet rho and do not affect Miller capacitance significantly. However, on figure 3b there is a clear impact of the applied additional plasma resist strip on the device performance plot for PMOS device. All plasma strip related data points in this plot show an increased scattering and a clear shift, twice as much as on n-channel MOS transistor. The $CF_4$ experiment results for the splits in figure 3b is not shown, because this dependence is out of scale.

60

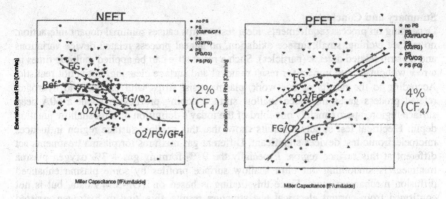

**Figure 3:** Silicon sheet resistance vs. electrical capacitance on device test structure for different plasma treatments

According to the results of the performance plots in figure 3 the $O_2$/FG/$CF_4$ split shows the most impact on sheet resistance of both NMOS and PMOS device test structures. However, test structures are designed very sensitive for a certain purpose. Since these test structures were designed to magnify the strip effects, actual device performance will be affected to a lesser extent. Therefore, the performance of a more representative test structure ring oscillator is plotted in figure 4 also. Best performing ring oscillators have high speed at low static current consumption (SIDD) and at lowest possible variation. Looking on figure 4a, the $CF_4$ related split degrades ring oscillator most. As can be seen in figure 4b, all other plasma treatments except for the forming gas treatment (FG) shift the ring oscillator device but to a much lesser extent. This is a surprising result, because just one additional plasma treatment was done, out of multiple plasma resist strips applied in a row for processing fully integrated wafers.

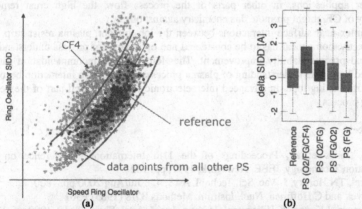

**Figure 4:** Impact of one additional plasma surface treatment post shallow transistor extension implants and pre dopant activation anneal on (a) ring oscillator device performance and (b) variation of static current consumption at reference speed.

## Summary and Conclusion

Depending on process requirements, ideal resist strips causes minimal dopant interaction, no surface etching, small surface oxidation, no anneal process related device variations and no strip related defects (particles). Such a resist strip can be applied multiple times in a row without device impact for resist removal and surface cleaning, but is not realistic. According to the results of this work, plasma resist strips with oxygen and hydrogen related process gases influence shallow surface doping profiles in about a 60A deep surface region, approximately one third of the today's definition of ultra shallow junction depth. Electrical test structure results show, that this shallow surface region influences microelectronic test devices significant. Different gas mixtures for plasma treatments act different at this surface region. Especially the 97% forming gas + 3% oxygen plasma treatment is smoothing out ultra shallow surface profiles by some plasma enhanced diffusion mechanism most. Since this finding is based on TOF-SIMS data, but is not confirmed from current electrical test structure results, this kind of hydrogen assisted plasma enhanced diffusion effects might become importance in next microelectronic technology generation and will be a matter of further investigation. Whenever oxygen is involved in the plasma resist strip processes of this work, surface oxidation, surface dopant profile variations and shift electrical device performance is observed simultaneously. Depending on the amount of oxygen and forming gas in the resist strip process, these effects can be balanced out depending on process requirements. Looking across all splits in this work, maintaining a certain amount of surface oxide above native oxide level of about 12Å before resist stripping can help to prevent surface material loss. Most likely this will not eliminate surface dopant effects, which penetrates through surface oxide level of 40Å and more.

From sheet resistance measurement on non structured wafers the $CF_4$ plasma resist strip causes a linear shift in sheet resistance vs. anneal temperature dependence, but does not influence dopant activation behavior. This effect is attributed to surface material loss therefore, caused by the $CF_4$ etching component. Because this surface material loss can not be tolerated at ultra shallow junctions from device performance point of view, $CF_4$ containing resist strip process are not recommended for ultra shallow junction processing. For other applications, in other parts of the process flow, the high crust removal capability of $CF_4$ might promote this chemistry against others.

The complex near surface interactions between the ion implant, plasma resist strip, and implant activation anneal has to be considered and optimized for process understanding, tuning and process stability improvement. Therefore no clear recommendation can be concluded here. The understanding of plasma processes and surface interaction becomes a highly interesting topic in advanced microelectronic technology and part of the device targeting process.

## References

[1] M. A. Jones et al., Proceedings of the 11th International Conference on Ion Implantation Technology, IEEE, New York, 1997, pp. 182
[2] AS Perel, TN Horsky, J. Vac. Sci. Technol. A 18, 4, Jul/Aug 2000 pp. 1800
[3] K J Orvek and C Huffman, Nucl. Instrum. Methods B7/8 (1985) P501
[4] S. Fujimura, J Konno, K Hikazutani, H Yano, Jpn, J. Appl. Phys. 28, 10, 1989 pp. 2130
[5] P. Gillespie, I. Berry, P. Sakthivel, Semiconductor International, October, 1999

[6] JI. McOmber K Ostrowski, M. Meloni, R. Eddy, P. Buccos, Nucl. Instrum. Methods B74 (1993) pp. 266-270

[7] K. Reinhardt, EG. Pavel, N. Fernandes, D. Neil, IBM Technical Symposium, France October 1999

[8] A Kirkpatrick, N Fernandes, T Uk, G Patrizi, MICRO:July/August 1998:

[9] K. T. Lee and S. Raghavan, Electrochemical and Solid-State Letters, 2 (4) 172-174, 1999

[10] D. M. Knotter, S.de Gendt, P.W. Mertens, and M. M. Heyns, Journal of The Electrochemical Society, 147 (2) 736-740, 2000

[11] L. Liu, Pey, K.L., P. Foo, Electron Devices Meeting, IEEE Hong Kong, pp17, June, 1996

[12] S. A. Vitale and B. A. Smith, J. Vac. Sci. Technol. B 21(5), Sep/Oct 2003

[13] K. Kim, et.al., J. Vac. Sci. Technol. B 14(4), Jul/Aug 1996

[14] E. G. Seebauer, Proc. MRS Spring Symposium, 2008

[15] F. Arnaud, et.al. Proceedings UCPSS Conference, Brussels Belgium, September 2004

[16] G. H. Buh, et.al. J. Vac. Sci. Technol. B, Vol. 24, No. 1, pp 449, Jan/Feb 2006

[17] M. Q. Huda, K. Sakamoto, H. Tanoue, Electrochemical and Solid-State Letters, 6 (10) G117-G118,2003

[18] K. Shibahara, H. Furumoto, K. Egusa, M. Koh, S. Yokohama, Mat. Res. Soc. Symp. Proc. 532, 23 (1998)

# Shallow Junction Contacting

Mater. Res. Soc. Symp. Proc. Vol. 1070 © 2008 Materials Research Society 1070-E02-01

# Ion implantation for low-resistive source/drain contacts in FinFET devices

Mark J. H. van Dal[1], Ray Duffy[1], Bartek J. Pawlak[1], Nadine Collaert[2], Malgorzata Jurczak[2], and Robert J. P. Lander[1]

[1]NXP-TSMC Research Center, Kapeldreef 75, Leuven, 3001, Belgium
[2]IMEC, Kapeldreef 75, Leuven, 3001, Belgium

## ABSTRACT

FinFET is one of the leading candidates to replace the classical planar MOSFET for future CMOS technologies due to the double-gate configuration of the device leading to an intrinsically superior short channel effect (SCE) control. A major challenge for FinFETs is the increase in parasitic source-drain resistance ($R_{sd}$) as the fin width is scaled. As fins must be narrow in order to control SCEs, $R_{sd}$ reduction is critical. This work will deal with the challenges faced in the use of ion implantation for the low-ohmic source-drain contacts. Firstly a new technique to characterize fin sidewall doping concentration will be introduced. We will have a closer look at the $R_{sd}$ dependency upon fin width for different fin implant conditions and investigate how the implant conditions affect FinFET device performance. It will be shown that the cause of the device degradation upon fin width scaling is related to the fundamental issues of silicon crystal integrity in thin-body Si after amorphizing implant and recrystallization during source-drain activation.

## INTRODUCTION

Multi-gate metal-oxide-semiconductor (MOS) devices and in particular Fin Field Effect Transistors (FinFETs) exhibit excellent SCE immunity due to the superior electrostatic gate coupling to the channel [1-5], enabling the aggressive scaling of CMOS technologies beyond the 32nm node. Furthermore because of the fully-depleted nature of the device no channel doping is needed to adjust the threshold voltages of the device. This eliminates dopant fluctuations, which dramatically improves variability compared to bulk CMOS.

Figure 1 shows a schematic of a FinFET device. A FinFET consists of a vertical standing Si body (fin) and the gate is wrapped around either side creating two channels on the sides and one on the top. An important advantage of the FinFET compared to alternative planar double gate architectures is that gates are self-aligned and can be fabricated with a single lithography and etch step. High-aspect-ratio trigate FinFETs with aggressively scaled fin widths (20 nm and narrower) are of particular interest as they combine excellent SCE immunity with high drivability per unit chip area.

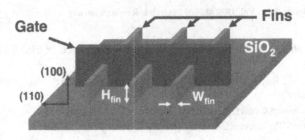

**Figure 1.** Schematic drawing of a Silicon-on-Insulator (SOI) FinFET with three fins in parallel.

As the device width of a FinFET is defined as the sum of the side and the top surface dimensions, optimization of the side surface (sidewall) doping for extension and Highly Doped Drains (HDDs) can add an extra degree of freedom for improved FinFET performance. However, the 3D configuration also imposes a challenge to form low-resistive contacts, especially for the extensions. A conformal deposition and in-diffusion methodology may produce equal doping on top and side surfaces, and therefore could be advantageous in this regard. In-diffusion [6,7], plasma doping [8] and metallic contacts [9] have been proposed as possible solutions but have yet to demonstrate an improved performance-leakage trade-off with respect to conventional implantation. Ion implantation remains a strong candidate as the means to introduce dopants into the fin, as it is an established and conventional technique.

Ion implantation for extensions and HDDs in FinFET integration can indeed yield high performance FinFETs [10]. However, recently it was shown that aggressively scaling the fin width in FinFETs leads to a better SCE control but is accompanied by a severe drive degradation [5]. The origin of the device degradation was found in the increase in access resistance. When fin width is scaled, the $R_{sd}$ extracted from FinFET devices increases drastically, and deviates from simulated trend where ideal dopant activation is assumed (Fig. 2).

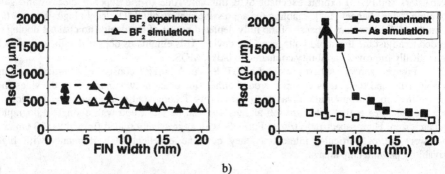

a)                                                    b)

**Figure 2.** $R_{sd}$ vs. fin width for pMOS FinFETs with a) $BF_2$ self-amorphizing extension implants and b) As self-amorphizing extension implants. Simulations match experiments for Fins wider than 10nm. Below 10nm, a clear deviation is observed.

This work will deal with the classical extension/HDD/silicide configuration for making low-ohmic contacts in highly scaled FinFET devices. Firstly we will demonstrate a new technique to characterize the fin side wall doping concentration after implantation, which can be used in optimizing implants for fins. Then the recrystallization after an amorphizing implant in ultra-thin Si body is examined. Finally, the relation between FinFET device performance degradation upon fin width scaling and the extension/HDD implant conditions will be discussed in more detail. It is concluded that complete amorphization of fins after ion implant can result in poor recrystallization. As a consequence extensions become highly resistive explaining the severe device degradation in aggressively scaled FinFETs.

## EXPERIMENTAL RESULTS

### Measuring fin side wall doping: The SIMS through fins concept

Due to the 3D nature of the FinFET, major challenges arise for junction characterization. Two orientations exist, and while they are not entirely independent, the top surface doping profile may differ significantly from that on the side. Thus, characterization of the sidewall doping becomes a pressing issue. Accessing the sidewall is a difficult task, and some recent work has done so through cross-sectioning the fin [11]. Here, we aim to apply a 1D characterization technique with fast turnaround time to a 2D system. Dedicated sample definition can enable us to use Secondary Ion Mass Spectrometry (SIMS) analysis, with special emphasis on the characterization of the dopants located on the fin sidewall.

The structure that is used for the SIMS analysis consists of arrays of long Si fins with an effective pitch of 600 nm. The fin array test structure is large enough to incorporate the entire SIMS crater. Experiments were performed on (100) p-type Si wafers, with bulk resistivity of 20 $\Omega \cdot$cm. Fin structures were patterned in the ‹110› direction by 193 nm lithography and an optimized reactive ion etching (RIE) process. In these *SIMS through fins* structures, the fin height was targeted at 200 nm to produce a clear portion in SIMS profile that could only come from the sidewalls. Implantation was then performed with splits on dose, energy and tilt angle which was succeeded by an activation anneal at 1050 °C. A 500 nm thick layer of undoped silicon was then deposited in order to fill the trenches between the fins. The test structure is shown schematically in Fig. 3. It was important to fill in the trenches with a material that had the same mass as silicon, and thus would have the same sputter rate during SIMS analysis. Finally, Chemical Mechanical Polishing (CMP) was performed to planarize the surface of the sample.

Representative cross-sectional TEM images of the samples are shown in Fig. 4. The crystalline fins are labeled, as well as the regions where the fill material is present. It can also be seen that the region in between the fins was not completely filled, as holes/voids (black triangular regions) are present in this layer. A close-up view in the magnified area in Fig. 4 reveals that the fill material between the fins is epitaxially grown Si with many faults parallel to {111}-crystal planes.

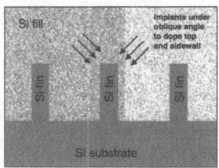

**Figure 3.** Schematic cross-section of the structure used for SIMS analysis of fin structures. After implant and anneal, the regions between fins are filled with Si, and CMP is used to planarize the top surface.

Before we proceed, some remarks need to be made on the implant angle and fin sidewall dose. Consider the system depicted schematically in Fig. 5a. The implant angle is $\alpha$ and the ion beam hits the sidewall at an angle $(90-\alpha)°$. When dose is specified by the user, it typically refers to the surface of a planar wafer, and thus corresponds to the plane of the top surface of the fin here. An important factor at this stage is the dose correction performed by an industrial implanter. As $\cos(\alpha)$ is <1 the user gets less dose than desired. An industrial implanter will automatically correct for implant angles so that the surface of the wafer receives the dose specified by the user, i.e. the dose received by the top of the fin equals specified dose (= dose in the beam $\times \cos(\alpha)$). As a result, the dose received by the fin side wall equals specified dose $\times \tan(\alpha)$ (= dose in the beam $\times \sin(\alpha)$). Thus for example if the process engineer specifies a 10° implant with $1\times10^{15}$ cm$^{-2}$ dose, the top surface will receive $1\times10^{15}$ cm$^{-2}$, and the sidewall will receive $1\times10^{15} \times \tan(10°)$ cm$^{-2} = 1.7\times10^{14}$ cm$^{-2}$.

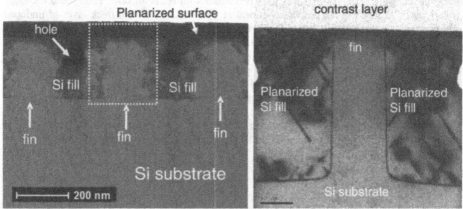

**Figure 4.** TEM image (left) depicting the fin arrays in overview. The dashed square indicates the position of the magnified area (right) which shows the fin in more detail.

$BF_2$ implants were performed at implant tilt angles of 45° and 10° and doses of $8\times10^{15}$ and $8\times10^{14}$ $cm^2$. Fig. 5b shows B concentration vs. depth from SIMS through fins analysis. Starting at a depth = 0 nm there is no concentration found, as some undoped deposited silicon remained on top of the fins after CMP. At a depth = 50 nm the signal from the top surface of the fins is detected. Note for a fixed implant dose, the dose correction done by the implanter means the concentration on the top surface is the same for both 45° and 10° implants.

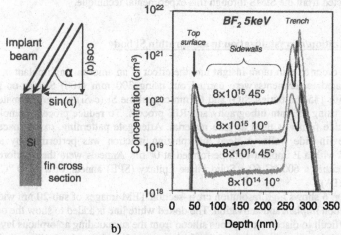

a)

b)

**Figure 5.** a) Schematic representation of a fin during ion implantation. The implant angle is α; b) B concentration vs. depth for different $BF_2$ implants. Signals are clearly discernible from top, sidewall, and trench surfaces.

The plateau in the SIMS profiles at depths ~75-250 nm corresponds to the sidewall doping. This is the concentration to be used to calculate the sidewall dose per fin and can be shown to be equal to the concentration extracted by SIMS × effective fin pitch. Finally there is another set of peaks in the SIMS profile at a depth ~250-300 nm. These come from the trenches in-between the fins and in-between the fin arrays. Multiple peaks are seen due to the incomplete filling by the deposited silicon between the fins in certain regions of the structure (Fig. 4).

A summary of the extracted sidewall doses is presented in Table 1 for the implants discussed here. In general we find that the experimental values are less than the theoretical value. This can be understood considering that ions may be lost during implant. For instance, ions can be lost through backscattering and return out of the sidewall they initially entered. Moreover, in case of very narrow fins ions can be lost out the opposite sidewall they initially entered. A second observation is that implanting at tilt angles (10°) produces relatively low sidewall doses. In recent work, it was shown that dose retention is very sensitive to implant angle [12]. These effects have to be taken into account when optimizing implant conditions for extensions in FinFETs.

71

| Species | Implant | Theoretical: Dose specified × tan(α) | SIMS: Sidewall dose per fin |
|---------|---------|--------------------------------------|------------------------------|
| $BF_2$ | 5 keV, $8\times10^{15}$ cm$^2$, 45° | $8\times10^{15}$ cm$^2$ | $5.1\times10^{15}$ cm$^2$ |
| $BF_2$ | 5 keV, $8\times10^{15}$ cm$^2$, 10° | $1.4\times10^{15}$ cm$^2$ | $1\times10^{15}$ cm$^2$ |
| $BF_2$ | 5 keV, $8\times10^{14}$ cm$^2$, 45° | $8\times10^{14}$ cm$^2$ | $6\times10^{14}$ cm$^2$ |
| $BF_2$ | 5 keV, $8\times10^{14}$ cm$^2$, 10° | $1.4\times10^{14}$ cm$^2$ | $1.3\times10^{14}$ cm$^2$ |

**Table 1.** Implant conditions and corresponding sidewall dose per fin, extracted from theory and extracted from the SIMS through fins experimental technique.

## Amorphization/recrystallization in an ultrathin Si body

In order to gain more insight into the effect of an amorphizing implant in ultra-thin body Si, dedicated experiments were carried out using 300 mm (100) Silicon-on-Insulator (SOI) wafers with 145 nm buried $SiO_2$ and 65 nm crystalline Si (c-Si) as starting material. Fins were patterned using 193 nm lithography and RIE process. To reduce process complexity, a 60 nm silicon oxide film was used as gate material. After gate patterning, oxide spacer were formed along the fin side walls. A deep amorphizing junction was performed by an As implant, combined with a P implant both performed at 0° tilt. Anneals were then performed in an inert ambient, either a 600 °C 60 s Solid Phase Epitaxy (SPE) anneal, or a 1050 °C rapid-thermal-anneal (RTA). Fig. 6 shows high resolution cross-section TEM images of sub-20 nm wide fins after the deep junction implant and activation. The dotted white line is added to show the outline of the fin as it is difficult to discern amorphous silicon from the surrounding amorphous layers. In Fig 6a it is observed that the top portion of the fin has expanded. Amorphous silicon is less dense than crystalline silicon [13], and with oxide spacers on either side of the fin, expansion of the silicon region only takes place at the top of the structure. Fig. 6b shows the same fin after a 600 °C 60 s anneal. Regrowth appears to be faster in the middle of the fin and suppressed at the sidewall surfaces. In bulk silicon the ‹100› regrowth rate is approximately 1 nm/s at 600 °C, so without the influence of the surfaces this structure would be completely recrystallized from the bottom to the top. Fig. 6c shows the silicon fin after the 1050 °C RTA. The bottom portion of the fin is defect-free mono-crystalline silicon. The middle part of the fin is crystalline but is highly defected. The dark lines at 45° to the side surfaces indicate the presence of {111} twin boundary defects. They originate from both right and left side surfaces. The top ~25 nm of the fin consists of poly-crystalline silicon grains. Thus despite the presence of a lattice template under the gate and at the bottom of the fin, SPE is repressed in these sub-20 nm wide fins.

Drosd and Washburn presented an atomistic model for the physics of recrystallization [15]. One requirement for a silicon atom to be incorporated in the growing lattice is that it should form 2 undistorted bonds with the crystal, i.e. bonds of characteristic angle and length as in crystalline silicon. In the presence of a surface the regular lattice is interrupted and the formation of 2 undistorted bonds becomes difficult, and SPE is retarded. Moreover in the FinFET the surfaces are less than ideal, as they may be terminated with native oxide and roughened either by the fin patterning or by the high dose ion implantations. The twin boundary defects observed in Fig. 6c are also known as stacking faults and are located in the {111} plane. On an atomistic level non-idealities such as roughness can hinder templated atom incorporation into the crystal lattice, and {111} defects may result [15].

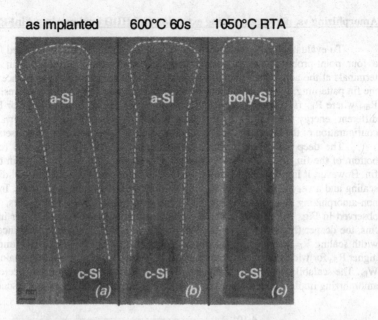

**Figure 6.** Cross-section TEM images showing a sub-20 nm wide silicon fin (a) after a deep implant which amorphizes the top 50 nm of the structure, where the top portion of the fin has expanded, (b) after implant and a 600 °C 60 s anneal where regrowth is incomplete, (c) after implant and a 1050 °C RTA anneal where regrowth is complete but yields many twin boundary defects, and the top ~25 nm has transformed into poly-silicon [14]. The dotted lines have been added to guide the eye.

As stated earlier the regrowth in these thin layers is retarded compared to the bulk silicon case. The poly-crystalline silicon formation observed in Fig. 6c is linked to untemplated recrystallization known as random nucleation and growth (RNG). In bulk silicon templated recrystallization (i.e. SPE) has an activation energy of ~2.7 eV, and advances rapidly even at low temperatures. RNG has an activation energy of ~4 eV, thus RNG may not have enough time to occur before all the amorphous silicon has been consumed by SPE. However if SPE is retarded by surface proximity, then RNG becomes more likely provided the thermal budget is high enough, as it is in the 1050 °C RTA case here.

Based on this study we can conclude that when fins are narrow and are fully-amorphized after ion implantation, poor recrystallization during activation anneal in the extension region can lead to the formation of poly-Si. In [16], it is demonstrated that in such a case carrier mobility is reduced dramatically leading to very high resistance. The strongest impact of poor crystallinity was observed for As doped layers, where a ~10 times $R_s$ increase was observed compared c-Si As doped thin films.

## Amorphizing vs. non-amorphizing extension and HDD implants in p-FinFETs

To evaluate the effect of fin width scaling on the resistance of a doped fin, we have used a four point-probe resistor structure consisting of a single 9-$\mu$m-long fin with four probe terminals at the active levels. These structures enable measuring fin resistance performing only the fin patterning, ion implantation and dopant activation. Fig. 7 shows the resistance of the fin, $R_{fin}$, where $R_{fin}$ ($\Omega\cdot\mu$m) = $R_{meas} \times (2H_{fin}+W_{fin})$ as a function of $W_{fin}$ for B or BF$_2$ implant with different energy and angle. These conditions are chosen in such a way that different configuration of the amorphous Si (a-Si) after implant exists in the fin (see insets in Fig. 7).

The deep self-amorphizing 0° BF$_2$ implant leaves a crystalline Si (c-Si) seed at the bottom of the fin, while for the 45° shallow BF$_2$ implant a c-Si core exists in the middle of the fin. However, it is important to realize that in the latter case the c-Si core will disappear upon fin scaling and a very narrow fin will be fully amorphous after such an implant. In contrast, for the non-amorphizing B implanted fins the Si integrity is preserved. The $R_{fin}$ vs. $W_{fin}$ dependency observed in Fig. 7 is related to the morphological condition of the fin after implant. For wide fins, the deeper (high energy) 0° implant for BF$_2$ result in the lowest resistance where upon fin width scaling $R_{fin}$ increases rather drastically. Although the 45° tilted BF$_2$ implanted fins have higher $R_{fin}$ for wide fins, these conditions lead to better scalability as $R_{fin}$ remains low for smaller $W_{fin}$. The scalability is further improved switching from BF$_2$ to B implant species. Hence the less amorphizing implant conditions lead to lower resistance for the aggressively scaled fins.

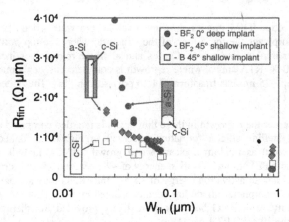

**Figure 7.** $R_{fin}$ as a function of fin width determined from a resistor short loop experiment for BF$_2$ and B doped fins.

The impact of amorphizing or non-amorphizing implant conditions for contact formation is further investigated in pMOS FinFETs devices. Fig. 8 shows the basic outline of FinFET integration flow [5,17]. SOI wafers as described earlier were used as starting material. Fins were patterned such that fin side walls are defined in the (110) Si planes. Extension and HDD conditions were chosen to be either amorphizing (using BF$_2$ implant species) or non-amorphizing (B). The effective device width is defined as $W_T = n(2H_{fin}+W_{fin})$, where n is the number of fins, $W_{fin}$ is the fin width and $H_{fin}$ is the fin height.

- Fin patterning (193 nm i-lithography)
- Gate stack deposition (HfSiO, TiN, poly-Si)
- Gate patterning
- Extensions
- Spacers
- Selective Epitaxial Si growth
- HDD
- NiSi (6 nm Ni: RTP1, selective etch, RTP2)

**Figure 8**: Process flow for fabrication of SOI FinFET devices used in the present work [5]

The improved SCE control for FinFETs when scaling fin widths can be seen when plotting Drain-Induced-Barrier-Lowering (DIBL) as a function of $L_g$ for different $W_{fin}$. Fig. 9 shows DIBL vs. $L_g$ for FinFETs with two extension/HDD configurations: $BF_2/BF_2$ and B/B. In both cases the DIBL starts to increase at shorter $L_g$ when $W_{fin}$ is scaled.

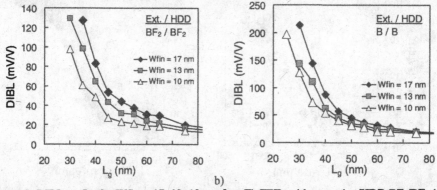

a)                                      b)

**Figure 9**. DIBL vs. Lg for Wfin = 17, 13, 10 nm for pFinFETs with extension/HDD $BF_2/BF_2$ (a) or B/B (b). Clear improvement in SCE control is seen when $W_{fin}$ is scaled.

In Fig. 10 a-c the $I_{on} - I_{off}$ characteristics of pMOS FinFETs are shown where either B or $BF_2$ implant was used for extension or HDD for three fin widths: 17 nm, 13 nm and 10 nm. In case of $BF_2$ extensions and $BF_2$ HDD, a clear degradation upon $W_{fin}$ scaling can be observed (Fig. 10a). Also the data seems to be rather scattered. If non-amorphizing B implant conditions are used for extensions, no $I_{on}$ degradation down to $W_{fin}=13$ nm is observed (Fig. 10b) and there is tighter data distribution. An even better scalability is seen when switching both extension and HDD to B implant as device performance is not compromised for $W_{fin} = 10$ nm (Fig. 10c). The effect of an amorphizing HDD implant was further investigated by implementing an extra amorphizing Ge implant accompanying the B HDD implant. Fig. 10d shows the $I_{on} - I_{off}$ of pFinFETs for a fixed fin width 17 nm comparing devices with B and B + Ge implant HDD. The increase in the data scatter in case of the B + Ge HDD is probably related to a problematic recrystallization of the S/D region leading to more scatter in the access resistance.

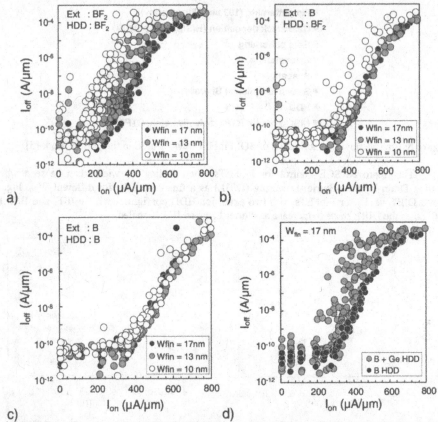

**Figure 10.** $I_{on}$ - $I_{off}$ plot for pMOS FinFETs measured at $|V_d|$ = $|V_{g,on}|$ = 1V. Figure a to c correspond to devices with $W_{fin}$ = 17, 13, 10 nm and different extension/HDD conditions: a) $BF_2/BF_2$, b) $B/BF_2$ and c) B/B. Figure d compares devices with $W_{fin}$ = 17 nm using either B or B+Ge HDD. Amorphizing implant in extension and HDD has impact on device performance, scalability and variability of FinFET devices.

The source/drain resistance $R_{sd}$ of these devices was extracted by plotting resistance R as function of physical gate length and extrapolating to $L_g$ = 0. R was measured at $V_d$ = 50 mV and at $V_g$ = 3V in order to obtain the gate-length-independent R. Fig. 11 shows access resistance $R_{sd}$ versus $W_{fin}$ for the different splits. In case of $BF_2$ extensions the $R_{sd}$ increases rather drastically upon fin width scaling in agreement with the device performance degradation (Fig. 11a). Low $R_{sd}$ for narrow fins is retained for B extension implants. In summary, the use of amorphizing implant conditions for both the HDD and extension result in a performance degradation due to an increase in access resistance and lead to more variability. The increase in fin resistance, in turn, originates from the problematic recrystallization after amorphizing implant and activation anneal in ultra-thin Si body.

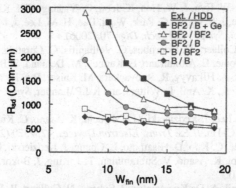

**Figure 11.** $R_{sd}$ as a function of $W_{fin}$ for pMOS FinFET devices with different implant species for extensions and HDD (ext./HDD). $R_{sd}$ was extracted by extrapolating the $R_{on}$ (measured at $V_{g,on}$ = 3 V) versus $L_g$ plot to $L_g = 0$.

## CONCLUSIONS

The present work deals with the use of ion implantation to achieve low-ohmic S/D contacts in FinFETs. The 3D configuration of a FinFET poses challenges not only for the device integration but also in material characterization, e.g. the fin top and sidewall doping concentration. In this work, the *SIMS through fins* concept is introduced in order to determine fin side wall doping concentration. Poor FinFET device performance and variability for aggressively scaled FinFETs is linked to problematic recrystallization after ion implantation, leading to defect formation and high resistance. As a conclusion we can state the use of doping techniques to form S/D contact in FinFET integration that fully amorphize the fin should be avoided.

## ACKNOWLEDGMENTS

The authors would like to acknowledge M. Kaiser, R. G. R. Weemaes, J. G. M. van Berkum and P. Breimer from Philips Research Laboratories Eindhoven, The Netherlands for TEM sample preparation, analysis and SIMS characterization. This work has been partially funded by the European PULLNANO Integrated Project (FP6 – IST-026828).

## REFERENCES

1. D. Hisamoto, T. Kaga, Y. Kawamoto, and E. Takeda, *IEDM Tech. Dig.*, 833 (1989).
2. F.-L. Yang, D.-H. Lee, H.-Y. Chen, C.-Y. Chang, S.-D. Liu, C.-C. Huang, T.-X. Chung, H.-W. Chen, C.-C. Huang, Y.-H. Liu, C.-C. Wu, C.-C. Chen, S.-C. Chen, Y.-T. Chen, Y.-H. Chen, C.-J. Chen, B.-W. Chan, P.-F. Hsu, J.-H. Shieh, H.-J. Tao, Y.-C. Yeo, Y. Li, J.-W. Lee, P. Chen, M.-S. Liang, and C. Hu., *Symp. VLSI Tech. Dig.*, 196 (2004).

3. B. Yu, L. Chang; S. Ahmed, W. Haihong, S. Bell, C.-Y. Yang, C. Tabery, H. Chau, X. Qi, T.-J. King, J. Bokor, C.-M. Hu, M.-R. Lin, and D. Kyser, *IEDM Tech. Dig.*, 251 (2002).

4. H. Lee, L.-E. Yu, S.-W. Ryu, J.-W. Han, K. Jeon, D.-Y. Jang, K.H. Kim, J. Lee, J.-H. Kim, S. C. Jeon, G. S. Lee, J. S. Oh, Y. C. Park, W. H. Bae, H. M. Lee, J. M. Yang, J. J. Yoo, S. I. Kim, and Y.-K. Choi, *Symp. VLSI Tech. Dig.*, 70 (2006).

5. M.J.H. van Dal, N. Collaert, G. Doornbos, G. Vellianitis, G. Curatola, B.J. Pawlak, R. Duffy, C. Jonville, B. Degroote, E. Altamirano, E. Kunnen, M. Demand, S. Beckx, T. Vandeweyer, C. Delvaux, F. Leys, A. Hikavyy, R. Rooyackers, M. Kaiser, R.G.R. Weemaes, S. Biesemans, M. Jurczak, K. Anil, L. Witters, and R.J.P. Lander, *Symp. VLSI Tech. Dig.*, 110 (2007)

6. D. Hisamoto, W.-C. Lee, J. Kedzierski, H. Takeuchi, K. Asano, C. Kuo, E. Anderson, T.-J. King, J. Bokor, and C. Hu, *IEEE Trans. Electron Devices*, 47, 2320 (2000).

7. X. Huang, W.-C. Lee, C. Kuo, D. Hisamoto, L. Chang, J. Kedzierski, E. Anderson, H. Takeuchi, Y.-K. Choi, K. Asano, V. Subramnian, T.-J. King, J. Bokor, and C. Hu, *IEDM Tech. Dig.*, 67 (1999).

8. D. Lenoble, K.G. Anil, A. De Keersgieter, P. Eybens, N. Collaert, R. Rooyackers, S. Brus, P. Zimmerman, M. Goodwin, D. Vanhaeren, W. Vandervorst, S. Radovanov, L. Godet, C. Cardinaud, S. Biesemans, T. Skotnicki, and M. Jurczak, *Symp. VLSI Tech. Dig.*, 212 (2006).

9. A. Kaneko, A. Yagishita, K. Yahasi, T. Kubota, M. Omura, K. Matsuo, I. Mizishima, K. Okano, H. Kawasaki, S. Inaba, T. Izumida, T. Kanemura, N. Aoki, K. Ishimura, H. Ishiuchi, K. Suguro, K. Eguchi, and Y. Tsunashima, *IEDM Tech. Dig.*, (2006).

10. G. Vellianitis, M. J. H. van Dal, L. Witters, G. Curatola, G. Doornbos, N. Collaert, C. Jonville, C. Torregiani, L. S. Lai, J. Petry, B.J. Pawlak, R. Duffy, M. Demand, S. Beckx, S. Mertens, A. Delabie, T. Vandeweyer, C. Delvaux, F. Leys, A. Hikavyy, R. Rooyackers, M. Kaiser, R.G.R. Weemaes, F. Voogt, H. Roberts, D. Donnet, S. Biesemans, M. Jurczak, and R. J. P. Lander, *IEDM Tech. Dig.*, 681 (2007).

11. A. A. Khajetoorians, J. Li, C. K. Shih, X.-D. Wang, D. Garcia-Gutierrez, M. Jose-Yacaman, D. Pham, H. Celio, and A. Diebold, *J. Appl. Phys.* **101**, 034505 (2007).

12. R. Duffy, G. Curatola, B. J. Pawlak, G. Doonbos, K. van der Tak, P. Breimer, J. G. M. vna Berkum, and F. Roozboom, *J. Vac. Sci. Technol.* B **26**, 402 (2008).

13. J. S. Custer, M. O. Thompson, D. C. Jacobson, J. M. Poate, S. Roorda, W. C. Sinke, and F. Spaepen, *Appl. Phys. Lett.* **64**, 437 (1994).

14. R. Duffy, M. J. H. van Dal, B. J. Pawlak, M. Kaiser, and R. G. R. Weemaes, *Appl. Phys. Lett.*, **90**, 241912 (2007).

15. R. Drosd and J. Washburn, *J. Appl. Phys.* **53**, 397 (1982).

16. B. J. Pawlak, R. Duffy, M. J. H. van Dal, F. C. Voogt, F. Roozeboom, R. G. R. Weemaes, P. Breimer, and P. Zalm, MRS Spring meeting, 2008, San Francisco.

17. M. J. H. van Dal, G. Vellianitis, R. Duffy, G. Doornbos, B. J. Pawlak, B. Duriez, L-S Lai, A. Hikavyy, T. Vandeweyer, M. Demand, E. Altamirano, R. Rooyackers, L. Witters, N. Collaert, M. Jurczak, M. Kaiser, R. G. R. Weemaes, and R. J. P. Lander, ECS Spring meeting 2008, Phoenix.

Mater. Res. Soc. Symp. Proc. Vol. 1070 © 2008 Materials Research Society  1070-E02-09

# Pt Segregation at the NiSi/Si Interface and a Relationship with the Microstructure of NiSi

Haruko Akutsu[1], Hiroshi Itokawa[1], Kazuhiko Nakamura[1], Toshihiko Iinuma[1], Kyoichi Suguro[1], Hiroshi Uchida[2], and Masanori Tada[2]

[1]Toshiba Corporation Semiconductor Company, 8, Shinsugita-Cho, Isogo-Ku, Yokohama, Japan
[2]Toshiba Nanoanalysis Corporation, Japan

## ABSTRACT

Platinum Segregation at the NiSi/Si Interface and a relationship with the microstructure of NiSi were studied. We investigated the Platinum distribution by using TEM, EDX and Atomprobe. We found Pt segregation at the grain boundary of NiSi. And we studied shapes of NiSi(without Pt) grains and Pt doped NiSi grains by using SEM. The SEM apparatus has an angle selective Backscattered electron detector. (ASB detector) The ASB detector was adjusted to get channeling contrast. We found that a Pt doped NiSi grain is smaller than a NiSi grain.

## INTRODUCTION

The purpose of using Nickel monosilicide (NiSi) is lowering the resistances of doped Si layers. However NiSi is not thermally stable. D. Mangelinck et al. found that Pt doped Ni-silicide has rather good tolerance of thermal agglomeration and morphology. [1] [2]  Several possible mechanisms have been reported to play a role in the improved thermal stability of the NiSi:(1) the addition of Pt appears to influence the texture of NiSi film, thus affecting the interface energy, (2) the expansion of the NiSi lattice due to the additioin of Pt may result in the transformation of the MnP-type structure into more stable NiAs-type structure, and (3) since Pt is insoluble in $NiSi_2$ , the nucleation of $NiSi_2$ will be more difficult because of the required expulsion of Pt from the nucleation region and the resulting change in entropy of mixing. [1] [3]

On another front, T.F.Kelly et al. analyze Ni(Pt)Si/Si(As doped) interface with Atom Probe. They found the As segregation at the NiSi/Si interface and the Pt segregation of the surface in the 1-D profile. [4] We first reported the Pt segregation was found at a NiSi/Si interface by AP analysis. [6]

In this study, we studied a precise Pt distribution and a difference between NiSi with Pt and NiSi without Pt. First we investigated the Pt distribution with Atomprobe, TEM and EDX. Second we observed NiSi and Ni(Pt)Si films by using a SEM with angle selective Backscattered electron detector. Finally we discuss that the Pt distribution and its role.

## EXPERIMENT

Ni-Pt alloy film of 8-12nm in thickness were deposited on As implanted (100)Si substrates. Some of the samples were annealed to form Ni silicide in $N_2$ after Ni sputtering. As deposited sample(sample A) and the sample after silicidation (sample B) were prepared for AP analysis using LEAP3000(IMAGO) with a special resolution of 0.4 nm×0.4 nm×0.2 nm , detection efficiency about 50 atomic % and sensitivity about 0.5atomic %. To complement TEM, FE-SEM observation was carried out.

Figure 1(a) shows the depth profile through the Ni(Pt)/Si interface analyzed by AP. Z=0nm is Ni =50% isosurface. Fig. 1(b) shows that of sample B. Z=0nm is Ni =25% isosurface.

There is a small peak in profile of Pt at a Ni/amorphous-NiSi interface as shown in Fig.1(a). There are three peaks in the profile of Pt as shown in Fig.1 (b). Peak 1 locates at the top surface, Peak 2 is about 6nm above the NiSi/Si interface and Peak 3 is at the NiSi/Si interface.

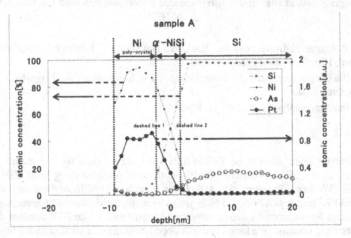

Figure 1(a) The Depth Profile of sample A (Z=0nm is Ni =50% isosurface)

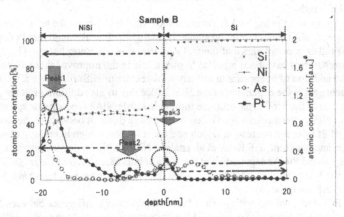

Figure. 1(b) The Depth Profile of sample B(Z=0nm is Ni =25% isosurface)

To analyze the Pt distribution in Sample B more properly, we processed the 3D image as follows: a cylindrical part is extracted from the 3D image; it is devided into 5nm thick slices; and 2D images depicting density-distribution of Pt and As in the slices are calculated as shown in Figs. 2(a) to 2(h). Fig. 2(b) shows that Pt atoms segregate around grains of NiSi. Fig. 2(d) shows that Pt atoms exist below the NiSi/Si interface, same as As. Fig. 2(f) also shows that As atoms segregate at grain boundary of NiSi.

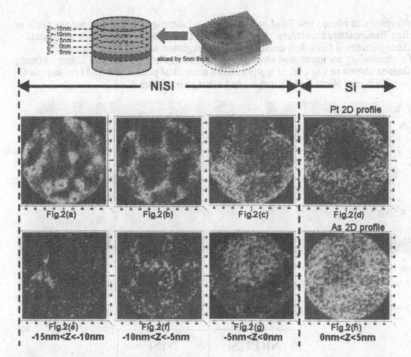

Figure2: 2D concentration of Pt,As in 5nm thick slices.(Z=0nm is Ni =25% isosurface. (a)~(d) indicate Pt distribution of the each slice. (e)~(h) indicate As distribution of the each slice.)

Fig.3(a) shows results of TEM & EDX on sample A , and Fig.3(b) shows a result of TEM on sample B. As shown in Fig. 3(a), sample A which is as sputtered sample has an amorphous layer on the substrate and a Ni poly-crystal layer on the amorphous layer. The thickness of the amorphous layer is about 4nm from the TEM observation. While in Fig.1(a), the dashed line 1 is placed at Ni 80% compared with the Ni-poly-crystal region. The dashed line 2 at Ni 20%. The thickness between these two lines is about 4nm and corresponds to the Fig.3(a). This region is an amorphous-NiSi layer. As shown in Fig. 3(b), NiSi grains grow by annealing. A thickness of the Nickel-silicide film from the TEM observation is 20nm on an average. It also corresponds to a thickness from the Atomprobe profile (Fig.1(b)).

| point | Si[atomic%] | Ni[atomic%] | Pt[a.u.] |
|-------|-------------|-------------|----------|
| 1 | 8.64 | 85.98 | 1.08 |
| 2 | 45.39 | 57.13 | 0.32 |

Figure 3:(a) The Results of TEM & EDX on Sample A

(b) The Result of TEM on Sample B

81

Fig.4 shows results of plane view SEM on sample B and sample C. These images are made up with channeling Backscattered electrons. A grain in Fig.4 corresponds with a grain of crystal. Contrast of these pictures is faint. It is due to that As segregation at the surface changes a trajectory of a channeling backscattered electron. The grain size of Ni(Pt)Si are 20nm∼60nm φ, average is 30nm as shown in Tig4.(a). Fig.4(b) shows a result of plane view SEM on sample C. The grain size of NiSi are 50nm φ ∼250nm φ, and average is 150nm φ.

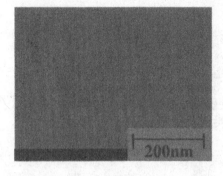 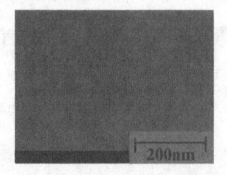

Figure4: (a)The Result of plane view SEM on Sample B    (b)The Result of plane view SEM on Sample C

Table 1. Ni(Pt)Si and NiSi grain size and shape

|  | Ni(Pt)Si | NiSi |
|---|---|---|
| **Max.[nm]** | 60 | 250 |
| **Min.[nm]** | 20 | 50 |
| **Ave.[nm]** | 30 | 150 |
| **shape** | polygonal | spindly |

DISCUSSION

(1)Fig. 1(a) shows that Pt atoms are localized at the interface between the Ni poly crystal layer and the amorphous layer in the as-sputtered sample (sample A). On the other hand, Fig. 1(b) shows that there are three peaks in the depth profile of Pt after the annealing in sample B. Figs. 2(a)∼(d) show Pt atoms distribution in detail. The 2D image of Fig. 2(a) corresponds to the depth of Peak 1 in Fig. 1(b). In view of Peak 1 which is at the top surface of the NiSi grains, Fig. 2(a) shows that Pt atoms are swept out from the NiSi grains during the silicidation. The 2D image of Fig .2(b) corresponds to the depth of Peak 2. In view of Peak2, Fig. 2(b) shows that Pt atoms are piled up at the side faces of the NiSi grains due to the segregation of Pt atoms. The 2D image of Figs. 2(c) and (d) correspond to the depth of Peak 3. In view of Peak 3, Figs. 2(c) and (d) show that Pt atoms are uniformly distributed at the bottom of the NiSi grains. From these results, it is evident that Pt atoms segregate around the NiSi grains. At the same way As atoms also segregate around the NiSi grains. And these results leads to the following assumption about Ni(Pt) silicidation scheme.(Fig.6)

(1) While Ni(Pt) atoms are sputtered on the Si substrate, Ni atoms diffuse much faster than Pt atoms in the Si matrix. [5] As the result, the Pt rich layer is formed at the top of αNiSi layer.

(2) In the annealing step, Pt atoms are unstable in NiSi grains so that Pt atoms are piled up at the grain boundaries of the NiSi grains.

(3) Consequently, as shown in Fig. 4(c), Pt segregate at the grain boundaries of the NiSi grains and Pt profile has 3 peaks in the NiSi. We also point out that there are few Pt atoms in a NiSi grain. Pt atoms exist around NiSi grain-boundary. It suggests that Pt is not able to change a character of a bulk-NiSi because Pt does not exist inside of a NiSi grain.

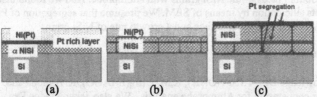

Figure5:Pt segregation scheme.( Figs. 6(a), (b) and (c) show after sputtering, during the silicidation , after the silicidation, respectively.)

Fig.4 and Table 1 show that grain size of NiSi without Pt is larger than Ni(Pt)Si.  It suggests that a crystal growth rate of NiSi is higher than Ni(Pt)Si. We suggest that Pt reduces a growth rate of NiSi grain as shown in Fig.7.

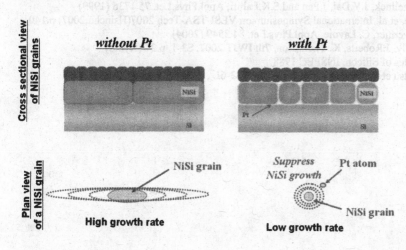

Figure6:NiSi grain growth scheme.(a)NiSi grain growth model without Pt (b) Ni(Pt)Si grain growth model with Pt

Pt atoms at the grain boundary of NiSi would be meta-stable and reduce interface energy. Pt has rather strong electron affinity compared with Ni and Si. An existence of Pt atoms at the NiSi/Si interface would deform a lattice structure of the interface. As a result Pt atoms work as barriers against the diffusion or reaction of Ni and Si atoms. In this way, we presume that the thermal agglomeration is improved.

## CONCLUSIONS

We found that Pt doped Ni-silicide has segregated Pt atoms at NiSi grain boundaries, both of side and bottom faces of the NiSi grains with Atomprobe. And we found that Pt segregation affects silicidation by means of SEM. We presume that segregation of Pt atoms changes the energy of Ni(Pt)Si/Si interface and a growth rate of NiSi grain.

## ACKNOWLEDGMENTS

We acknowledge Drs. T.Iba(NOAH), R.M. Ulfig(IMAGO), K.Tomoda and K.Mori(TNA) for cooperation for Atomprobe analysis. We also acknowledge Drs. Y.Tsunashima, Y.Saito, T.Yamauchi, A.Shimazaki, Y.Toyoshima, K.Ohuchi, S.Kawanaka, A.Hokazono, T.Sonehara , I.Mizushima for fruitful discussion.

## REFERENCES

1. D.Mangelinck, J.Y.Dai, J.Pan and S.K.Lahiri, Appl.Phys.Let. 75 1736 (1999)
2. J.Strane, et al. International Symposium on VLSI-TSA-Tech 2007(Hsinchu,2007) p.140
3. C.Detavernier, C. Lavoie, Appl.Phys.Let. 84 3549 (2004)
4. T.F.Kelly, J.Roberts, K. Thompson, 7th IWJT 2007, S3-1, p.27 (Kyoto)
5. Properties of Silicon, INSPEC 1988
6, H. Akutsu et al., Ext. Abstacts of SSDM, F-3-6L , p. 296 (Tsukuba, 2007).

Mater. Res. Soc. Symp. Proc. Vol. 1070 © 2008 Materials Research Society                    1070-E02-10

# Investigation of Platinum Silicide Schottky Barrier Height Modulation using a Dopant Segregation Approach

Nicolas Breil[1,2], Aomar Halimaoui[1], Emmanuel Dubois[2], Evelyne Lampin[2], Guilhem Larrieu[2], Ludovic Godet[3], George Papasouliotis[3], and Thomas Skotnicki[1]

[1]STMicroelectronics, 850, rue Jean Monnet, Crolles, 38920, France
[2]IEMN-ISEN, UMR CNRS 8520, Cité Scientifique, Avenue Poincaré, Villeneuve d'Ascq, 59652, France
[3]Varian Semiconductor Equipment Associates, Inc., 35, Dory Rd., Gloucester, MA, 01930

## ABSTRACT

The role of the dopant activation on the segregation efficiency during the formation of platinum silicide (PtSi) is investigated in this paper. Using an implant before silicidation technique, we first demonstrate an important Schottky Barrier Height (SBH) modulation for As and B segregation. In the case of As, we highlight that an activation of the dopants before the silicidation does not impact the SBH modulation. On the contrary, an important impact of the dopant crystalline position is evidenced for Boron. Also, a comparison of conventional implant versus a PLAsma Doping (PLAD) highlights the suitability of the latter implantation tool for the SBH modulation. Those results are interpreted on the basis of SIMS depth profiling.

## INTRODUCTION

As CMOS technology scaling continues, the contact resistance between the silicide and the doped junctions appears as a severe issue. This contact resistance shows an exponential dependence to the ratio between $\phi_b$, the Schottky Barrier Height (SBH), and $N_d^{1/2}$ the square root of the underlying silicon doping concentration [1]. As this latter parameter almost reached the dopant solubility limit, the modulation of $\phi_b$ appears as a promising performance booster. Using dopant segregation approaches, such SBH-engineered silicides can be integrated in a standard integration scheme, as well as in a more aggressive Schottky-Barrier transistor (SBMOS) approach [2].

Recently, severe issues appeared since the 65nm node concerning the stability of the conventional nickel silicide. Abnormal penetration under the gate and a poor thermal stability are two of the main drawbacks associated to this material. To leverage this issue, an increasing amount of platinum added during the nickel metal deposition is known to increase the process stability [3]. In this study, we investigate the potential of pure platinum as a successor to nickel. Furthermore, an original method for the selective etching of platinum versus platinum silicide was recently demonstrated from a morphological [4] and an electrical point of view [5], which makes this material a promising candidate for the future technology nodes.

Recently, some authors have shown that dopant segregation techniques are an efficient way for the SBH engineering [6]. This study focuses on the impact of the dopant crystalline position in Si on the SBH modulation efficiency.

## EXPERIMENT

The test structures used in this study were patterned in a thermally grown 120nm thick $SiO_2$ layer on p-type Si substrates. They are composed of two $100x100\mu m^2$ openings separated by a micrometric Si gap. Dopant implantations were realized after the $SiO_2$ test structures fabrication, and before the silicide formation.

The segregation of n-type dopants was investigated using an As implantation at 1keV and a dose of $10^{15}$ cm$^{-2}$, with a Ultra-Low Energy beamline implanter. Some of the samples received a spike annealing at 1080°C.

Regarding the segregation of Boron, the samples were given to a beamline $BF_2$ implantation at an energy of 1keV and a dose of $10^{15}$ cm$^{-2}$. A Solid Phase Epitaxial Regrowth (SPER) annealing at 700°C for 2min was realized on some of the samples, in order to activate B atoms while limiting diffusion. Sheet resistance measurements, as well as SIMS depth profiling have confirmed that dopants were activated without evidence of measurable diffusion. An alternative route was also investigated on some samples, using a $BF_3$ Plasma DOPing (PLAD) implant at a bias of 1keV with doses of $10^{15}$ cm$^{-2}$ or $5x10^{15}$ cm$^{-2}$.

After the implantation step, a de-oxidation of the Si surfaces by hydrofluoric dip was realized, and the samples were immediately loaded in an e-beam evaporation system. Once a base pressure of $1x10^{-7}$mbar was reached, an Ar plasma etching at 60eV for 30sec was realized, in order to remove any surface contamination. The samples were then transferred in the evaporation chamber, where a 20nm thick layer of Pt was deposited. After unloading, the samples received a Rapid Thermal Annealing (RTA) at temperatures ranging from 200 to 600°C for 4min in a forming gas ($N_2$:$H_2$-95:5). Unreacted Pt was then selectively removed using a standard Aqua Regia etching.

In this study, the SBHs were measured using a back-to-back diode structures. Current-Voltage characteristics as a function of the temperature were used to plot the results in the Arrhenius $I/T^2$ vs. $1000/T$ format. Coupling the experimental data with a numerical modeling already described in [7], we are able to extract the SBH as well as other parameters such as the surface injection, and the doping level of the underneath doped silicon.

## RESULTS

### Arsenic Segregation

The SBH measurement of the As (1keV ; $10^{15}$ cm$^{-2}$) implanted samples is shown in Fig. 1. As expected, a large SBH of 0.60eV is measured for a silicidation annealing temperature of 200°C. However, as the annealing temperature increases, the SBH decreases down to 0.15eV for an annealing temperature of 600°C.

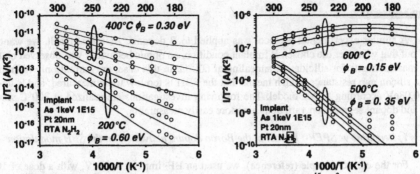

Fig. 1 *Schottky Barrier Height measurement of the As 1keV $10^{15}cm^{-2}$ implanted samples. The silicidation temperature of the 20nm thick layer of Pt is indicated.*

SIMS depth profiling analysis was performed on the samples annealed at 300°C and 600°C (Fig. 2). At 300°C, a segregation of As is observed at the surface and at the PtSi/Si interface. This segregation increases at 600°C, with a maximum concentration of $2x10^{21}$at./cm$^3$ at the PtSi/Si interface. Thus, the SBH modulation observed in Fig. 1 is attributed to the presence of dopants, segregated at the interface.

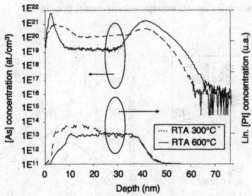

Fig. 2 *SIMS concentration profiles of Pt and As after a silicidation annealing of 300°C or 600°C.*

SBH measurements were also performed on samples that received a Spike anneal at 1080°C after the implantation step. Sheet resistance measurements confirmed that the dopants were activated after the annealing, thus moving from interstitial to substitutional position. However, no major change is observed on the SBH modulation as a function of the temperature. Thus, we conclude that the dopant position (substitutional versus interstitial) has a minor impact on the segregation process, and thus on SBH modulation.

## Boron Segregation

A similar experimental protocol was applied to B dopants. As a first result, we want to point out that whatever the chosen B implant condition, an ohmic behaviour was observed on the SBH measurements for silicidation annealing at 600°C. In this case, the current only is limited by the silicon gap resistance, which means that the SBH is too low to be measured, and is lower than 0.08eV according to our model. The following investigations are realized at 300°C where SBH and dopant segregation variations are more easily measurable.

### 1) Beamline vs SPER : Impact of the Boron dopant activation on the SBH modulation

For the control sample (reference), we used an $BF_2$ implant at 1keV, with a dose of $10^{15}$ cm$^{-2}$. After a RTA silicidation annealing at 300°C, a SBH of 0.20eV is measured (Fig.3a), which is the standard value for PtSi [8]. The SIMS profile (Fig. 3b) shows that the initial implantation profile is only slightly impacted by the PtSi formation.

Using the same experimental conditions, a SPER annealing at 700°C for 2min was realized just after the implantation step. A huge impact is observed on the electrical characteristics of the contact in Fig. 3a), where a SBH as low as 0.055eV is measured. It is worth noting that the surface injection also decreases. The SIMS depth profiling evidences a dopant segregation at the PtSi/Si interface. Using a SPER annealing, the segregated B concentration is increased by more than a factor 2, as shown in Fig. 3b). This observation evidences that the B segregation is enhanced when B dopants occupy substitutional positions.

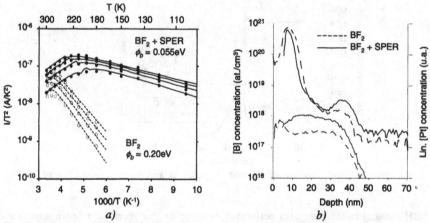

Fig. 3 Arrhenius plot a) and SIMS depth profiling b) of a $BF_2$ beamline implant at 1keV and $10^{15}$cm$^{-2}$, with or without a SPER annealing, and after a silicidation annealing at 300°C.

### 2) PLAD implant : Impact of the Boron dose

Some samples have been implanted using PLAsma Doping technique (PLAD). We should note that, due to the ion collisions in the plasma ionic sheath, a PLAD implant results in a non-Gaussian B profile. Furthermore, from a throughput point of view, a PLAD tool allows substantially higher doses than a standard beamline implanter. Using an implantation energy of 1kV, our samples have received doses of $10^{15}$ or $5\times10^{15}$ cm$^{-2}$.

The sample implanted at a dose of $10^{15}$ cm$^{-2}$ is shown in Fig. 4. A SBH of 0.22eV is measured (Fig. 4a). When the implanted dose is increased to $5\times10^{15}$ cm$^{-2}$, a substantial decrease of the SBH is measured at 0.08eV, as shown in Fig. 4a. This SBH modulation can be related to the important concentration of B at the PtSi/Si interface of $2\times10^{19}$ at./cm$^3$, measured in Fig. 4b.

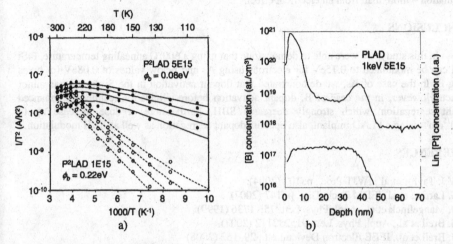

*Fig. 4 Arrhenius plot a) and SIM depth profiling b) of a PLAD BF$_x$ implant at 1keV and $10^{15}$ cm$^{-2}$ or $5\times10^{15}$ cm$^{-2}$, after a silicidation annealing at 400°C.*

## DISCUSSION

Very different behaviors are observed depending on the type of segregated dopants. Whereas As is not sensitive to the dopant activation, this parameter has a main impact on the segregation of B. Also for B, the increase of the implanted dose to a value of $5\times10^{15}$ cm$^{-2}$ using PLAD increases the concentration of dopants at the interface. It is interesting to compare the results of Fig. 3b and Fig. 4b. Whereas, the segregated B concentration is higher with the PLAD implant, the impact on the SBH is more important in the case of a beamline implant with a SPER annealing. However, it is difficult to compare those two results because of the specificity of the PLAD implant as compared to a conventional beamline implant, considering the dopant concentration profile, as well as the incorporated fluorine dose.

Actually, many mechanisms are involved in the segregation of dopants. First of all, the well known snow-plough effect pushes dopants at the front of the silicide interface during its formation. This effect is directly related to the limit solid solubility of the dopants in the silicide. However, in the case of PtSi, the presence of an intermediate phase Pt$_2$Si in the formation sequence is well known [9]. Pt is the main diffusing specie for the formation of Pt$_2$Si, whereas Si is the one for the formation of PtSi. The limit solid solubility of dopants being a function of the phase and of the temperature, the understanding of the snow-plough appears as being a complex process.

Finally, it is generally considered that segregated dopants have an electrical effect on the SBH modulation. However, we would like to point out that some authors have shown, using thermodynamical considerations, that in the case of highly doped Si layers the formation of Pt-Dopant compounds is sometimes energetically favorable [9]. In this case, the SBH modulation would originate from a chemical effect – related to the workfunction of the Pt-Dopant compound modulation – more than from an electrical effect.

## CONCLUSIONS

In this study, we were able to demonstrate that using a 600°C annealing temperature, PtSi SBH can be modulated to 0.15eV for electrons using As and below values of 0.08eV for holes using B. In the case of As, we demonstrated that dopant activation in the Si crystal has minor impact. However, in the case of B, dopant activation before silicidation resulted in enhanced dopant segregation, which strongly decreases SBH. Furthermore, a dose increase up to $5 \times 10^{15} \mathrm{cm}^{-2}$ using a PLAD implant, also benefits dopant segregation as well as SBH modulation.

## REFERENCES

1. W.J. Taylor et al., IWJT Proc., p. 107 (2004).
2. G. Larrieu et al, IEDM Tech. Dig., p. 147 (2007)
3. D. Mangelinck et al., Appl. Phys. Lett. 75, 1736 (1999).
4. N. Breil et al., Appl. Phys. Lett. 91, 232112 (2007).
5. N. Breil et al., IEEE Electron Device Lett., 29, 152 (2008).
6. Z. Zhang et al., IEEE Electron Device Lett., 28, 565 (2007).
7. E. Dubois et al., J. Appl. Phys. 96, 729 (2004).
8. K. Maex and M. van Rossum, Properties of Metal Silicides. London, U.K.: INSPEC, (1995).
9. K. Maex et al., J. Appl. Phys. 66, p. 5327 (1989).

# Poster Session

Mater. Res. Soc. Symp. Proc. Vol. 1070 © 2008 Materials Research Society        1070-E03-01

# A Comprehensive Atomistic Kinetic Monte Carlo Model for Amorphization/Recrystallization and its Effects on Dopants

Nikolas Zographos[1], and Ignacio Martin-Bragado[2]
[1]Synopsys Switzerland LLC, Affolternstrasse 52, Zürich, 8050, Switzerland
[2]Synopsys Inc., 700 East Middlefield Road, Mountain View, CA, 94043

## ABSTRACT

This work shows a comprehensive atomistic model to describe amorphization and recrystallization, and its different effects on dopants in silicon. We begin by describing the physical basis of the model used, based on the transformation of ion-implanted dopants and generated point defects into amorphous pockets of different sizes. The growth and dissolution of amorphous pockets is simulated by the capture and recombination of point defects with different activation energies. In some cases, this growth leads to the formation of amorphous layers. These layers, composed of a set of amorphous elements, have an activation energy to be recrystallized. The recrystallization velocity is modeled not only depending on temperature, but also on dopant concentration. During the recrystallization, dopants can move with the recrystallization front, leading to dopant redistribution during solid phase epitaxial regrowth (SPER). At the edge of the amorphous/crystalline (A/C) interface, the remaining damage forms end-of-range (EOR) defects.

Once the model is explained, we discuss the calibration methodology used to reproduce several A/C experiments, including the dependencies of the A/C transition temperature on dose rate and ion mass, and the A/C depth on ion implant energy.

This calibrated model allows us to explore the redistribution of several dopants, including B, As, F, and In, during SPER. Experimental results for all these dopants are compared with relevant simulations.

## INTRODUCTION

The semiconductor industry is systematically reducing the thermal budget allowed for diffusion of dopants while using preamorphization and SPER as a main strategy to increase active dopant concentration and to better control the doping profile on current devices. As the scenario shifts from a diffusion-controlled paradigm to a diffusion-less one, the role played by the redistribution of dopants during SPER gets more important. This optimist scenario that avoids the difficulties of modeling diffusion conveys other important problems: SPER produces dramatic changes in dopants and impurities profiles. It is the aim of this work to explore and propose a comprehensive model to predict the final doping profiles after SPER has taken place.

## PHYSICAL MODEL

Our simulator [1], using the kinetic Monte Carlo (KMC) technique [2,3], gets the coordinates of all the particles for each collision cascade from a Binary Collision Approximation Monte Carlo simulator. Instead of undergoing immediate IV recombination, amorphous pockets (APs) are formed, simulating disordered regions. The amorphous pockets shrink by recombining internal IV pairs with a frequency given by [4,5]

$$\alpha s^{\beta} \times \exp\left(-E_{act}^{IV}(s)/k_B T\right), \tag{1}$$

$\alpha$, $\beta$ and $E_{act}^{IV}$ being calibration parameters, and s being the cluster size, i.e., the number of IV pairs included in the AP. The activation energy dependence on size implements a mechanism to allow small APs to dissolve faster than bigger, more stable ones [6].

During the implant, the dynamic annealing of the APs can yield to two different scenarios. a) Recombination is faster than the damage formation, or b) recombination cannot cope with damage formation. In this last scenario, APs grow and, eventually, may overlap and form amorphous layers. The simulator detects the presence of an amorphous layer when the concentration of damage (I plus V) exceeds an amorphization threshold.

A SPER process is also simulated converting the amorphous layers back into crystalline silicon. This recrystallization is implemented by splitting the amorphous layer into different amorphous elements, each of them with a recrystallization frequency depending on the number of neighboring amorphous elements, as given by

$$\Delta L v \times \exp\left(-[E_{act}^{SPER}(s) + c]/k_B T\right) \tag{2}$$

being $\Delta L$ the element length in the recrystallization direction, c a correction term, and v and $E_{act}^{SPER}(s)$ the recrystallization velocity prefactor and activation energy. $E_{act}^{SPER}(s)$ is made dependent on the number of neighboring amorphous elements, to simulate the faster recrystallization of isolated amorphous island, and c and v are explained in more detail below.

The growth dependency on Fermi-level and impurity concentration [7] is also included. The Fermi-level dependency is modeled by defining the velocity prefactor, v as

$$v = v_0\left(1 + |K \times C|\right), \tag{3}$$

being $v_0$ the recrystallization velocity in intrinsic conditions, K a calibration parameter, and C the local amorphous element doping.

The SPER dependency on impurities and not just dopants is including modeling of the correction term c as $c = \sum_i C_i^{\delta_i} / 5e22 \times (E_i - E_0)$ where $C_i$ are the dopant concentrations, $\delta_i$ and $E_i$ are calibration parameters, and $E_0$ is the activation energy for recrystallization of a planar front without impurity effects, accounted as 2.68 eV [7].

Finally, SPER redistribution is simulated using a simple atomistic model. When an amorphous element is recrystallizing, it has a probability P to deposit the impurity in the recrystallized area. This P is different for each impurity being computed using a prefactor and activation energy.

## CALIBRATION PROCESS

Following the calibration methodology stated in Ref. [5], the amorphization model is calibrated to explain a) the amorphous layer thickness as a function of Ge preamorphization, represented in Fig. 1a, and b) the A/C transition temperatures as a function of dose rate for (100) Si implanted with C, Si, and Ge as reported by [8], see Fig. 1b.

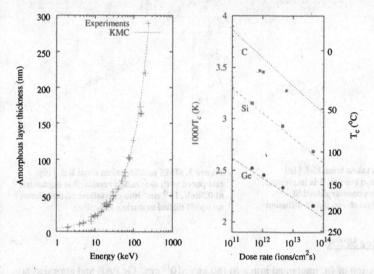

**Figure 1a (left). Amorphization depth predicted by KMC (line) compared with different experimental results (symbols). Figure 1b (right). Simulations (lines) compared to experimental data (symbols) taken from [8] for A/C transition temperature as a function of dose rate, for (100) Si implanted with 80keV ions, $1 \times 10^{15}$ cm$^{-2}$ for Si and Ge, and $2 \times 10^{15}$ cm$^{-2}$ for C.**

After this calibration, the amorphous threshold was found to be $1.0 \times 10^{22}$ cm$^{-3}$, the prefactor $\alpha$ equal to $4 \times 10^4$ cm$^2$ s$^{-1}$, the exponent $\beta$ equals to 1, and the activation energy for IV pair recombination linearly dependent on the size, with 0.7 eV for $s=1$ and 2.2 eV for $s=199$.

## RESULTS AND DISCUSSION

### Boron uphill diffusion

B uphill diffusion is a recently observed phenomenon, where B at high concentrations diffuses preferentially towards the surface after preamorphization implantation (PAI) [9]. The implant damage under amorphization conditions is the driving force behind it [10]. Fig. 2 shows experimental results for uphill diffusion, taken from Ref. [10] and compared with our simulation results. The agreement is excellent. The conclusion is that a proper modeling of damage accumulation and recrystallization, together with "standard" EOR formation, B diffusion and boron-interstitial cluster (BIC) formation satisfactorily explains this phenomenon without the necessity of including new models or mechanisms such as surface-near traps. Please note that no B is redistributed by the recrystallization front [11]. This is an example of how atomistic simulations can explain relatively new phenomena as the result of having several known physical mechanisms working together at the same time. Finally, without PAI, there is no up-hill diffusion (see Fig. 3)

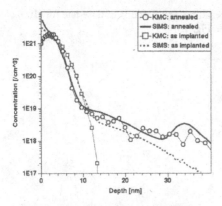
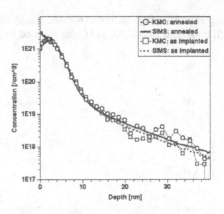

Figure 2. SIMS profiles taken from Ref. [10] compared with simulation results. B is implanted at 0.5keV, $10^{15}$ cm$^{-2}$ into preamorphized Si ($1.2 \times 10^{15}$ cm$^{-2}$ 15keV), and shows uphill diffusion during annealing.

Figure 3. SIMS profiles taken from Ref. [10] compared with simulation results. B is implanted at 0.5keV, $10^{15}$ cm$^{-2}$ into crystalline Si, and shows no uphill diffusion during annealing.

## Redistribution during SPER

The redistribution of In, implanted into a-Si (80 keV, $10^{15}$ cm$^{-2}$ Ge PAI) and annealed at 600C for 1 and 2 minutes is shown in Fig. 4 and 5 respectively. The experimental data have been taken from [11]. The swept of In towards the surface during SPER is clearly visible. It can be reproduced by the model presented in this work assuming a 20% In deposition probability during SPER and enhanced recrystallization velocity in p-type doped regions.

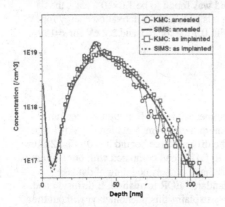
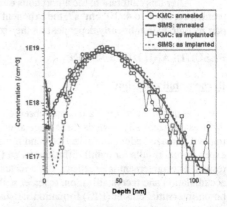

Figure 4. In 90 keV $4 \times 10^{13}$ cm$^{-2}$ implanted into a-Si (Ge PAI) and annealed at 600C for 60s. The simulation predicts the dopant peak due to the impurity swept during recrystallization.

Figure 5. In 90 keVm $4 \times 10^{13}$ cm$^{-2}$ implanted into a-Si (Ge PAI) and annealed at 600C for 120s. The simulation produces the right redistribution of dopants close to the surface.

96

An example of As redistribution during SPER can we found at Ref. [12]. In this experiment (shown in Fig. 6), As is implanted (3 keV, $10^{15}$ cm$^{-2}$) and annealed for 10 seconds at 600C. The dopant peak moves towards the surface due to the redistribution of As atoms during SPER. Using a deposit probability of 45%, enhanced recrystallization velocity in n-type doped regions, and reduced recrystallization velocity in presence of high As concentration, our model shows an excellent agreement with the experiment.

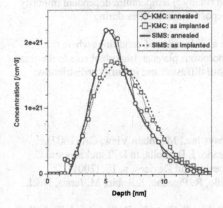 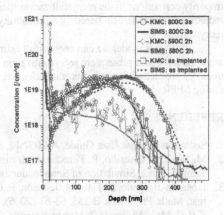

Figure 6. As redistribution profiles during SPER, with SIMS taken from Ref. [12]. Our model correctly simulates the peak movement when using a partial deposit probability during SPER.

Figure 7. F concentration profiles of the as-implanted sample ($4\times10^{14}$ cm$^{-2}$ F at 100keV) and after SPER at different temperatures: 580C for 120min., 800C for 3s. SIMS data from Ref. [13].

Another drastic case of redistribution happening during SPER is reported in Ref. [13]. In this cited work, the authors show very important differences in the final shape of F profiles by changing the temperature at which the SPER takes place, and allowing the annealing enough time to complete it. More in particular, Fig. 7 shows the profiles of an implanted sample (F $4\times10^{14}$ cm$^{-2}$ at 100keV) in preamorphized Si after annealing at different temperatures: 580C for 120 minutes and 800C for 3 seconds. These profiles are created by the competition between regrowth and F diffusion in amorphous. At low temperatures, the SPER (slowed down by the F concentration) is much slower than F diffusion in amorphous, allowing for the F in the sample to get to the interface and dissolve there. For higher temperatures, the diffusion in the amorphous region cannot catch up with the F swept by the recrystallization front, and more F is captured inside the sample. The agreement is good once the probability of deposit for F is made temperature dependent (~1% at 580C, ~20% at 800C). In addition, the presence of F slows down the SPER velocity, which is simulated by the model's impurity effect on SPER.

Finally, we need to add that the SPER redistribution is very sensitive to recrystallization speed. Consequently the previous examples are also valid to show the overall agreement of our model to predict the right recrystallization speed, and also the necessity to include small correction to the the velocity activation energy and prefactor (as explained in the model section of this paper) dependent on the particular impurity present. Although this model is able to reproduce the SPER effects of most dopants, some of them, such as the As "hysteresis" effect reported in Ref. [7], may need extra physics embedded in the model.

## CONCLUSIONS

We have presented a comprehensive model to accurately simulate several important effects produced on the doping distribution of impurities after SPER. This model simulates a) damage accumulation and dynamic annealing to predict amorphization conditions, b) accurate amorphization depth, c) SPER regrowth accounting for d) Fermi-level corrections and e) impurity corrections to the recrystallization speed and finally f) temperature dependent impurity redistribution mechanism based on a probability to deposit the impurities during recrystallization.

With this model we can successfully simulate a wide range of conditions where the amorphization and subsequent recrystallization of amorphous plays an important role in the final impurity profile, while paying special attention to uphill diffusion and impurity redistribution during SPER.

## REFERENCES

1. Sentaurus Process User Guide, A-2007.12, Synopsys Inc., Mountain View, CA (2007).
2. M. Jaraiz, P. Castrillo, R. Pinacho, I. Martin-Bragado, J. Barbolla, in D. Tsoukalas and C. Tsamis (Eds.), Simulation of Semiconductor Processes and Devices, p. 10. (2001).
3. I. Martin-Bragado, S. Tian, M. Johnson, P. Castrillo, R. Pinacho, J. Rubio, M. Jaraiz, Nucl. Inst. Meth. Phys. Res. B 253, 63-67 (2006).
4. K.R.C. Mok, M. Jaraiz, I. Martin-Bragado, J. E. Rubio, P. Castrillo, R. Pinacho, J. Barbolla, M. P. Srinivasan, J. Appl. Phys. 98, 046104 (2005).
5. K.R.C. Mok, B. Colombeau, F. Benistant, R. S. Teo, S. H. Yeong, B. Yang, M. Jaraiz, Shao-Fu Sanford Chu, IEEE Trans. Elec. Dev. 54 (9) (2007).
6. L. Pelaz, L.A. Marqués, J. Barbolla, J. Appl. Phys. 96, 11, 5947 (2004).
7. G.L. Olson and J.A. Roth, Materials Science Reports 3, 1-78 (1988).
8. R.D. Goldberg, J.S. Williams, R.G. Elliman, Nucl. Inst. Meth. Phys. Res. B 106, 242-247 (1995).
9. H.C.H. Wang, C.C. Wang, C.S. Chang, T. Wang, P.B. Griffin, C.H. Diaz, IEEE Elec. Dev. Lett. 22 (2), 65 (2001).
10. R. Duffy, V.C. Venezia, A. Heringa, T.W.T. Husken, M.J.P. Hopstaken, N.E.B. Cowern, P.B. Griffin, C.C. Wang, Appl. Phys. Lett. 82, 3647 (2003).
11. V.C. Venezia, R. Duffy, L. Pelaz, M. Aboy, A. Heringa, P.B. Griffin, C.C. Wang, M.J.P. Hopstaken, Y. Tamminga, T. Dao, B.J. Pawlak, F. Roozeboom, IEDM Tech. Digest, 489-492 (2003).
12. K. Suzuki, Y. Kataoka, S. Nagayama, C.W. Magee, T.H. Büyüklimanli, T. Nagayama, IEEE Trans. Elec. Dev. 54 (2) (2007).
13. G. Impellizari, S. Mirabella, F. Priolo, E. Napolitani, A. Carnera, J. Appl. Phys. 99, 103510 (2006).

Mater. Res. Soc. Symp. Proc. Vol. 1070 © 2008 Materials Research Society                    1070-E03-02

# Performance Characteristics of 65nm PFETs Using Molecular Implant Species for Source and Drain Extensions

C. F. Tan[1], L. W. Teo[1], C-S. Yin[1], J. G. Lee[1], J. Liu[1], A. See[1], M. S. Zhou[1], E. Quek[1], S. Chu[1], C. Hatem[2], N. Variam[2], E. Arevalo[2], A. Gupta[2], and S. Mehta[2]

[1]Technology Development, Chartered Semiconductor Manufacturing, 60 Woodlands Industrial Park D, Singapore

[2]Varian Semiconductor Equipment Associates, Gloucester, MA, 01930

## ABSTRACT

We investigated the performance of 65nm pFETs whereby the source and drain extensions (SDE) were implanted with Carborane, $(C_2B_{10}H_{12})$ a novel form of molecular species. The high atomic mass of this molecule (146 a.m.u.) and the number of boron atoms transported per ion enables the productivity at low energy required for manufacturing of ultra shallow junctions for advanced scaling. In this investigation, Carborane was implanted at 13 keV to produce a Boron profile near equivalent to that produced by the reference $BF_2$ implant. Results of electrical measurements did not exhibit any compromise in the I-V characteristics in terms of $I_d$-$V_g$ and $I_d$-$V_d$ and $I_{on}$-$I_{off}$. External resistance and $V_t$ roll-off shifted slightly with respect to the reference devices. This is attributed to a deeper junction with Carborane due to slight offset in the profile matching. It will be shown that with fully matched profiles, a perfect match of the device characteristics can be achieved.

## I. INTRODUCTION

Suppression of short channel effects in advanced MOSFET devices is driving the SDE junction depths shallower, leading to significant drop in associated implant energies. These energies are now reaching the sub keV scale for Boron [1,2]. The consequential drop in productivity of the implant process is resulting in a high level of interest in molecular forms of implantation to extend the energy range below 500 eV for devices beyond the 45 nm node [3].

**Figure 1.** Molecular structure of Carborane. [1]

Carborane, a form of molecular species (referred to as CBH in the text of this paper), shown in Figure (1) has been recently [4] demonstrated as a means to form junctions as shallow as 10 nm at 5E18 to meet the 32nm node requirements. Nevertheless, for the successful integration of CBH into the fabrication process, a better understanding of the performance characteristics of devices fabricated with this molecular species is necessary. This work is the first investigation of electrical characteristics of pFETs, whereby implantation of CBH was employed to dope the SDE regions for a 65 nm process.

## II. DEVICE FABRICATION

The fabrication of the pFET transistor follows a typical flow as shown in the flowchart in Figure 2. SDE splits between the reference $BF_2$ and CBH were performed on device wafers from the same lot. In both cases a standard Ge pre-amorphization implant was done before the SDE implant. SIMS analysis was used to match the as-implanted B profiles from CBH to that of the reference $BF_2$ implant. Implant energy of 13 keV was used for CBH to achieve a close match as depicted in Figure 3. With the exception of the SDE implant, all other processing steps were identical for both sets of device splits. In addition to SIMS analysis, cross-sectional transmission electron microscopy (XTEM) measurements were also used to characterize the amorphous layer. Results depicted in Figure 4 show that the amorphous layer thickness is ~ 233A and is identical for CBH and $BF_2$ implants indicating that the thickness of the layer is determined by the Ge PAI implant.

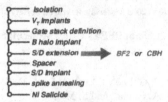

- Isolation
- $V_T$ Implants
- Gate stack definition
- B halo implant
- S/D extension ➡ BF2 or CBH
- Spacer
- S/D Implant
- spike annealing
- Ni Salicide

**Figure 2.** Fabrication flow chart of the pFET transistors

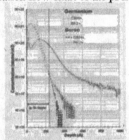

**Figure 3.** As- implanted SIMS profile of the $BF_2$ and CBH used in the experiment.

**Figure 4.** XTEM images compare the damage profile of the as-implanted (a) CBH and (b) $BF_2$

## III. RESULTS AND DISCUSSION

Figure 5 shows the sheet resistance ($R_s$) versus junction depth ($X_j$) of boron implanted into bare Si. It can be observed that B from the CBH conforms well to the behavior of the $BF_2$ junctions, thus indicating that there is no deviation in the junction activation with CBH. As expected, the CBH species shows a favorable shift in the universal curve when laser annealing was used.

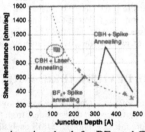

**Figure 5.** Sheet resistance versus junction depth for $BF_2$ and CBH implanted into bare silicon.

$I_{on}$-$I_{off}$ characteristics of the pFETs are presented in Figure 6. While the CBH implanted pFETs exhibited higher $I_{on}$, it remains along the same technology curve. No performance degradation in the devices was observed.

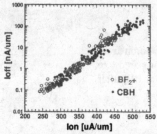

**Figure 6.** $I_{on}$ - $I_{off}$ behavior showing strong alignment between CBH and $BF_2$ implanted pFETs

External resistance ($R_{ext}$) measurements performed on the devices as depicted in Figure 7(a) show that the $R_{ext}$ with CBH is lower by 5%. A comparison of the Vt roll-off characteristics is shown in Figure 7(b). Slight degradation in the roll-off with CBH suggests that the SDE junction is deeper in the CBH implanted device.

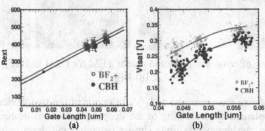

**Figure 7.** (a) $R_{ext}$ versus $L_{poly}$ measurement for CBH and $BF_2$ devices (b) $V_{tsat}$ roll off characteristics compared for CBH and $BF_2$ implanted device

SIMS measurements made on the annealed wafers show that the CBH profile is deeper than the $BF_2$ profile by ~ 10nm and is attributed to the slight offset in energy matching. (see Figure 8). This difference in the depth of Boron leads to the observed differences in the electrical parameters mentioned above.

Also noted in Figure 8 is the formation of a second boron peak at ~8 nm depth with CBH. This peak is attributed to the segregation of B due to the presence of shallow carbon inherent to the CBH implant. To confirm this hypothesis, the $BF_2$ sample was co-implanted with carbon at the same depth as inherent to CBH. The presence of the shallow B peak in this sample is identical to the one with CBH as shown in Figure 9 hence confirming the hypothesis.

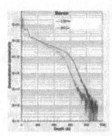 **Figure 8.** SIMS profile of boron from CBH compared to boron from BF₂ after annealing.

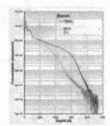 **Figure 9.** SIMS profile of boron from CBH compared to boron from BF₂ co-implanted with a shallow C implant

Cross-sectional and planar TEM measurements were made on annealed samples for both CBH and BF₂. Results depicted in Figure 10 show no difference between the 2 cases. No evidence of defects was observed.

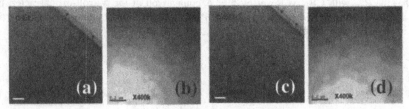

**Figure 10.** Comparison of cross-sectional and planar TEM for CBH and BF₂. (a) & (b) CBH. (c) & (d) BF₂

Figure 11(a) shows a comparison of $I_d$-$V_g$ behavior of the two devices. There is a shift in $V_{tsat}$ for the CBH implanted device owing to the deeper junction, as discussed earlier. The sub-threshold slope of the two devices exhibits similar behavior and drain induced barrier lowering (DIBL) is also similar. The $I_d$-$V_d$ characteristics are compared in Figure 11(b). Results show that for a constant gate overdrive, the transistor characteristics of the two devices are almost identical. Both curves indicate that the transistors fabricated with the CBH species exhibit excellent I-V characteristics.

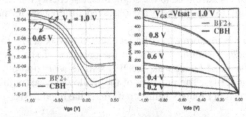

**Figure 11.** (a) $I_d$-$V_g$ and (b) $I_d$-$V_d$ plots comparing the performance of the two pFETs.

Based on the above results, another split of devices was implanted with CBH profiles matched to those of the reference wafer. Excellent matching of the transistor I-V characteristics was achieved as shown in Figure 12. The $V_{tsat}$ roll-off characteristics are also completely matched to the reference device as depicted in Figure 13, indicating identical short channel behavior.

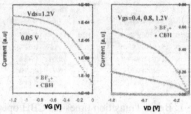

**Figure 12.** (a) $I_d$-$V_g$ (b) Id-Vd curve comparing the pFET devices fabricated with a matched SDE profile.

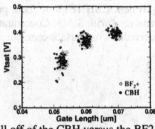

**Figure 13.** Vtsat roll off of the CBH versus the BF2 for a matched device.

## IV. UNIVERSALITY OF CBH

While molecular implantation enables production worthy beam currents at low energy required for advanced nodes, the molecules proposed to date have shown poor thermal stability requiring a specialize cold source. This specialized source in turn has reduced productivity for typical high current implants with ions such as $BF_2^+$ and $P^+$.

VSEA has developed a solution with a thermally stable boron molecule that ionizes at higher temperatures and can be run in a standard ion source and does not affect normal operation. The high thermal stability can be seen in figure 14.

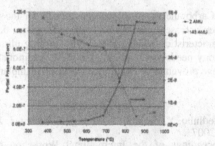

**Figure 14.** RGA data shows no significant decomposition to >700°C.

The ability to transition from one ionic species to another is necessary from a manufacturing perspective. Table 15 shows ion beam transition-times from preliminary tests for transition from boron to carborane and vice versa. These transition times will be reduced substantially with optimization of software control.

103

| Transition | Tune Time | | Uniformity |
|---|---|---|---|
| | Source | Total | |
| 20KeV B to 200eV CBH | 11:28 | 14:14 | 0.44% |
| 20KeV B to 500eV CBH | 11:22 | 13:56 | 0.53% |
| 500eV CBH to 20KeV B | 4:26 | 6:18 | 0.48% |
| 500eV CBH to 200eV CBH | 3:17 | 5:31 | 0.34% |
| 200eV CBH to 500eV CBH | 3:16 | 5:24 | 0.44% |

**Table 1:** Beam Transition times for CBH operation.

Specie cross contamination is also of interest from a manufacturing perspective to avoid counter doping. For this purpose cross contamination of $As_2^+$ in CBH was analyzed. A wafer each was implanted with 500eV $B^+$ equivalent CBH $1E15cm^{-2}$ both prior to after a continuous 4 hour run of $As^{2+}$. These wafers were analyzed with SIMS. Concentration profiles from these tests are shown in Figure 15 and shows sputtered arsenic surface contamination of 0.84%.

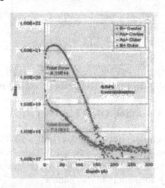

**Figure 15.** SIMS analysis showing less than 1% $As_2^+$ contamination.

## V. CONCLUSIONS

The feasibility of Carborane as a molecular approach for the formation of ultra shallow junctions in pFET has been assessed using 65nm high performance devices. The devices exhibit excellent performance characteristics. This approach to forming ultra shallow junctions is well suited to meet the technology needs for 32nm and beyond. Additionally, the thermal stability of CBH and its ease of use have proven to be an attracted molecular implant solution.

## V. REFERENCES

[1] H.-J. Gossmann, G. Redinbo, Y. Erokhin, T. Romig, K. Elshot, J. Xu, and J. McComb, Solid State Technology, July 2007.
[2] H.-J. Gossmann, Proceedings of the International Workshop on Semiconductor Device Fabrication, Metrology, and Modeling (INSIGT-2007).
[3] D. Jacobson et al, Nucl. Instr. & Meth. in Phys. B 237 (2005) 406-410.
[4] A. Renau, IWJT 2007, p. 107.
[5] N. Variam et al, 15th International Conference on Advanced Thermal Processing of Semiconductors, RTP 2007, p. 291.

Mater. Res. Soc. Symp. Proc. Vol. 1070 © 2008 Materials Research Society          1070-E03-03

# Phosphorus diffusion and activation in silicon: Process simulation based on *ab initio* calculations

Beat Sahli[1], Kilian Vollenweider[1], Nikolas Zographos[2], Christoph Zechner[2], and Kunihiro Suzuki[3]

[1]Integrated Systems Laboratory, ETH Zurich, Zurich, 8092, Switzerland
[2]Synopsys Switzerland LLC, Zurich, 8050, Switzerland
[3]Fujitsu Laboratories Ltd, Atsugi, 243-0197, Japan

## ABSTRACT

We present the results of extensive *ab initio* calculations for phosphorus clustering and diffusion in silicon and the application of these results in a state-of-the-art process simulator. The specific defects and the parameters that are investigated are selected according to the needs of diffusion and activation models, taking into account the availability of experimental data, the capabilities of current *ab initio* methods and the requirements for advanced technology development. The calculated formation energies, binding energies and migration barriers are used to determine a good starting point for the calibration of a diffusion and clustering model implemented in an atomistic process simulator. The defect species V, I, P, PV, PI, PI$_2$, P$_2$, P$_2$V, P$_2$I, P$_3$, P$_3$V, P$_3$I and P$_4$V are considered in all relevant charge states. The *ab initio* results are discussed as well as the transfer of this information into the process simulation model and the impact on model quality.

## INTRODUCTION

Phosphorus is used for n-type doping of source/drain regions of nMOSFETS due to its high activation. However, phosphorus suffers strong transient enhanced diffusion (TED), and therefore a good knowledge of its diffusion mechanism is needed, but yet not fully established. Generally, the need for increasingly accurate and predictive modeling of dopant diffusion and activation leads to a move away from phenomenological models to increasingly detailed physics based models, e.g. atomistic simulations based on the kinetic Monte Carlo (KMC) approach or continuum based simulations with a large number of defect species. However, such models also require a large number of physical model parameters, for many of which there is no specific experimental data available. The process of calibrating an initial model using a database of dopant profiles from varying processing conditions is well developed. However, the development of the initial model including the large number of physical model parameters is still a very challenging task.

The aim of this work is to explore the use of systematic *ab initio* calculations to support model development. The *ab initio* calculations are performed according to the needs of model development, taking into account the capabilities and limitations of current *ab initio* methods and the availability of experimental data. The *ab initio* results are used together with experimental data to determine the initial parameter values for the model, which is then calibrated with the traditional approach. This is done for the example of phosphorus diffusion and clustering, but the methodology and the required software tools are developed in such a way that they are also useful for other dopants, co-dopants or impurities.

## *AB INITIO* CALCULATIONS

We performed *ab initio* calculations for eleven defect species containing one or several phosphorus atoms, as shown in Table I. In addition, we performed calculations for the self-interstitial, the vacancy and for perfect silicon, which are needed for the determination of binding and formation energies. All simulations were performed with supercells charged from -4e to +4e (with a neutralizing homogeneous background charge). In order to increase the confidence that the lowest energy configuration for each defect is found, several different initial configurations were used as starting points of total energy minimizations (relaxations) for each defect. The barrier to PV migration via the ring mechanism was investigated with the nudged elastic band (NEB) method.

The *ab initio* calculations based on density functional theory (DFT) were performed with the Vienna *Ab Initio* Simulation Package (VASP) [1, 2]. Cubic 216-atom supercells were used, with the lattice constant fixed at 5.47 Å, as determined with VASP using the same parameter settings as for the calculations with the defects. The generalized gradient approximation was used (GGA, Perdew-Wang 91), together with the projector-augmented wave (PAW) method. The plane-wave cutoff energy was set to 191.312eV. The Brillouin zone was sampled only at the Γ-point. The conjugate gradient algorithm was used for the relaxations, stopping when the energy difference between two steps was smaller that 1e-3eV.

## DATABASE OF *AB INITIO* RESULTS FOR MODEL DEVELOPMENT

The above described VASP parameters lead to relatively low precision calculations. However, to perform several relaxations for all defects in all charge states, roughly 900 individual VASP runs were required, already using about 80,000 CPUh. Setting up, running and analyzing such a large number of simulations would be a bottleneck just as restrictive as the large computational effort. Therefore, we implemented a framework of software tools to automate the most frequent tasks to a large degree. With these tools, we can now conveniently set up, run and analyze additional calculations, and we are essentially restricted only by the available computational resources. Additional calculations can be made for additional defects, for additional initial configurations for relaxations, or for other VASP parameters. Higher-precision calculations would require a computational effort which can be larger by one order of magnitude or more.

Since the development and refinement of diffusion and clustering models is a continually ongoing effort, there must be a continual exchange of *ab initio* results and feedback from model development. The *ab initio* results presented in this paper are therefore just a snapshot in time of a continually growing database of *ab initio* results, with new calculations being added according to the current needs of model development. For example, a sensitivity analysis of the current model can give valuable input on which parameters to calculate with increased precision settings. It is not desirable to have an *ab initio* database with very few selected results calculated with maximum precision. Instead, it is preferable to have a large set of consistent *ab initio* results calculated with sufficient precision to be useful as initial values for calibration.

We will repeat a selected set of the calculations with VASP parameter settings for higher precision, i.e. calculations with a higher plane-wave cutoff energy, more k-points, and corrections for charged supercells. From test calculations with the PV pair we expect that changing these parameters will usually not change the location of the local energy minima. Therefore, the minimum energy configuration found with the presented calculations will serve as good initial configurations for the relaxations with higher precision. The impact of the improved VASP parameter settings on the calculated binding and formation energies will be taken into account in the decision on further calculations, together with the results from the calibration procedure.

## *AB INITIO* RESULTS

For each defect species and each charge state, the lowest energy configuration was determined. Due to restrictions in paper length, it is not possible to discuss here all initial configurations and the lowest energy configurations for all defects (in all charge states). Furthermore, it is planned to perform analogous calculations for a much larger number of cluster types. Therefore, tools for an interactive exploration of the configurations included in the database of *ab initio* simulations are required. As an example, Figures 1a), 1b) and 1c) show a selection from the various initial configurations for the relaxations of the $P_2I$ cluster. The atoms have been randomly moved by a small distance from their high symmetry positions. The relaxation starting from the configuration in Figure 1a) leads to the lowest energy configuration found, as shown in Figure 1d). |Several further initial configurations have been considered for the $P_2I$ cluster. This example illustrates that even for a relatively small and simple cluster like $P_2I$ there is already a considerable number of plausible initial configurations that must be considered.

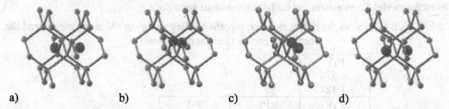

a)                    b)                    c)                    d)

Figure 1: a, b, c) Selection of three initial configurations for the relaxations for the $P_2I$ cluster. d) The lowest energy configuration found after relaxation. The bonds from a perfect reference lattice are shown in gray.

The formation energies $E_F$ of the defects were calculated according to

$$E_F = E_{Def} + (n_P(n_L - 1) - (n_L - n_P + n_I - n_V))\frac{E_{Perf}}{n_L} - n_P E_P$$

where $E_{Def}$ is the total energy of the supercell with the respective defect, $E_{Perf}$ is the total energy of a supercell without any defects (a perfect silicon lattice), $E_P$ is the total energy of a supercell with one substitutional phosphorus atom, $n_L$ is the number of lattice sites in the supercells (216 in this case), and $n_P$, $n_I$ and $n_V$ are the numbers phosphorus atoms, interstitials and vacancies in the defect. (For each defect at most one of $n_I$ and $n_V$ is nonzero.) The binding energies were calculated according to

$$E_B = (E_{Def} + (n_P + n_I + n_V - 1)E_{Perf}) - (n_P E_P + n_I E_I + n_V E_V)$$

where $E_I$ and $E_V$ are the total energies of supercells with one self-interstitial or one vacancy, respectively. This definition of binding energy is consistent with the definition of total binding or potential energy used in the KMC approach, as described below. A negative binding energy means that the defect configuration is energetically favorable compared to the reference situation with separated substitutional phosphorus atoms and self-interstitials or vacancies. The results for all defects in the neutral charge state are shown in Table I. Ionization energies were calculated from the differences in total energy for the differently charged supercells for each defect. For each charge state, the defect configuration with the lowest total energy was selected. Therefore these ionization energies contain also energy difference contributions from configurational changes. We observed the well known underestimation of the band gap with DFT/GGA.

The results of the NEB calculations for PV diffusion via the ring mechanism indicate that the dominant energy barrier for diffusion is the one for moving the phosphorus atom into the vacancy on the first nearest neighbor position (exchange of the P and V position). This finding is consistent with the results of previous *ab initio* studies of PV diffusion [3, 4]. These energy barriers are 0.94eV, 1.12eV and 1.19eV for the singly negative, the neutral and the singly positive charge state. For negative charge states, the results indicate that the diffusion mechanism is not the basic ring mechanism with the vacancy moving on first, second and third nearest neighbor lattice sites. Further relaxations and NEB calculations would be necessary to determine the diffusion mechanism in more detail. However, since currently only the dominant barrier is of interest for model development, no further calculations were done.

Table I: Each cell shows the defect name on top, the binding energy in eV in the middle and the formation energy in eV at the bottom.

| $P_4V$ | | | |
|---|---|---|---|
| -4.674 | | | |
| -1.121 | | | |
| $P_3V$ | $P_3$ | $P_3I$ | |
| -3.328 | 0.127 | -2.034 | |
| 0.225 | 0.127 | 1.419 | |
| $P_2V$ | $P_2$ | $P_2I$ | |
| -2.488 | 0.066 | -2.037 | |
| 1.064 | 0.066 | 1.416 | |
| PV | P | PI | $PI_2$ |
| -1.207 | - | -0.852 | -2.583 |
| 2.346 | - | 2.601 | 4.323 |
| V | No defect | I | |
| - | - | - | |
| 3.553 | - | 3.453 | |

## KMC SIMULATIONS

To verify that the *ab initio* results are consistent with the experimental data, we have used the energetics from the first principles calculations in a KMC simulator [5]. Based on a completed calibration for damage accumulation, point defects and extended defects [6], the phosphorus parameters were adopted as shown in the following. While the binding and migration parameters are transferred from *ab initio* as is, the ionization levels have some uncertainty due to the underestimated band gap. Assuming a realistic band gap and taking the measured ionization energy of $P^+$ of 0.045eV from the conduction band [7] as a reference, we scale the ionization energy differences from *ab initio* to get the ionization energies of the PI and PV pairs of positive and negative charge for KMC. From the valence band, the first donor and acceptor level of PI are determined to be +0.08eV and +0.18eV, and of PV to be +0.28eV and +0.76eV, respectively. Based on these ionization energies, the ionization energies of point defects, and the binding energies of neutral PI and PV pairs given by *ab initio*, the binding energy of $PI^+$ is -1.77eV and the one of $PV^+$ is -1.53eV according to the reactions illustrated by Martin-Bragado [8]. For interstitial mediated diffusion, we take the macroscopic diffusivity of phosphorus [9]

$$D_P = 0.453 \cdot \exp\left(-\frac{3.482eV}{kT}\right) cm^2/s + \frac{n}{n_i} \cdot 1.61 \cdot \exp\left(-\frac{3.647eV}{kT}\right) cm^2/s$$

and translate it to microscopic parameters following again Martin-Bragado [8]. Taking into account the binding and ionization energies from above, this leads to migration prefactors and migration energies (1.82eV) for the neutral PI pair and (1.25eV) for the positive charged PI pair. In contrast, to simulate the vacancy mediated diffusion, we apply the migration energies for positive, neutral and negative charged PV pairs given by *ab initio*. The corresponding migration prefactors remain subject to calibration. Finally, the complete P-I and P-V cluster family with total binding energies of clusters in neutral charge states from *ab initio* were used for the KMC simulations.

Based on SIMS after ion implantation and thermal anneals, the PV migration prefactors were calibrated, along with parameters for P segregation at the Si/SiO$_2$ interface and P activation after recrystallization. To allow for the formation of kink and tail in diffused P profiles (see Figure 2 and Figure 3), our model assumes that mobile PV pairs form at high P concentration and enhance the diffusion in these regions. A relatively high migration prefactor for $PV^-$ compared to neutral $PV^0$ and $PV^+$ results in qualitatively good agreement of KMC simulations with experiments. The corresponding macroscopic, vacancy mediated diffusivity is

$$D_{PV} = 8 \cdot 10^{-5} \cdot \exp\left(-\frac{3.408eV}{kT}\right) cm^2/s + \frac{n}{n_i} \cdot 8 \cdot 10^{-5} \cdot \exp\left(-\frac{3.128eV}{kT}\right) cm^2/s$$
$$+ \left(\frac{n}{n_i}\right)^2 \cdot 0.781 \cdot \exp\left(-\frac{3.622eV}{kT}\right) cm^2/s$$

However, the dose loss at high concentrations is currently underestimated, and therefore, further calibration is needed to improve the agreement quantitatively.

At higher phosphorus concentrations, KMC predicts the formation of phosphorus clusters with the dominating metastable cluster type P$_2$, which can be interpreted as the precipitation of a SiP phase. Since the KMC simulation results are sensitive to the binding energy of P$_2$, *ab initio* simulations with higher precision for this particular cluster type will be performed.

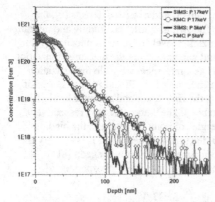 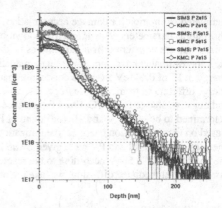

Figure 2: Comparison of SIMS [10] and KMC for P 2e15 5keV/17keV implant and anneal at 1025C for 3sec.

Figure 3: Comparison of SIMS [10] and KMC for P 2e15/5e15/7e15 17keV implant and anneal at 1025C for 3sec

## CONCLUSIONS

The presented preliminary results indicate that the approach of building a systematic *ab initio* database of model parameters to use as initial values for calibration is very promising. The close collaboration between the model developers and the researchers performing the *ab initio* calculations is essential.

## REFERENCES

1. G. Kresse, and J. Hafner, *Phys. Rev. B* **47**, R558 (1993)
2. G. Kresse, and J. Furthmüller, *Phys. Rev. B* **54**, 11169 (1996)
3. X. Liu, W. Windl, K. Beardmore, and M. P. Masquelier, *Appl. Phys. Lett.* **82**, 1839 (2003)
4. M. G. Ganechenkova, V. A. Borodin, and R. M. Nieminen, *Nucl. Instrum. Meth. B* **228**, 218 (2005)
5. *Sentaurus Process User Guide*, Version A-2007.12, Synopsys Inc., Mountain View, CA (2007).
6. *Advanced Calibration User Guide*, Version A-2007.12, Synopsys Inc., Mountain View, CA (2007).
7. S.M. Sze, *Physics of Semiconductors Devices*, 2nd ed., Wiley, New York, 1981.
8. I. Martin-Bragado, P. Castrillo, M. Jaraiz, R. Pinacho, J. E. Rubio, J.Barbolla, *Phys. Rev. B* **72**, 035202 (2005)
9. P. Pichler, *Intrinsic Point Defects, Impurities and Their Diffusion in Silicon*, Springer 2004, Vienna
10. K. Suzuki, H. Tashiro, Y. Tada, and Y. Kataoka, *IEEE Transactions on Electron Devices* **49** (11), (2002)

Mater. Res. Soc. Symp. Proc. Vol. 1070 © 2008 Materials Research Society          1070-E03-04

# Infrared Semiconductor Laser Annealing Used for Formation of Shallow Junction

Toshiyuki Sameshima[1], Yuta Mizutani[1], Naoki Sano[2], and Masao Naito[3]
[1]Tokyo A&T Univ., Koganei, 184-8588, Japan
[2]Hightec Systems Corporation, Yokohama, 222-0033, Japan
[3]Nissin Ion Equipment Co., Ltd., Koka, Japan

## ABSTRACT

We report continuous-wave (CW) infrared-semiconductor laser annealing of silicon implanted with boron atoms with assistance of diamond-like carbon (DLC) films as optical absorption layer in order to form shallow junctions. Boron ions were implanted at 10 keV at concentrations $5 \times 10^{14}$, $1 \times 10^{15}$ and $2 \times 10^{15}$ cm$^{-2}$. Boron clusters were also implanted at 30 and 6 keV at $1 \times 10^{15}$ cm$^{-2}$. The samples were coated with 200-nm-thick DLC films and annealed by irradiation with a 940 nm CW laser at 70 kW/cm$^2$ with a beam diameter of 180 μm for a dwell time of 2.6 ms. The boron in-depth concentration profiles hardly changed from the initial profiles because of rapid heat treatment. The effective doping depths at $1 \times 10^{18}$ cm$^{-3}$ were 203, 45 and 7 nm for implantation of samples implanted with boron atoms at 10 keV, boron clusters at 30 and 6 keV, respectively. The free carrier photo absorption analysis revealed complete activation of boron atoms implanted with boron clusters. Shallow doping was achieved by the present method.

## INTRODUCTION

Rapid heating is important for activating semiconductor materials implanted with impurity atoms. A high activation ratio and no marked impurity diffusion are required to fabricate extremely shallow source/drain extension (SDE) region with depth of 10 nm order in metal-oxide-semiconductor (MOS) transistor devices for 45-nm-node and beyond, which cannot be realized by conventional rapid thermal annealing (RTA) [1-4]. Flash lamp annealing (FLA) or laser spike annealing (LSA) for several milliseconds and excimer laser annealing for the order of nanoseconds have been developed [2-4]. We have also developed an annealing method on the order from $10^{-5}$ to $10^{-3}$ s using infrared lasers [5-7]. Infrared semiconductor lasers are an attractive light source because they can have a high power ~10 kW, a high conversion efficiency ~50% and a stable emission. We use a carbon optical absorption layer in order to solve the problem of the low optical absorbance in infrared regions for silicon. Black carbon layers have a high optical absorbance in an infrared range because of a high extinction coefficient and a low refractive index giving a low reflection loss. A high thermal durability to a temperature around 5000 K enables the carbon layers to act as a heat source to the underlying silicon substrates to be annealed at a high temperature [8]. In this paper, we report annealing of silicon implanted with boron atoms using infrared semiconductor laser. We demonstrate an effective activation and a low electrical resistance by the free carrier optical absorption analysis. We also report that initial boron in-depth profiles are not significantly changed by laser annealing.

## EXPERIMENTAL PROCEDURE

The ion implantation was conducted for n-type silicon substrates with a resistivity of 10-50 $\Omega$cm. Boron atoms were implanted at an acceleration energies of 10 keV with concentrations of $5 \times 10^{14}$, $1 \times 10^{15}$ and $2 \times 10^{15}$ cm$^{-2}$. Boron clusters were also implanted at 30 and 6 keV with a concentration of $1 \times 10^{15}$ cm$^{-2}$ for shallow doping. DLC films with a thickness of 200 nm were formed on the silicon surface by the sputtering method at room temperature with Ar gas. Optical measurement revealed that the optical absorbance was 85 % at 940 nm, which was the wavelength of our laser light. Samples were normally irradiated with a fiber-coupled continuous wave (CW) laser diode with a wavelength of 940 nm and a power of 20 W in air at room temperature. The diameter of the core and the numerical aperture (NA) of the fiber were 400 μm and 0.22, respectively. The diverged beam was concentrated on the sample surface using a combination of six aspherical lenses for 2:1 image formation. The power distribution of the beam was Gaussian like. The size of the beam spot was 180 μm at the full width at half maximum (FWHM) of the laser power distribution. The peak power density was 70 kW/cm$^2$ at the sample surface. Samples were mounted on an X-Y stage driven by linear motors at a constant velocity of 7 cm/s in the Y direction. The dwell time of the laser beam, which was defined as (beam spot size)/(scanning velocity), was 2.6 ms. The stage was also moved with a step of 100 μm in X direction. After laser irradiation, the carbon layer was removed by oxygen plasma treatment. Our previous study of numerical heat flow simulation has indicated that silicon surface was heated to a high temperature of 1390°C just below the melting point [9]. Boron clusters implantation caused serious amorphization of the silicon surface. Measurement of optical reflectivity spectra in the ultra-violet region and Raman scattering revealed that the surface region was completely recrystallized by the laser annealing [9]. Rapid heating at a high temperature in solid phase is achieved. The boron concentration in-depth profiles were measured by secondary ion mass spectroscopy (SIMS) analysis. The free carrier optical absorption effect was analyzed in order to investigate the hole carrier concentration and its mobility. Optical transmissibity spectra were measured by conventional Fourier transform infrared (FTIR) spectrometry after polishing of the rear surface of the samples. The optical transmissivity spectra were analyzed using a numerical calculation program, which was constructed with the optical interference effect using a model of air/60-layered doped Si/Si substrate structures [10]. The optical transmissivity of the sample surface is governed by the real refractive indexes and extinction coefficients. The free carrier in Si causes changes in the real refractive index and extinction coefficient. These changes depend on the carrier density and carrier mobility. The most possible in-depth distributions of the electron carrier density and carrier mobility were obtained by fitting the calculated reflectivity spectra to the experimental reflectivity spectra. The sheet resistance was also investigated using a four-point-probe measurement system.

## RESULTS AND DISCUSSION

Figure 1 shows the boron atoms in-depth profiles of the samples $2 \times 10^{15}$-cm$^{-2}$-as-boron implanted at 10 keV and laser-annealed (a), $1 \times 10^{15}$-cm$^{-2}$-as-boron-clusters implanted at 30 keV and laser annealed (b), $1 \times 10^{15}$-cm$^{-2}$-as-boron-clusters-implanted at 6 keV and laser annealed (c). The peak boron concentration depths were 50 nm for the sample as-boron implanted at 10 keV. The boron concentration gradually decreased to $1 \times 10^{18}$ cm$^{-3}$ as the depth increased to 200 nm. No marked change in the boron concentration profile was observed after laser annealing. After

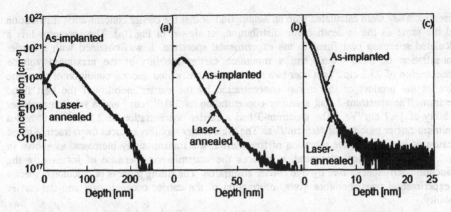

Fig.1: Boron atoms in-depth profiles of the samples $2\times10^{15}$-cm$^{-2}$-as-boron implanted at 10 keV and laser-annealed (a), $1\times10^{15}$-cm$^{-2}$-as-boron-clusters implanted at 30 keV and laser annealed (b), $1\times10^{15}$-cm$^{-2}$-as-boron-clusters-implanted at 6 keV and laser annealed (c).

laser annealing, the peak boron concentration was also observed at 50 nm and the depth at $1\times10^{18}$ cm$^{-3}$ were 203 nm. In the case of boron-clusters implantation at 30 keV, the peak concentration appeared at 6 nm from the surface for the sample as-implanted. The depth at $1\times10^{18}$ cm$^{-3}$ was 45

nm. The in-depth boron concentration profiles after laser annealing was almost same as that as-implanted within a resolution of 2 nm. Boron clusters implantation at 6 keV resulted in the peak concentration at 1 nm just below the surface. The depth of $1\times10^{18}$ cm$^{-3}$ was 9 nm. Laser annealing resulted in a very steep in-depth boron concentration profile with depths of 1 nm for the peak concentration and 7 nm for the concentration of $1\times10^{18}$ cm$^{-3}$. Those results mean that silicon surface regions were heated to a high temperature by laser annealing, and that the heating duration was too short for boron atoms to diffuse into the silicon substrate significantly.

Figure 2 is a demonstration of fitting of calculated spectra to the experimental FTIR spectrum for samples with $5\times10^{14}$ cm$^{-2}$-boron implantation and laser annealing. Three calculated spectra are

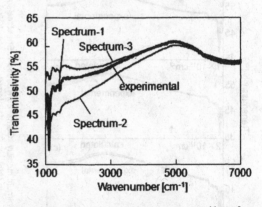

Fig.2: Experimental spectrum for $5\times10^{14}$-cm$^{-2}$-boron implantation and laser annealing. Spectra are calculated with carrier concentrations of $5\times10^{14}$ cm$^{-2}$ and minimum carrier mobilities of 23.3 cm$^2$/Vs (spectrum-1), $1\times10^{15}$ cm$^{-2}$ and 16.7 cm$^2$/Vs (spectrum-2), and $2.5\times10^{14}$ cm$^{-2}$ and 46.6 cm$^2$/Vs (spectrum-3)

presented. They were calculated with an assumption of that the carrier concentration distribution had the same as the in-depth boron distribution, as shown in Fig.1(a). The spectrum-1 is a calculated spectrum best fitted to the experimental spectrum. It was obtained with a carrier concentration of $5 \times 10^{14}$ cm$^{-2}$ and a minimum carrier mobility at the maximum volume concentration of 23.3 cm$^2$/Vs. Other two spectra were calculated under a condition of the same value of the product of the carrier concentration by the carrier mobility of the best fitted spectrum. The spectrum-2 had a carrier concentration of $1 \times 10^{15}$ cm$^{-2}$ and a minimum carrier mobility of 16.7 cm$^2$/Vs. The spectrum-3 had a carrier concentration of $2.5 \times 10^{14}$ cm$^{-2}$ and a minimum carrier mobility of 46.6 cm$^2$/Vs. The high carrier mobility reduces the refractive index because of high anti-phase vibration of free carrier, the transmissivity increases, as shown in Fig.2. The high carrier concentration reduces the transmissivity because of increase in the extinction coefficient caused by free carrier absorption. The fitting process of calculated spectra to experimental ones therefore gives information of the carrier concentration and the carrier mobility.

Figure 3 shows transmittance spectra for 10-keV-boron implanted samples annealed by laser irradiation in the cases of boron concentration of $5 \times 10^{14}$ cm$^{-2}$ (a), $1 \times 10^{15}$ cm$^{-2}$ (b) and $2 \times 10^{15}$ cm$^{-2}$ (c). Figure 3 also represents calculated spectra best fitted to the experimental spectra. The transmissivity decreased as the wavenumber decreased because free carrier absorption was effective for low wavenumber regions. The transmissivity also decreased as the doping concentration increased especially in the low wavenumber region because of high free carrier effect. Figure 4 shows sheet resistance, carrier concentration and the minimum carrier mobility

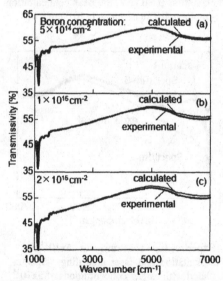

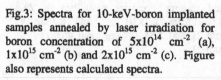

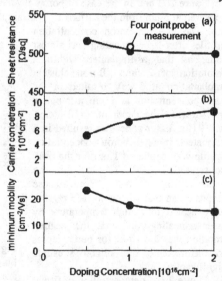

Fig.3: Spectra for 10-keV-boron implanted samples annealed by laser irradiation for boron concentration of $5 \times 10^{14}$ cm$^{-2}$ (a), $1 \times 10^{15}$ cm$^{-2}$ (b) and $2 \times 10^{15}$ cm$^{-2}$ (c). Figure also represents calculated spectra.

Fig.4: Sheet resistance (a), carrier concentration (b), and the minimum carrier mobility (c) obtained from analysis of spectra shown in Fig.3. A sheet resistance by four point probe measurement is also presented.

obtained from analysis of FTIR spectra shown in Fig.3 as a function of doping concentration. The sheet resistance gradually decreased from 515 to 505 $\Omega$/sq. Four point probe measurement gave a good agreement with the sheet resistance obtained by free carrier absorption analysis. The boron atoms was completely activated and hole carriers with a high concentration were generated for samples with $5 \times 10^{14}$ cm$^{-2}$-implantation. On the other hand the carrier generation ratio decreased as the doping concentration increased, as shown in Fig.4 (b). This probably results from decrease in the boron ionization probability according to the high doping concentration doping. The carrier mobility decreased as the doping concentration increased because of increase in the impurity scattering.

Figure 5 shows the transmittance spectra for samples implanted with boron clusters at 30 and 6 keV and annealed by laser irradiation. Figure 5 also represents calculated spectra best fitted to the experimental spectra. They were calculated with an assumption of that the carrier concentration distributions had the same as the in-depth boron distributions, as shown in Fig.1 (b) and (c). Although the doping concentration was set at $1 \times 10^{15}$ cm$^{-2}$, SIMS measurement revealed that total boron concentration implanted in silicon was $7.1 \times 10^{14}$ and $4.4 \times 10^{14}$ cm$^{-2}$ for samples implanted at 30 and 6 keV, respectively, as shown in Table I. Good agreements between experimental and calculated spectra resulted in a high carrier generation ratio almost 1, although the carrier mobility was very low, as shown in Table I. The very high volume concentration caused by shallow doping shown in Fig. 1 probably induced change in atom bonding configuration so that the impurity band was formed. The high carrier density was probably generated by placing the Fermi level in the impurity band.

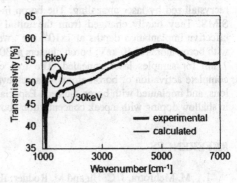

Fig.5: Spectra for samples implanted with boron clusters at 30 and 6 keV and annealed by laser irradiation. Figure also represents calculated spectra.

TABLE I.   Doping concentration obtained by SIMS, carrier concentration, sheet resistance and the minimum carrier mobility obtained from analysis of spectra shown in Fig.5.

| Energy (keV) | Doping concentration (cm$^{-2}$) | Carrier concentration (cm$^{-2}$) | Sheet resistance ($\Omega$/sq) | Minimum mobility (cm$^2$/Vs) |
|---|---|---|---|---|
| 30 | $7.1 \times 10^{14}$ | $7.1 \times 10^{14}$ | 232 | 15 |
| 6 | $4.3 \times 10^{14}$ | $4.3 \times 10^{14}$ | 564 | 10 |

## CONCLUSIONS

CW infrared-semiconductor laser annealing was investigated for activating silicon implanted with boron atoms with assistance of DLC films as optical absorption. Boron ions were implanted at 10 keV at concentrations $5 \times 10^{14}$, $1 \times 10^{15}$ and $2 \times 10^{15}$ cm$^{-2}$. Boron clusters were also implanted at 30 and 6 keV at $1 \times 10^{15}$ cm$^{-2}$. The samples were coated with 200-nm-thick DLC films. the optical absorbance was 0.85 at 940 nm. Samples were annealed by irradiation with a 940 nm CW laser at 70 kW/cm$^2$ with a beam diameter of 180 μm for a dwell time of 2.6 ms. Amorphous surface regions induced by implantation of boron clusters were completely recrystallized by laser annealing. The boron in-depth concentration profiles were measured by SIMS. They hardly changed from the initial profiles because of rapid heat treatment. The effective implantation depths at $1 \times 10^{18}$ cm$^{-3}$ were 200, 45 and 9 nm for samples as-implanted with boron atoms at 10 keV, boron clusters at 30 and 6 keV, respectively. They were 203, 45 and 7 nm for samples laser annealed. The free carrier photo absorption analysis revealed that complete activation of boron atoms was achieved for samples implanted at 10 keV with boron ions, and implanted with boron clusters. Especially boron clusters implantation at 6 keV resulted in shallow doping with a peak concentration above $1 \times 10^{21}$ cm$^{-3}$.

## REFERENCES

1. M. Mehrotra, J. C. Hu and M. Rodder: IEDM Tech. Dig. 1999, p.419.
2. T. Ito, K. Suguro, M. Tamura, T. Taniguchi, Y. Ushiku, T. Iinuma, T. Itani, M. Yoshioka, T. Owada, Y. Imaoka, H. Murayama, and T. Kusuda, Ext. Abstr. 3rd Int. Workshop on Junction Technol., 2002, p.23.
3. A. Shima and A. Hiraiwa: Jpn. J. Appl. Phys. 45 5708 (2006).
4. K. Goto, T. Yamamoto, T. Kubo, M. Kase, Y. Wang, T. Lin, S. Talwar and T. Sugii, IEDM. Tech. Dig. 1999, p. 931.
5. T. Sameshima and N. Andoh: Jpn. J. Appl. Phys. 44 7305 (2005).
6. N. Sano, M. Maki, N. Andoh and T. Sameshima: AM-FPD Digest of Tech. Papers (2006, Tokyo) 329.
7. N. Sano, M. Maki, N. Andoh and T. Sameshima: Jpn. J. Appl. Phys. 46 (2007) 1254.
8. T. Sameshima and N. Andoh, Proc. in Mat. Res. Soc. Symp. 849 (2004) KK9.5.
9. T. Sameshima, M. Maki, M. Takiuchi, N. Andoh, N. Sano, Y. Matsuda and Y. Andoh, Jpn. J.Appl. Phys. 46 6474 (2007).
10. T. Sameshima, K. Saitoh, N. Aoyama, S. Higashi, M. Kondo and A. Matsuda, Jpn. J. Appl. Phys. 38 1892(1999).

Mater. Res. Soc. Symp. Proc. Vol. 1070 © 2008 Materials Research Society 1070-E03-05

# Modelling of Sb Activation in Ultra-shallow Junction Regions in Bulk and Strained Si

Yan Lai, Nicolas Cordero, and James C Greer
Tyndall National Institute, Cork, Ireland

## ABSTRACT

The activation behaviour of dopants in ultra-shallow junctions on strained silicon is investigated from a simulation vantage point. Process models available in commercial simulation tools are typically developed for junctions formed with high implantation energies (> 50 keV) and for long anneal times. Hence the question arises as to whether these models and parameter sets can accurately predict the active profile for highly doped, ultra-shallow junctions formed thin strained silicon layers using short rapid thermal anneals (RTA, <10 seconds) at temperatures below 800 °C. By incorporating the results from experimental data, we develop modified models allowing for improved predictions of antimony activation within both bulk and strained silicon.

## INTRODUCTION

For advanced CMOS technologies, highly conductive, abrupt, ultra-shallow regions are required as source and drain extensions. These ultra-shallow junctions require very high active dopant concentrations. Another feature of advanced CMOS is the use of strained silicon ($\sigma$-Si) to take advantage of the improved carrier mobility relative to relaxed silicon. When both technologies are combined, due to the limited thermal budget (annealing temperatures below 800 °C to avoid stress relaxation) and the very thin active layer (less than 20 nm), the activation of these ultra-shallow dopants poses a challenge. Under these conditions, antimony (Sb) appears as a promising alternative to arsenic, as it shows higher activation in strained silicon than in bulk silicon [1]. However, the dopants exhibit anomalous activation/diffusion behaviour in the ultra-shallow regions. This paper focuses on the numerical modelling of these effects using common process models incorporated within commercial simulation tools.

## SIMULATIONS

### Sb activation in bulk Si

It has been reported that rapid thermal annealing of Sb implanted silicon at 700 °C, an active concentration of $4.5 \times 10^{20}$ cm$^{-3}$ can be achieved [2]. This far exceeds the solid solubility reported by Trumbore ($7 \times 10^{19}$ cm$^{-3}$) [3]. At higher temperature anneals, this supersaturation effect is reduced. As we will demonstrate, a naïve application of the models available in commercial process modelling software (Synopsys TCAD [4] and Silvaco VWF [5]), the supersaturation phenomena for Sb in silicon is not accurately captured (Figure 1, Table 1).

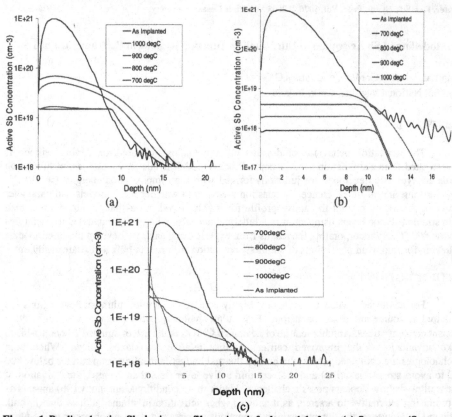

**Figure 1.** Predicted active Sb doping profiles using default models from (a) Sentaurus (Synopsys TCAD) and (b) TSuprem4-Athena (Silvaco VWF) (c) TSuprem4 (Synopsys TCAD). Sb were implanted at 2 keV with a dosage of $3.5 \times 10^{14}$ cm$^{-3}$ followed by a 10 seconds RTA.

| Temperature | 700 °C | 800 °C | 900 °C | 1000 °C |
|---|---|---|---|---|
| Solid Solubility in Sentaurus(1/cm$^3$) | $5.089 \times 10^{19}$ | $7.937 \times 10^{19}$ | $1.148 \times 10^{20}$ | $1.566 \times 10^{20}$ |
| Solid Solubility in TSuprem4 (1/cm$^3$) | $1.90 \times 10^{19}$ | $2.30 \times 10^{19}$ | $3.10 \times 10^{19}$ | $4.00 \times 10^{19}$ |

**Table 1.** Solid solubility limit at different temperatures as implemented with Sentaurus and TSuprem4

At present, experimental work has shown that Sb tends to form clusters prior to precipitation for very heavy doping [6]. For short RTA times and low annealing temperature (< 800 °C), clustering is believed to be the dominant deactivation mechanism for Sb. The supersatuarated levels of Sb are a result of metastability, and a dramatic drop in activation levels upon further annealing occurs until clustering restores thermal equilibrium.

To describe this behaviour, the transient activation model was adopted in both Sentaurus and TSuprem4 (Synopsys TCAD and Silvaco VWF) to simulate this scenario. The low activation profiles given by default models are caused by a low initial activation and high clustering rate. By enhancing the initial activation conditions and decreasing the clustering rate, the activation level of Sb can be elevated as shown in Figure 2. In TSuprem4-Athena, the initial activation concentration plays a more important role than other factors for Sb doping. The post-implant and pre-diffuse initialisations are more complex with TSuprem4-Synopsys TCAD. The default models in TSuprem4-Synopsys TCAD assume the Sb dopants in the amorphous regions are all initially clustered. Si becomes amorphous after implantation within the first few nanometers of the surface and this region overlaps with the peak region of the Sb concentration profile. The full clustering assumption in this case results in a very low activation in the active concentration profiles.

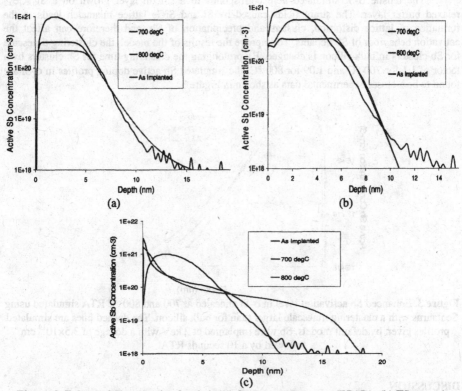

**Figure 2.** Enhanced Sb activation levels in (a) Sentaurus (Synopsys TCAD), (b) TSuprem4 (Silvaco VWF) and (c) TSuprem4 (Synopsys TCAD).

Comparing the different models, the results given by TSuprem4-Athena (Figure 2.b) are restrained by the equilibrium solid solubility limits assumed for Sb, and contradict the fact that Sb achieves its maximum activation concentration at 700 °C for short RTA times. Results from

TSuprem4-Synopsys TCAD yield too much out- and in-diffusion at 800 °C. For this example Sentaurus (Figure 2.a) provides the best dopant-cluster models for transient activation of Sb dopants in light of experimental findings.

## Sb activation in strained Si

It has been demonstrated experimentally that compressive biaxial strain enhances Sb diffusion while tensile strain retards diffusion compared to the diffusion in the strain relaxed system [7]. Recent experiments [1] have shown that the activation of Sb is enhanced in strained silicon compared to bulk silicon for the low annealing temperatures ($\leq$ 800 °C) and short RTA times (10 seconds) required for formation of ultra-shallow junctions.

The tensile σ-Si simulated here corresponds to a silicon layer grown on a $Si_{0.8}Ge_{0.2}$ relaxed buffer layer. The strain fields caused by Si and SiGe lattice mismatch leads to the formation of lattice distortion, clusters and precipitation of Sb, and therefore can affect the activation behaviour of Sb dopants. To improve the results of the model, the clustering timescale for Sb clusters in bulk silicon is changed. By prolonging the clustering time for Sb clusters by a factor of 1.1 for 700 °C and 1.09 for 800 °C, the simulated Sb active doping profiles in σ-Si are found to be a closer experimental data as shown in Figure 3.

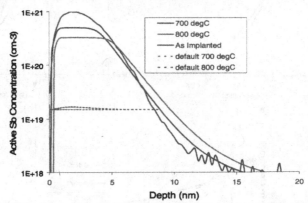

**Figure 3.** Enhanced Sb activation level in σ-Si annealed at 700 and 800 °C RTA simulated using Sentaurus with a clustering timescale larger than for bulk silicon. The dashed lines are simulated profiles given by default models. Sb were implanted at 2 keV with a dosage of $3.5\times10^{14}$ cm$^{-3}$ followed by a 10 seconds RTA.

## DISCUSSION

Simulated active Sb profiles annealed at 700 °C in both bulk and σ-Si in comparison with experimental differential Hall measurements are shown in Figure 4. Though more experimental data is needed to determine the full active Sb dopant profile, the simulated enhancement of activation level in strained Si compared with bulk Si is consistent overall with the differential

Hall measurement results. Note that the model results near the interface may overpredict active doping levels due to the fact that the formation of an amorphous layer upon implantation and recrystallization are not adequately treated.

From the modified models, it can be concluded that post-implant initialisation non-equilibrium models are essential to simulate the Sb clustering kinetics for short anneal times. However, the dopant diffusion and activation are not only time, but also temperature and concentration dependant processes. Different diffusion/activation mechanisms take place under different process conditions, and there is no single existing model or parameter set that governs all the scenarios. Therefore, the modified models here only work for the activation of Sb at low temperature and for short anneal times, but are not transferable to activation or diffusion for higher temperatures, or for longer anneal times.

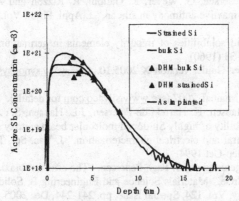

**Figure 4.** Simulated active Sb profiles (10 seconds of RTA at 700 °C) in comparison with differential Hall measurement

## CONCLUSIONS

We have simulated the supersaturated effects of Sb activation using both Sentaurus and TSuprem4 by modifying the non-equilibrium clustering models. Sentaurus yields the best agreement with experiment of the models considered due to its superior transient dopant-cluster models. The activation of Sb in strained Si has also simulated using Sentaurus, and results show enhanced activation level relative to bulk Si, in agreement with differential Hall measurements. The models were modified exclusively for short anneal times and low RTA temperatures, and are not transferable to Sb activation in other scenarios. Insight into the critical parameters governing the formation of metastable Sb doping levels above the solid solid solubility limit has been obtained. The parameter set extracted represents an improvement for the simulation of ultra-shallow Sb junction formation for advanced CMOS technologies.

## ACKNOWLEDGMENTS

This work has been supported by Science Foundation Ireland. The authors would like to acknowledge Nick S. Bennett (Univ. of Newcastle and formerly Univ. of Surrey) for providing us with experimental data from differential Hall measurements.

## REFERENCES

1. N.S. Bennett, N.E.B. Cowern, A.J. Smith, R.M. Gwilliam, B.J. Sealy, L. O'Reilly and P.J. McNally, "Highly conductive Sb-doped layer in strained Si," Applied Physics Letters, Vol 89, No. 18, Article 182122 (Oct 2006).
2. A.N. Larsen, F.T. Pedersen, G. Weyer, R. Galloni, R. Rizzoli and A. Armigliato, "The nature of electrically inactive antimony in silicon," J. Appl. Phys., Vol.59, No. 6, pp. 1908-17, Mar 1986.
3. F.A Trumbore, "Solid solubilities of impurity elements in germanium and silicon," Bell Syst. Tech. J. 39:205-33 (1960)
4. Synopsys TCAD User Guide, version X-2005.10. (see www.synopsys.com/products/tcad/tcad.html for details).
5. Silvaco Athena User Manual, 2007 (see www.silvaco.com for details).
6. E.V. Thomsen, O. Hansen, K. Harrekilde-Petersen, J.L. Hansen, S.Y. Shiryaev and A.N. Larsen, "Thermal stability of highly Sb-doped molecular beam epitaxy silicon grown at low temperatures: Structural and electrical characterization," J. Vac. Sci. Technol. B. Vol. 12, No. 5, pp. 3016-22, Sep-Oct 1994.
7. A.N. Larsen, N. Zangenberg and J. Fage-Pedersen, "The effect of biaxial strain on impurity diffusion in Si and SiGe," Materials Science and Engineering B, Solid State Materials for Advanced Technology, Vol. 124, Special Issue, pp. 241-244, Dec 2005.

Mater. Res. Soc. Symp. Proc. Vol. 1070 © 2008 Materials Research Society

## Study on the Effect of RTA Ambient to Shallow N+/P Junction Formation using PH3 Plasma Doping

Seung-woo Do[1], Byung-Ho Song[1], Ho Jung[1], Seong-Ho Kong[1], Jae-Geun Oh[2], Jin-Ku Lee[2], Min-Ae Ju[2], Seung-Joon Jeon[2], Ja-Chun Ku[2], and Yong-Hyun Lee[1]
[1]School of Electrical Engineering and Computer Science, Kyungpook National University, Daegu, 702-701, Korea, Republic of
[2]Hynix Semiconductor Inc., Kyoungki-do, 467-701, Korea, Republic of

## ABSTRACT

Plasma doping (PLAD) process utilizing $PH_3$ plasma to fabricate n-type junction with supplied bias of $-1$ kV and doping time of 60 sec under the room temperature is presented. The RTA process is performed at 900 °C for 10 sec. A defect-free surface is corroborated by TEM and DXRD analyses, and examined SIMS profiles reveal that shallow $n^+$ junctions are formed with surface doping concentration of $10^{21}$ atoms/cm$^3$. The junction depth increases in proportion to the $O_2$ gas flow when the $N_2$ flow is fixed during the RTA process, resulting in a decreased sheet resistance. Measured doping profiles and the sheet resistance confirm that the $n^+$ junction depth less than 52 nm and minimum sheet resistance of 313 $\Omega/\square$ are feasible.

## INTRODUCTION

Transistor doping challenges for advanced technology nodes, illustrated by the recent international technology roadmap for semiconductor (ITRS 2005), include high dose and low energy implants for gate, source/drain and source/drain extension dopings[1]. As the scaling down of metal oxide semiconductor (MOS) device continues, MOS device requires ultra shallow source/drain junction depth to reduce the short channel effect and enhance the performance of the devices[2,3]. Among many doping techniques to form a ultra shallow junction, the plasma doping (PLAD) has attracted much attention due to the high-dose ion doping with low ion energy, high activation efficiency, reduced damage and improved doping profile[4,5]. PLAD is a surface treatment technique characterized by the implantation of energetic ions that are generated by immersing the substrate into plasma and applying a pulsed negative voltage to the substrate [6].

In this study, $n^+$ shallow junctions are formed using PLAD technique and rapid thermal annealing (RTA) method. The PLAD is followed by RTA process performed either in $N_2$, $O_2$ or $O_2+N_2$ ambient.

## EXPERIMENT

The PLAD is performed on (100) p-type silicon wafer using a plasma doping system. A standard wafer cleaning process is performed followed by a diluted HF dip to remove native oxide prior to loading the wafer in the plasma doping system. Plasma doping source is gas mixture of 10 % $PH_3$ diluted with He. The argon gas is injected to the process chamber to help the first ignition of plasma and stopped just before the bias voltage applying. The wafer is placed in the plasma generated with 200 W and a negative DC bias (–1 kV) is applied to the substrate for 60 sec under no substrate heating. The flow rate of the diluted $PH_3$ and the process pressure are 100 sccm and 10 mTorr, respectively. After PLAD, RTA process performed either in $N_2$, $O_2$ or $O_2+N_2$ ambient at 900 °C for 10 sec.

The shallow $n^+$ doping profile, crystalline defect and sheet resistance are examined after PLAD and following RTA by secondary ion mass spectrometry (SIMS), transmission electron microscopy (TEM), double crystal X-ray diffraction (DXRD) and 4-point probe analysis methods, respectively.

## DISCUSSION

As shown in Figure 1, we performed transmission electron microscopy (TEM) analysis to confirm the defect structure on the surface of the plasma-doped sample. TEM images reveal no crystalline defect even directly after the plasma doping. For a precise defect investigation after plasma doping on the substrate surface, double crystal X-ray diffraction (DXRD) measurement is also accomplished and shown in Figure 2. The DXRD curves of as-doped sample and annealed one after PLAD have a wider peak than bare silicon wafer. It means that doped samples have minor crystalline defects. However it is clear that RTA-processed sample, which has a sharper peak, results in a fewer defect than as-doped one.

Figure 1. TEM images of the PH₃ plasma-doped samples at (a) non-annealed and (b) RTAed at 900 °C for 10 sec.

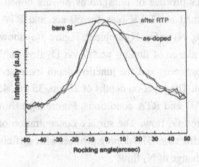

Figure 2. Double crystal X-ray diffraction (DXRD).

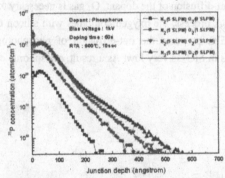

(a) SIMS profiles as a function of O₂ flow during RTA.

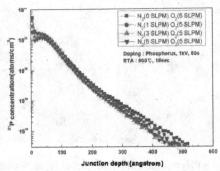

(b) SIMS profiles as a function of $N_2$ flow during RTA.

Figure 3. SIMS profiles of samples annealed gas mixture ambient at 900 °C for 10 sec.

Figure 3 shows SIMS profiles of phosphorus plasma doped and annealed in different ambient. PLAD was performed at −1 kV bias for 60 sec and RTA process was followed at 900 °C for 10 sec either in $O_2$, $N_2$ or $O_2+N_2$ ambient. Figure 3(a) shows the junction depth of the doped sample and the dependence of doping profile on $O_2$ flow with fixed $N_2$ flow. Both the surface concentration of phosphorus and the junction depth increased in proportion to $O_2$ flow compared to $N_2$-only ambient. The junction depths of 25 nm, 35 nm, 45 nm and 52 nm have been achieved with proposed PLAD and RTA conditions. Figure 3(b) shows the doping profile as a function of $N_2$ flow with fixed $O_2$ flow. The surface concentration of phosphorus and junction depth did not show a large difference. The doping profiles with sufficient $O_2$ flow above 3 SLPM were less dependent on the change of $N_2$ flow.

The RTA process performed in $N_2$-only ambient showed a relatively low concentration of phosphorus. It seems that phosphorus is out diffused in $N_2$-only ambient during the RTA process. In order to prevent the out-diffusion of the dopant, $O_2$ gas is necessary to be flowed into the RTA chamber. During the RTA process, the oxygen combines with silicon to form $SiO_2$ thin film at the surface and $SiO_2$ film prevents the out-diffusion of phosphorus. It is known that the diffusivity of phosphorus in $SiO_2$ is very low. As a result, phosphorus concentration increases in the surface.

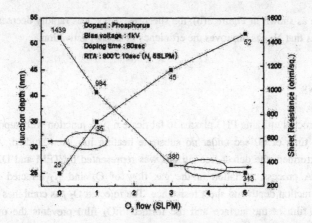

Figure 4(a). The variations of sheet resistance and junction depth as a function of $O_2$ flow.

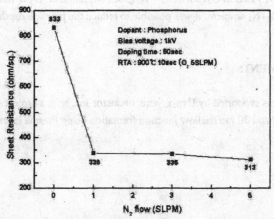

Figure 4(b). The variations of sheet resistance as a function of $N_2$ flow.

Figure 4 shows the variations of sheet resistance and junction depth as a function of gas flow. For both cases of $O_2$ and $N_2$ injection during the RTA process, the sheet resistance decreases and junction depth increases. As can be seen in Figure 4(a) and 4(b), $O_2$ flow forms $SiO_2$ thin film and gives less out-diffusion of implanted dopants, resulting in deeper junction and smaller sheet resistance. Although there was no dominant difference in the doping profile with

N$_2$–flow change as shown in Figure 3(b), the sheet resistance was rapidly decreased at 1 SLPM N$_2$ flow. It seems that N$_2$ gas improves the efficiency of dopant activation.


## CONCLUSIONS

PLAD process utilizing PH3 plasma to fabricate n-type junction with supplied bias of –1 kV and doping time of 60 sec under no substrate heating has been studied. After the RTA process was performed, the defect-free surface was represented by TEM and DXRD analyses. During the RTA process, variation in the gas flow of O$_2$ and N$_2$ effected to the surface concentration, junction depth and sheet resistance. The injected O$_2$ gas combines with silicon to form SiO$_2$ thin film at the surface and the formed SiO$_2$ film prevents the out-diffusion of phosphorus. As a result, surface concentration and junction depth increase so that the sheet resistance decreases.

A shallow n$^+$ junction depth with phosphorus, which surface concentration is about $10^{21}$ atoms/cm$^3$, could be controlled by modifying the ambient condition of RTA. The sheet resistance decreased down to 313 $\Omega/\square$ in O$_2$ (5SLPM) + N$_2$ (5SLPM) ambient. By optimizing the annealing condition in O$_2$ or O$_2$+N$_2$ ambient, it was possible to reduce the junction depth below 30 nm.


## ACKNOWLEDGMENTS

This work was supported by Hynix Semiconductor Inc. with a project entitled, 'N+/P+ poly counter doping and 30 nm shallow junction formation using Plasma Immersion Ion Doping (PI$^2$D)'.

## REFERENCES

1. International Technology Roadmap for Semiconductors, Front End Processes, 23, (2005).
2. M. Y. Tsai and B. G. Streetman, J. Appl. Phys., 50, p188, (1979).
3. H. Jin, S. K. Oh, H. J. Kang, S. W. Lee, Y. S. Lee and K. Y. Lim, J. Korean Phys. Soc. 46, S52, (2005)
4. Paul K Chu, Plasma Phys. Control Fusion, 45, (2005)
5. F. Lallement et al., Symp. VLSI Tech. Dig., 178, (2004)
6. A. Anders, Handbook of Plasma Immersion Ion Implantation and Deposition, (2000)

Mater. Res. Soc. Symp. Proc. Vol. 1070 © 2008 Materials Research Society

# Modeling of Effect of Stress on C Diffusion/Clustering in Si

Hsiu-Wu Guo[1], Chihak Ahn[2], and Scott T Dunham[1]

[1]Electrical Engineering, University of Washington, Seattle, WA, 98195

[2]Physics, University of Washington, Seattle, WA, 98195

## ABSTRACT

Extensive *ab-initio* calculations were performed to find formation energies of stable C complex configurations in silicon as function of stress. The results indicate that substitutional C is the lowest energy state, while the <100> split interstitial is the dominant mobile species. Investigation of small carbon/interstitial clustering suggests that these clusters are only significant under a substantial interstitial supersaturation. We studied the diffusion path for neutral C including the impact of stress. Through KLMC analysis of stress effect on diffusivity, we found that tensile biaxial strain enhances the effective C diffusivity, with a stronger stress dependence for C diffusivity in the out-of-plane direction.

## INTRODUCTION

The formation of very shallow and abrupt dopant profiles with high electrical activation is required for the exponential downscaling of metal-oxide-semiconductor field-effect transistors (MOSFETs). Carbon co-implantation can reduce B transient enhanced diffusion (TED) [1] and several experiments have shown that C provides a highly efficiency sink for excess interstitials during annealing [2-4]. Napolitani *et al.* [3] also found no detrimental effects on B electrical activation with the incorporation of C. Since strain engineering to enhance carrier mobility is widely used in CMOS technology, it is important for the behavior of impurity diffusion and activation under stress to be thorough understood. To that end, we perform *ab-initio* calculations for a range of carbon complex configurations and find the diffusion path of carbon interstitial in neutral charge statel. Using Kinetic Lattice Monte Carlo (KLMC) simulations, we predict the impact of stress on carbon diffusion.

## CARBON COMPEX FORMATION

Carbon is most stable as a substitutional impurity $C_s$ in the neutral charge state. Carbon interstitials (CI) can be generated through kick-out or Frank-Turnbull mechanisms [5-7].

$$C_s + I \Leftrightarrow CI \qquad (1)$$
$$CI + V \Leftrightarrow C_s \qquad (2)$$

Experimental results [5, 7] and theoretical study [8] suggest the <100> split carbon interstitial as the ground-state CI configuration. Our DFT calculations of formation energies of carbon interstitials in different configurations (see TABLE I) confirmed the <100> split as the most stable CI structure. In this configuration, the C and Si atoms are displaced along a <100> direction sharing a single lattice site. For all the calculations in this work, the density function theory code VASP [9, 10] was used with 64-atom supercell, energy cutoff of 340eV, and $2^3$ Monkhorst $\vec{k}$-point sampling method within generalized gradient approximation (GGA).

A possible clustering reaction to reduce the silicon interstitial concentration is CI-$C_s$ pairing (Eq.3), which forms a stable and immobile carbon complex [11].

$$CI + C_s \Leftrightarrow C_2I \qquad (3)$$

In equilibrium,

$$C_{C_2I} = \frac{C_{C_s}^2}{C_{Si}} \exp\left(-\frac{\Delta G_f^{C_2I}}{kT}\right), \quad (4)$$

where $\Delta G_f^{C_2I}$ is the free energy of formation for $C_2I$, and $C_{Si}$ is the density of lattice sites.

|  | Split <100> | Split <110> | Hex | Tet |  |
|---|---|---|---|---|---|
| $E_f$ (eV) | 1.97 | 2.96 | 3.11 | 4.22 |  |
|  | $C_2I$ <100> | $C_2I$ (A) | $C_2I$ (B) | $C_2I_2$ | $CI_2$ |
| $E_f$ (eV) | 0.51 | 0.98 | 0.95 | 2.16 | 3.44 |

TABLE I: Formation energies of various carbon complexes.

TABLE I also shows the calculation results for formation energies of CI, $C_2I$, and other small carbon complexes. Our calculations confirmed the results from Liu *et al.* [12], where the most stable $C_2I$ configuration was identified as <100> split (dumbbell), as shown in Fig. 1[a]. Leary *et al.* [13] and Capaz *et al.* [14] identified two other $C_2I$ configurations in Si. The type "A" configuration (see Fig. 1[b]) consists of a <100> split CI, slightly perturbed by neighboring substitutional carbon. In the type "B" configuration (see Fig. 1[c]), a Si interstitial is accommodated between two substitutional carbon atoms. Mattoni et al. [15] reported the type "A" configuration to be favored over the type "B" by 0.4 eV, and the formation energy for a <100> split CI to be 0.2 eV higher than the type "A" configuration. However, we found the energy of $C_2I$ in <100> split configuration to be lower than the two Capaz structures by 0.4 eV or more, which is consistent with the results from Liu *et al.* [12]. A possible reason for this discrepancy could be the different setup for VASP calculations. Mattoni *et al.* [15] used lower cutoff energy (160 eV), which might not be sufficient in this application.

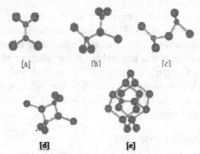

Figure 1: Various small carbon clusters. $C_2I$ with different configurations are shown in the top row: [a] <100> split, [b] type "A", and [c] type "B". [d] and [e] are $C_2I_2$ and $CI_2$. Blue and brown atoms are Si and C respectively.

Mattoni *et al.* [15] further suggested two other possible stable carbon complexes $C_2I_2$ and $CI_2$, as shown in Figs. 1[d] and [e]. We also performed calculations to obtain their formation energies (TABLE I). Note that the formation energies for all C complexes are positive, which implies that in the absence of strong interstitial supersaturation, the concentrations of all these complexes are expected to be well below that of the substitutional carbon $C_s$. The equilibrium concentration of $C_nI_m$ can be defined as

$$C_{C_nI_m}^* = \theta_{C_nI_m} C_{Si} \left(\frac{C_{C_s}}{C_{Si}}\right)^n \left(\frac{C_I}{C_I^*}\right)^m \exp\left(-\frac{\varepsilon_f^{C_nI_m}}{kT}\right), \quad (5)$$

where $E_f^{C_n I_m}$ is the formation energy for $C_n I_m$, $C_{Si}$ is the silicon lattice site density, and $\theta_{C_n I_m}$ is the number of possible cluster configurations per lattice site. Considering only three species ($C_s$, CI, and $C_2I$), the total carbon concentration becomes a function of $C_{C_s}$, $E_f$, and $C_I$. Fig. 2 illustrates the relationship at 1000°C with $C_I/C_I^* = 1$, which suggests that the cluster concentrations are much lower than $C_{C_s}$. This indicates that small C/I clustering is only significant in the presence of a large interstitial supersaturation.

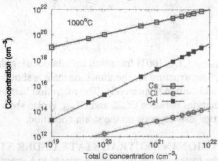

Figure 2: The local equilibrium concentrations of $C_s$, CI and $C_2I$ as a function of total C concentration at 1000°C at $C_I/C_I^* = 1$.

## C DIFFUSION IN NEUTRAL CHARGE STATE

The measured activation energies for the migration of interstitial carbon range from 0.73 to 0.87 eV [20, 21]. Capaz et al. [8] identified this migration path with an energy barrier of 0.51 eV in silicon by *ab-initio* calculations. Using a 65-atom supercell, Zhu [16] confirmed that the intermediate configuration is 0.5 eV higher in energy than the C interstitial (<100>split). The difference between the measured and calculated values can be due to the small supercell size. In our calculation, the nudged elastic band (NEB) method [17-19] was used to find the migration path. The initial state is [100] split, while the final state has the [001] split orientation. Fig. 3 shows the migration path with energy barrier ($E_m$) of 0.53 eV, which is comparable to the results from both Capaz et al. [8] (0.51 eV) and Zhu [16] (0.5 eV). For $C_s$ to diffuse, first CI (<100> split) needs to be formed and overcome the transition barrier ($E_m$) by passing through transition state (CI$_{tran}$). In this migration process, the Si atom remains in its original lattice site while the C atom moves to the neighbor site with different orientation. Thus, the CI diffusion does not affect Si self-diffusion. Note that CI reorientation is completed after each migration step.

| | $C_s$ | | | CI$_{split}$ [001] | | | CI$_{trans}$ | | | $C_2I$ <100> | | |
|---|---|---|---|---|---|---|---|---|---|---|---|---|
| $\Delta\vec{\varepsilon}$ | -0.42 | | | [-0.25 | -0.25 | 0.90] | [ 0.30 | -0.46 | 0.30] | [-0.50 | -0.50 | 0.63] |
| | | | | | | | | $C_2I_2$ | | | $CI_2$ | |
| $\Delta\vec{\varepsilon}$ | [-0.67 | -0.20 | -0.07] | [-0.31 | 0.55 | -0.39] | [-0.41 | -0.57 | -0.62] | [0.39 | 0.53 | 0.59] |
| | 0.20 | -0.67 | 0.07 | 0.55 | -0.31 | -0.39 | -0.57 | 1.12 | 0.56 | 0.52 | 0.39 | 0.59 |
| | -0.07 | 0.07 | 0.53] | -0.39 | -0.39 | -0.55] | -0.62 | 0.56 | -0.41] | 0.59 | 0.59 | 0.69] |

**TABLE II: Induced strains for various carbon complexes.**

131

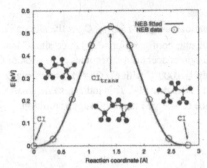

Figure 3: $CI_{split}$ [100] $\rightarrow$ $CI_{trans}$ $\rightarrow$ $CI_{split}$ [001] transition calculated using the NEB method [17-19] in unstrained silicon. The structure for the transition state is shown in the inset. Blue and brown atoms are silicon and carbon respectively. The migration barrier ($\sim 0.53$ eV), the difference of formation energies between $CI_{trans}$ and $CI_{split}$, is also shown in this figure. Notice that the migration barrier also depends on the strain condition.

## PREDICTION OF C DIFFUSION IN NEUTRAL STATE UNDER STRESS

The general form of the formation energy change related to a stress state can be written as

$$\Delta E_f(\vec{\sigma}) = -\Omega_0 \Delta \vec{\epsilon} \cdot \vec{\sigma}, \qquad (6)$$

where $\vec{\sigma}$ is the stress field, and $\Delta \vec{\epsilon}$ is the induced strain in a system with volume $\Omega_0$. The existence of induced strains is because the equilibrium lattice constant is different from the pure Si system. TABLE II lists the induced strains for various carbon complexes. The system with $C_s$ reduces the lattice constant by 0.42% in comparison to Si, while the system with a $CI_{split}$ shows the opposite behavior (a 0.13% increase in lattice constant). This indicates that the $C_s$ equilibrium concentration $C_{C_s}^i$ (solubility) increases in the presence of compressive strain, whereas the $CI_{split}$ equilibrium concentration $C_{CI_{split}}^i$ decreases. Tensile strain will reverse the effect; increasing $C_{CI_{split}}^i$ and decreasing $C_{C_s}^i$.

Based on harmonic transition state theory [22], the transition rate can be written as

$$\Gamma_i(\vec{\epsilon}) = \Gamma_0 \exp\left(-\frac{E_m^i(0) + \Delta E_m^i(\vec{\epsilon})}{kT}\right) \qquad (7)$$

where $\Gamma_0$ is the attempt frequency, and $i$ represents the different possible hopping directions. Here we include the stress effect under a given strain condition ($\vec{\epsilon}$). $\Delta E_m^i(\vec{\epsilon})$ is the change of the migration barrier under stress condition for a given hopping direction, and can be written as

$$\Delta E_m^i(\vec{\epsilon}) = [\Delta E_f^{trans}(\vec{\epsilon}) - \Delta E_f^{CI}(\vec{\epsilon})] \qquad (8)$$

In KLMC simulation, a lattice is used to define system states in a crystalline solid. Eq. 7 describes the transition rate for a neutral CI (<100> split) to migrate from one lattice site to the next. For a given orientation and position of CI, only 4 neighboring sites are accessible via the low transition barrier. The hopping probabilities, $p_i(\vec{\epsilon})$, in the presence of stress in different directions are defined as

$$p_i(\vec{\epsilon}) = \frac{\Gamma_i(\vec{\epsilon})}{\sum_{j=1}^{4} \Gamma_j(\vec{\epsilon})}. \qquad (9)$$

The diffusivity of CI ($d_{CI}$) and transition rate $\Gamma$ are directly linked by considering the diffusion as a random walk process at the atomic level. To keep track of the actual physical time, the average time for each migration step can be calculated as

$$\Delta t(\bar{\epsilon}) = \frac{1}{\sum_{j=1}^{A} r_j(\bar{\epsilon})} \qquad (10)$$

In one dimension, the following relation holds for CI diffusivity ($d_{CI}$):

$$d_{CI}(\bar{\epsilon}) = \frac{1}{2t(\bar{\epsilon})} \langle \Delta x(\bar{\epsilon})^2 \rangle, \qquad (11)$$

where $\langle \Delta x(\bar{\epsilon})^2 \rangle$ is the average of the square of displacement in $x$-direction after an N-step random walk process. $t$ is the physical system time and can be calculated by summing the time for each migration process (Eq. 10). The prediction of the relative $d_{CI}$ change under biaxial strain is shown in Fig. 4[left]. In the case of biaxial strain, the stress dependence of the equilibrium CI concentration can be expressed as

$$\frac{c_{CI}^{*}(\bar{\epsilon})}{c_{CI}^{*}(0)} = \frac{2}{3} \exp\left[-\frac{\Delta E_f^{CI(in)}(\bar{\epsilon})}{kT}\right] + \frac{1}{3} \exp\left[-\frac{\Delta E_f^{CI(out)}(\bar{\epsilon})}{kT}\right], \qquad (12)$$

where $\Delta E_f^{CI(in)}$ is the change in energy for the in-plane CI, whereas $\Delta E_f^{CI(out)}(\bar{\epsilon})$ is the change for out-of-plane components. With the multiplication of Eqs. 11 and 12, one can predict the effective diffusivity change ($D_C(\bar{\epsilon})/D_C(0)$) of C under biaxial stress, which is shown in Fig. 4[right]. Using the same approach, we can also predict the effective diffusivity change of C under uniaxial stress.

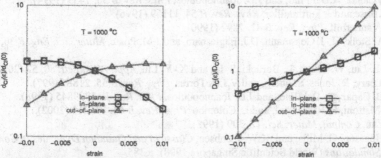

Figure 4: The relative change of C diffusivity ($d_{CI}$ and $D_C$) under biaxial strain at 1000°C determined by the KLMC analysis. [Left]: CI in-plane diffusivity $d_{CI}$ is reduced under tensile strain, but has a weak dependence under compressive strain. The out-of-plane diffusivity for CI shows the opposite behavior. [Right]: both in-plane and out-of-plane C diffusivities $D_C$ are enhanced under tensile strain and reduced under compressive strain. Note that out-of-plane diffusivity for C shows a stronger dependence on the strain condition.

## CONCLUSIONS

In this work, we have confirmed the migration path for C diffusion, which is consistent with findings from Capaz et al. [8] and Zhu [16]. *Ab-initio* calculations further suggested that carbon/interstitial clustering is only significant under a strong interstitial supersaturation. The results from these calculations were also used to obtain induced strains in order to describe the effects of stress on C diffusion/clustering. C clustering was found to be enhanced under tensile stress/strain. Through KLMC simulations, we found that tensile stress enhances C diffusion under biaxial strain, while compressive stress has the opposite impact. With the incorporation of models for excess point defect evolutions [23] and boron-interstitial clustering (BICs) under stress, these models provide insight into ultra-shallow-junction formation process for future devices which reduce TED effects.

## ACKNOWLEDGMENTS

This work was supported by the Semiconductor Research Corporation (SRC). Calculations were conducted on computing clusters donated by Intel and AMD.

## REFERENCES

1. S. Nishikawa, A. Tanaka, and T. Yamaji, *Appl. Phys. Lett.* 60, 2270 (1992).
2. P.A. Stolk, D.J. Eaglesham, H.-J. Gossmann, and J. M. Poate, *Appl. Phys. Lett.* 66, 1370 (1995).
3. E. Napolitani, A. Coati, D. De Salvador, A. Carnera, S. Mirabella, S. Scalese, and F. Priolo, *Appl. Phys. Lett.* 79, 4145 (2001).
4. B.J. Pawlak, T. Janssens, B. Brijs, W. Vandervorst, E.J.H. Collart, S.B. Felch, and N.E.B. Cowern , *Appl. Phys. Lett.* 89, 062110 (2006).
5. G.D. Watkins and K.L. Brower, *Phys. Rev. Lett.* 36, 1329 (1976).
6. F. Rollert, N.A. Stolwijk, and H. Mehrer, *Mater. Sci. Forum* 38-41, 753 (1989).
7. L.W. Song and G.D. Watkins, *Phys. Rev. B* 42, 5759 (1990).
8. R.B. Capaz, A.Dal Pino, and J.D. Joannopoulos, *Phys. Rev. B* 50, 7439 (1994).
9. G. Kresse and J. Furthmüller, *Phys. Rev. B* 54, 11169 (1996).
10. D. Vanderbilt, *Phys. Rev. B* 41, 7892 (1990)
11. P.A. Stolk, H.-J. Gossmann, D.J. Eaglesham, and J.M. Poate, *Mater. Sci. Eng. B* 36, 275 (1996).
12. C.-L. Liu, W. Windl, L. Borucki, S. Lu and X.-Y. Liu, *Appl. Phys. Lett.* 80, 52 (2002).
13. P. Leary, R. Jones, S. öberg, and V.J.B. Torres, *Phys. Rev. B* 55, 2188 (1997).
14. R.B. Capaz, A. Dal Pino, and J.D. Joannopoulos, *Phys. Rev. B* 58, 9845 (1998).
15. A. Mattoni, F. Bernardini, and L. Colombo, *Phys. Rev. B* 66, 195214 (2002).
16. J. Zhu, *Comput. Mater. Sci.* 12, 309 (1998).
17. H. Jönsson, G. Mills, and ,K.W. Jacobsen, *Classical and Quantum Dynamics in Condensed Phase Simulations* (World Scientific,Singapore,1998), p.385.
18. G. Henkelman and H Jönsson, *J. Chem. Phys.* 113, 9978 (2000).
19. G. Henkelman, B.P. Uberuaga, and H Jönsson, *J. Chem. Phys.* 113, 9901 (2000).
20. A.C. Tipping and R.C. Newman, *Semicond. Sci. Technol.* 2, 315 (1987).
21. L.W Song and G.D Wakins, *Phys. Rev. B* 42, 5759 (1990).
22. M.E. Glicksman, *Diffusion in solids: Field theory, solid-state principles, and applications*, (John Wiley, 2000).
23. H.-W Guo, S.T. Dunham, C-L. Shih, and C. Ahn, *Proceedings of IEEE 2006 International Conference on Simulation of Semiconductor Processes and Devices* (SISPAD, Monterey,CA, 2006), p.71.

Mater. Res. Soc. Symp. Proc. Vol. 1070 © 2008 Materials Research Society

# Atomistic Simulations of Epitaxial Regrowth of As-doped Silicon

Joo Chul Yoon, and Scott Dunham
Electrical Engineering, University of Washington, Seattle, WA, 98195

## ABSTRACT

We conducted molecular dynamics (MD) simulations of solid phase epitaxial growth of As-doped Si using a Tersoff potential characterized via comparison to density functional theory (DFT) calculations, including energies of $As_nV$ clusters. The Si:As systems were initialized by amorphizing the surface region of crystalline silicon via Si ion implantation. The remaining crystalline region provides dual function of controlling temperature in system without perturbing regrowth and providing seed for recrystallization. After recrystallization, isolated As atoms occupy substitutional sites, with the average number of nearest neighbors for As changing from about 3.3 in amorphous Si to 4 after crystallization. We observe V incorporation associated with high As concentrations, primarily at sites with multiple As neighbors. These observations are consistent with our previous model developed to explain kinetics of As shallow junction formation which assumed V incorporation at sites with 2 or more As nearest neighbors to account for experimental data.

## INTRODUCTION

Amorphization has become a ubiquitous part of doping technology as it enables sharper doping profiles (via reduction of channeling and TED) and high metastable dopant activation. The state of the material following regrowth determines to a large extent the final profile broadening and activation level even when SIMS shows no apparent change in the dopant distribution. This is because purely local rearrangements can lead to clustering and point defect incorporation. A key example of this is diffusion of high concentration As implants. Our work has indicated that formation of As-vacancy complexes during regrowth has a major impact on diffusion and activation in ultra-shallow junctions even when followed by a high temperature RTA [1]. The impact of processes in the amorphous material becomes even more critical as thermal budgets are reduced. A promising approach for ultra-shallow junctions is to limit annealing (RTA and flash) to just enough to allow recrystallization. For such processes, redistribution and activation in amorphous material and near amorphous/crystalline interface fully controls final structure.

In contrast to crystalline regions, for which there are a small number of equivalent sites, for amorphous (or just heavily disordered) regions, every microscopic region is different. Thus, one cannot use a small number of ab-initio (DFT) calculations to identify rate limiting processes and parameters and then apply the resulting model at each site. Also, because it is necessary to look at relatively large regions of the material in order to predict properties, ab-initio techniques cannot practically consider sufficiently large systems to capture the behavior of amorphous materials and amorphous/crystalline interfaces. Thus, we apply classical molecular dynamics methods in this work.

## ARSENIC IN SILCON

There are several features of high concentration As diffusion that must be considered in a comprehensive model [1]. At high doping levels (>$10^{20}$ cm$^{-3}$), As diffusion increases very rapidly with donor concentration [2]. High As concentrations (~$10^{21}$ cm$^{-3}$) deactivate rapidly even at low temperatures [3,4], injecting interstitials as evidenced by the enhanced diffusion of B buried layers [4]. Positron annihilation shows increased V concentration in laser-annealed Si:As [5]. At the same time, modeling suggests grown-in vacancies are present after regrowth of highly As-doped Si (Fastenko et al. [1]). The effect of grown-in vacancies is shown in Figure 1. Only by considering grown-in vacancies can the same model be used to predict both enhanced diffusion during annealing of highly As-doped laser-recrystallized Si and also As shallow junction profiles. The simulations shown use a simple model in which sites with multiple As neighbors regrow as vacancies.

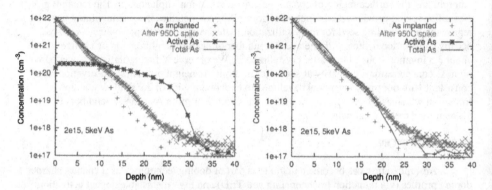

**Figure 1.** Comparison of models to measured junction profiles [6] following As implant and 950°C spike anneal. Without grown-in vacancies (left) I injection due to As deactivation leads to much more diffusion than seen experimentally. Including grown-in V for sites with 2 or more As neighbors leads to match with experiment (right).

## MOLECULAR DYNAMICS SIMULATIONS

To simulate epitaxial regrowth, we used simulation cells that were periodic in the lateral directions (Figure 2). The top surface was left free, while the bottom layer was fixed to bulk silicon lattice coordinates. To control temperature for annealing without disturbing the epitaxial regrowth process, a temperature control (TC) region was established in bottom few layers.

As foundation for potential for Si:As system, we took Si-Si and As-As parameters from Tersoff potentials calibrated to pure Si [5] and GaAs [6] systems. Due to the importance of As/V clusters to the targeted simulations, we used DFT calculations of the binding energy of As$_n$V clusters (Table I) as the basis for the Si-As interactions used.

**Table I.** $As_nV$ binding energies from DFT calculations used to establish As/Si interactions for Tersoff potential.

|       | Binding Energy (eV) |
|-------|---------------------|
| AsV   | 1.41                |
| $As_2V$ | 2.86              |
| $As_3V$ | 4.30              |
| $As_4V$ | 5.67              |

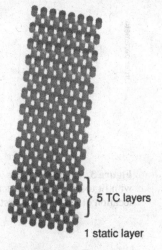

Figure 2. Example of molecular dynamics simulation cell.

To initialize the systems, As atoms were located randomly in substitutional sites at fixed concentrations ranging from $2 \times 10^{20}$ to $5 \times 10^{21}$ cm$^{-3}$. The top surface regions were amorphized via low energy Si implants simulated via molecular dynamics. The systems were then cooled to below room temperature before the temperature was raised to high temperature to give rapid (~ns) epitaxial regrowth.

## RESULTS

During regrowth, the amorphous crystalline interface was tracked via an order parameter based on alignment of atoms with ideal silicon lattice. This order parameter was fit to a Fermi-Dirac function in order to extract the interface location (threshold of 0.5). Results are shown in Figure 3.

We also studied the diffusion of As atoms in amorphous silicon by tracking the displacements of As atoms. We find that the diffusivity of As in amorphous Si is much higher than in crystalline material. However, the diffusivity in amorphous material does not seem to be high enough to show substantial redistribution prior to regrowth due to the very small thermal budgets involved. Figure 4 tracks the location of two As atoms in regrown region. Prior to recrystallization, significant diffusion is seen, which stops once the region in which the sit crystallizes.

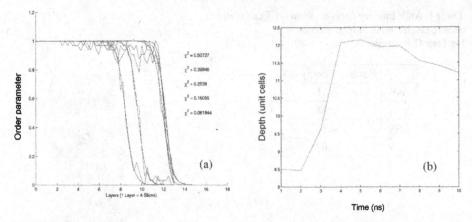

**Figure 3.** Order parameter versus vertical distance (unit cells) over a series of regrowth times, with fit to Fermi-Dirac function (a). Tracking the location of a given order threshold (0.5) gives location of amorphous-crystalline interface as function of time (b).

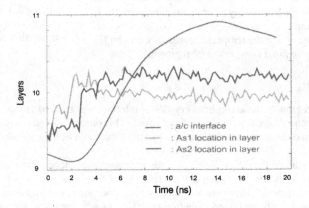

**Figure 4:** Vertical location of two As atoms tracked along with location of amorphous crystalline interface.

We also tracked the (smoothed) coordination number of As atoms, which revealed that As atoms have (on average) slightly more than 3 nearest neighbors in amorphous Si (as expected for valence 5 element), but shift to a coordination of 4 in crystalline material as they sit on substitutional Si sites (Figure 5).

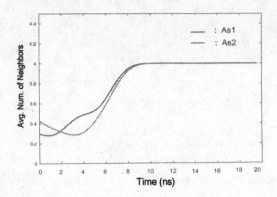

**Figure 5:** Average number of nearest neighbors for two As atoms during recrystallization.

After regrowth, we examined the recrystallized regions to check for the presence of point defects. In undoped Si, we found no point defect incorporation. However, for heavily As-doped material, we found V incorporation in the form of As/V clusters, primarily $As_2V$ structures.

## CONCLUSION

We conducted classical molecular dynamics simulations of Si:As regrowth with a Tersoff-type potential. The simulations support the formation of $As_nV$ clusters during regrowth. These grown-in defects can have a substantial effect on subsequent diffusion and activation of shallow junctions.

## REFERENCES

[1] P. Fastenko, PhD thesis, Univ. of Washington (2002).
[2] A.N. Larsen, P.E. Andersen, P. Gaiduk, K.K. Larsen, *Mater. Sci. Eng. B* **4**, 107 (1989).
[3] S. Luning, P.M. Rousseau, P.B. Griffin, P.G. Carey, J.D. Plummer, IEDM Tech. Dig. 1992, p 457-60.
[4] P.M. Rousseau, P.B. Griffin, W.T. Fang, J.D. Plummer, J.D, *J. Appl. Phys.* **84**, 3593 (1998).
[5] D.W. Lawther, U. Myler, P.J. Simpson, P.M. Rousseau, P.B. Griffin, and J.D. Plummer, *Appl. Phys. Lett.* **67**, 3575-7 (1995).
[6] A. Jain, Texas Instruments, private communication.
[7] J. Tersoff,. *Phys. Rev. B* **37**, 6991 (1988).
[8] P.A. Ashu, J.H. Jefferson, A.G. Cullis, W.E. Hagston, and C.R. Whitehouse, *J. of Crystal Growth* **150**, 176 (1995).

# Ultra Shallow Junctions II

Mater. Res. Soc. Symp. Proc. Vol. 1070 © 2008 Materials Research Society 1070-E04-01

# Source-Drain Engineering for Channel-Limited PMOS Device Performance: Advances in Understanding of Amorphization-Based Implant Techniques

Nick E. Cowern

School of Electrical, Electronic and Computer Engineering, University of Newcastle upon Tyne, Merz Court, Newcastle upon Tyne, NE1 7RU, United Kingdom

## ABSTRACT

This paper discusses the role of amorphisation and residual end-of-range defects in p-channel source/drain engineering. A comparison between preamorphisation and molecular implant approaches shows up some important common features of electrical activation, diffusion, and junction leakage, related to the formation and location of boron-interstitial and self-interstitial clusters. The success of these techniques depends on confining 'end-of-range' defects – whether TEM-visible defects or sub-microscopic clusters – within the narrow region between the boron implant peak and the source-drain/halo depletion region. This observation points to significant improvements that can still be made in implantation processing for ultrashallow junctions.

## INTRODUCTION

Formation of highly-doped, ultrashallow junctions with low junction leakage is essential for future CMOS technology generations. As channel engineering techniques continue to advance, performance scaling requires comparable improvements in source/drain series resistance, while maintaining very shallow and low leakage junctions. Here we focus on understanding the similarities and differences in two key techniques for achieving this goal.

P-channel MOS devices present particular difficulties owing to the tendency of boron to form electrically inactive interstitial clusters at high concentrations after implantation and annealing. To minimize this problem a variety of solutions have been attempted, including flash or laser annealing, with or without the use of preamorphization or molecular dopant ion implantation (which form amorphous layers prior to annealing).

This paper looks at the physical processes of defect formation, dopant activation and diffusion that play a part in ultrashallow junction formation using these approaches.

Traditionally, p-channel source-drain extensions have been created by ion implantation of B-containing ion species, either direct implantation of the doping ion into crystalline silicon, or pre-amorphization with heavy ions such as Ge to form an amorphous layer to limit boron channelling. Recently increasing emphasis has been placed on implantation of molecular species, because of the relatively high implantation energy that can be used even for ultrashallow junction formation, and because, empirically, reduced junction leakage is observed under appropriate annealing conditions [1] Since the benefits of pre-amorphisation arise from the fact that dopant is implanted into a 'pre-fabricated' amorphous layer in crystalline silicon, it is interesting to consider the similarities and differences between (a) pre-amorphization and dopant implantation, (b) direct amorphization during implantation of a heavy molecular dopant species. By combining knowledge about these two approaches we may hope to achieve a better basic understanding of both, and derive optimal processing opportunities which may have been missed up to now in 'design of experiments' analyses.

## PREAMORPHIZATION IMPLANTS

A range of preamorphization conditions is illustrated in Fig. 1, where frames (a-c) illustrate the impact of varying the energy/mass ratio of a pre-amorphizing ion species, keeping subsequent dopant implant and anneal conditions constant. In case (a) the interface is located within the doped region, in case (b) it is located in the region which will form a depletion layer between the source-drain extension and the substrate or halo implant, and in case (c) it is located deeper than this depletion layer.

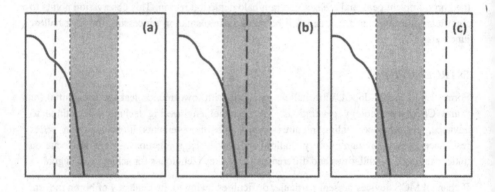

Fig. 1: Location of amorphous/crystalline (a/c) interface relative to the device depletion region (shown in gray). (a) Shallow a/c interface (vertical dashed line), located within the doping profile (solid curve) minimizing source-drain junction leakage. (b) Intermediate a/c interface depth causes junction leakage as a result of carrier generation at deep levels. (c) Deep a/c interface, again minimizing junction leakage, but potentially causing problems elsewhere in device. Changes in boron profile shape as a result of changing a/c interface depth have been neglected, for simplicity.

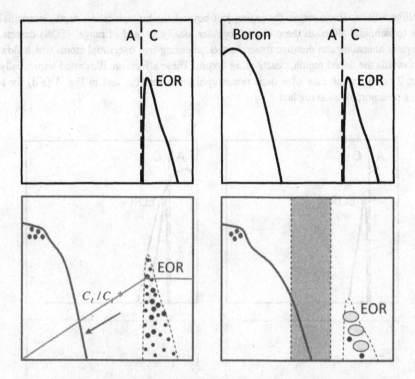

Fig. 2: Nanoscale events during fabrication of ultrashallow source-drain
extension by deep preamorphization (PAI). (a) PAI implant forms amorphous
layer and generates excess interstitial point defects in non-amorphized 'end-of-
range (EOR) region beyond a/c interface. (b) Boron implanted into amorphous
region. (c) Initial stages of annealing cause epitaxial regrowth of amorphous
layer. Excess interstitials and small mobile clusters in EOR region coalesce into
stable clusters. As clusters ripening into larger defects, sustained emission of free
interstitial atoms occurs. Surface acts as interstitial sink, so a gradient is set up
and interstitials flow towards surface. Interstitials reaching boron-doped region
catalyse crystalline-phase diffusion and clustering of boron. (d) After annealing,
a band of larger EOR defects formed by ripening may remain. These do not
contribute to junction leakage if outside depletion region (gray area).

Boron implanted within the preamorphized layer incorporates to a relatively high
concentration (a few times $10^{20}$cm$^{-3}$ depending on the anneal temperature and heating rate)
compared to that achieved in crystalline silicon, where boron clusters readily to form inactive
boron-interstitial complexes. The increased activation level is largely a result of the removal
of interstitial defects by the regrowth of amorphous silicon into crystalline silicon during
annealing, thus eliminating most of the self interstitial atoms introduced by implantation into

crystalline silicon. However, in the region just beyond the amorphization depth, interstitial atoms remain, and although these cluster together locally as 'end of range' (EOR) defects, subsequent annealing can dissolve these clusters, releasing free interstitial atoms which flow back towards the doped region, deactivating boron. These effects are illustrated sequentially in Fig. 2 (a-d) for the case of a deep preamorphization implant, and in Fig. 3 (a-d) for a shallow preamorphization implant.

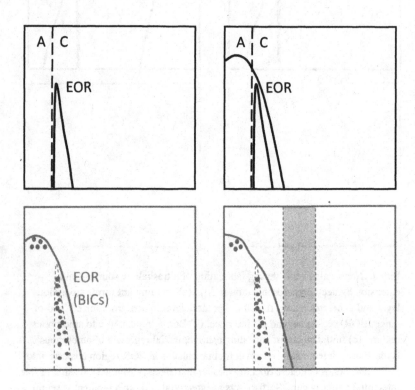

Fig. 3: Physical stages in fabrication of ultrashallow source-drain extension using shallow PAI. In this case EOR defects nucleate within the boron profile in the form of a band of boron-interstitial clusters (BICs). The higher stability of BICs compared to self-interstitial defects reduces the emission of free interstitials, thus limiting the deactivation of boron by long-range transport and trapping of interstitials during subsequent annealing. Transient enhanced diffusion of boron is likewise reduced. Careful choice of preamorphization depth relative to the boron implant profile ensures that the EOR defects do not overlap the source-drain/channel depletion region.

In the case of shallow preamorphisation, where the EOR defect band is located within the ultrashallow boron profile, the amorphous layer occupies only part of the boron profile, and the EOR defects also lie within the boron profile. Excess interstitial atoms formed in a region of high boron concentration react strongly with boron atoms, forming boron-interstitial clusters. It can therefore be expected that EOR defects in this case will consist of boron-interstitial clusters, and contribute to deactivation of part of the boron profile. If they are located close to the maximum of the boron profile, the clusters so formed will be rich in boron atoms and will thus deactivate a substantial amount of boron. If they are located deeper in the boron profile 'tail', they will contain relatively few boron atoms and the amorphous layer will be thicker, thus enabling higher activation, but there is a risk that the deeper portion of the defect band may overlap the source-drain/channel depletion region during device operation, causing an unacceptable level of junction leakage [2], unless annealing is used to drive the boron source-drain profile beyond the depth of these defects. The positioning of the EOR defect band within the source-drain region is therefore a critical parameter for device performance, and a sensitive function of the implantation and annealing conditions.

## MOLECULAR IMPLANTS

In the case of implantation of heavy boron-containing molecular ions, the location of the EOR band and the depth of the boron implant profile are closely related. Fig. 4 illustrates qualitatively the impact of changing the mass of the molecular dopant implant species, keeping its energy/mass constant to maintain an essentially unchanged dopant implant depth distribution. The heavier implant species creates a deeper amorphous layer, occupying a larger fraction of the boron implant profile.

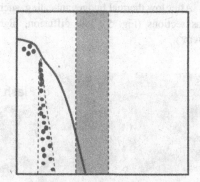 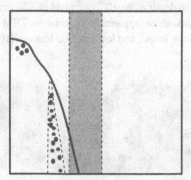

Fig. 4: Impact of amorphous layer depth on activation of shallow molecular boron implant. A heavier molecular ion produces a deeper amorphous layer and thus better activation, but there is closer proximity between EOR defects and the source-drain junction depletion region.

147

This leads to increased activation of the implant, since the regrown layer is thicker and the EOR defects are located in a less heavily boron doped region of the implant. The position of the EOR band relative to the source-drain depletion region is set by the ion implant energy and the device operating conditions (how far the depletion region penetrates into the boron profile during device operation). Since the only process parameters that can change this relative positioning are the ion mass and the annealing conditions, one can select an appropriate ion species by purely experimental means. Thus, although there may be a very narrow (~nm) physical window for the relative placement of the amorphous/crystalline interface within the dopant profile, the corresponding implant *process* window may be quite wide, and able to be determined by other process constraints.

## COMPARISON OF RESULTS

One of the most commonly used molecular dopant ions used for p-channel source-drain extensions is octadecaborane. This gives a favourable activation level, low levels of diffusion, no TEM-visible EOR defects, and low junction leakage when implanted and thermally processed by low-temperature SPE, flash or laser annealing [3]. It is a matter of considerable interest as to why this species gives such favourable results. To address this question we compare results obtained using octadecaborane with results obtained using elemental boron implantation into amorphous layers generated by low energy preamorphisation implants. It turns out that the favourable result with octadecaborane is not unique to this species, but is a result of a special set of process conditions that can also be created using preamorphization.

At 500 eV incident energy and $3 \times 10^{14}$ B/cm$^2$ dose, octadecaborane (B$_{18}$H$_{22}$) implants create ~8 nm thick amorphous layers, leaving end-of-range defects at a depth of ~10-12 nm, and a junction depth at the $10^{18}$/cm$^3$ level of ~20 nm [1]. After low thermal budget annealing, such implants show apparently defect-free TEM cross sections (Fig. 5), low diffusion, high activation levels, and low junction leakage ($10^{-7}$ A/cm$^2$).

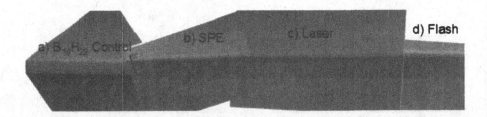

Fig. 5: TEM cross sections of B$_{18}$H$_{22}$ implanted silicon before and after annealing (low temperature SPE, laser, or flash anneal). No TEM-visible defects are present after annealing [3].

The same study showed poorer results using preamorphization implants, although the comparison was somewhat approximate because the preamorphization energies used produced deeper amorphous layers than the $B_{18}H_{22}$ implants. In order to make a direct comparison with the molecular implant results, we now revisit recently reported observations using preamorphization implants with a range of incident energies [4].

Ge preamorphization with 2, 5, 10 keV Ge was used to create amorphous layers with thicknesses of 8, 15, 23 nm, respectively. The 8 nm amorphous layer was closely comparable to that of the $B_{18}H_{22}$ in Ref. [2]. Subsequently elemental boron with an energy of 0.5 keV, the same energy/mass as that of the $B_{18}H_{22}$ implant in Ref. [2], was implanted into the amorphized layers. Plan view TEM data for the three preamorphization implant conditions are shown in Fig. 6.

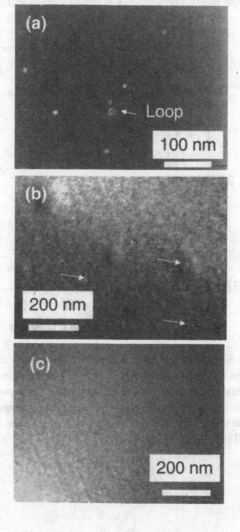

Fig. 6: Plan view TEM measurements taken in weak-beam dark field (g,3g) condition with g=[224], to show all EOR dislocation loops that are present (Ref. [4]). In case (a), 10 keV Ge, a significant number of loops are present after laser anneal. In case (b), 5 keV Ge, only a few small loops are present, and in case (c), 2 keV Ge, no TEM-visible defects are found. In case (c), the EOR defects are located within the boron implant profile and almost certainly consist of sub-microscopic boron-interstitial clusters.

## RELATIVE STABILITY OF SELF- AND BORON-INTERSTITIAL CLUSTERS

The reason why no defects are observed when the EOR band is located within the boron profile is that the incorporation of boron atoms into the defects increases their stability, slowing down the process of Ostwald ripening. The closer proximity of the defects to the silicon surface also allows interstitials lost from the defects to flow more quickly to the surface, thus allowing the defects to dissolve in preference to recapturing interstitials emitted from other defects within the band.

The stability of self-interstitial and boron-interstitial clusters as a function of size, expressed as differential formation energy [5] is shown in Fig. 7. Boron-interstitial clusters have lower formation energy, corresponding to greater stability [6,7].

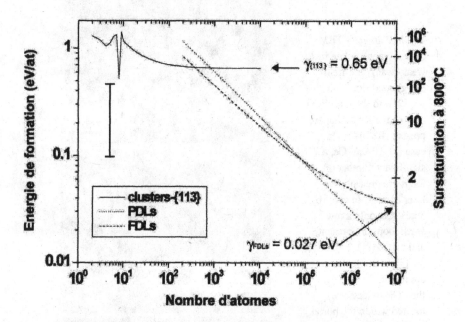

Fig. 7: Stability of self interstitial and boron-interstitial defects as a function of the number of atoms contained in the defect, expressed as differential formation energy. Boron-interstitial defects have lower formation energy, corresponding to greater stability. The error bar is an approximate estimate of the energy range of more stable (thus populated) boron-interstitial clusters with respect to dissociation, based on approximate inference from theory [8].

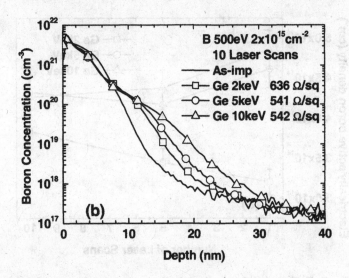

Fig. 8: Diffusion of ultrashallow boron implant during scanning laser anneal, from ref. [9]. Transient enhanced diffusion is reduced when EOR defects are located within (2 keV Ge) or close to (5 keV Ge) the boron-doped region.

## CONSEQUENCES FOR BORON DIFFUSION AND ACTIVATION

The impact that more stable EOR defects have on diffusion can be seen from boron atomic depth profiles, measured by SIMS [4], shown in Fig. 8. Boron implanted into a 2 keV Ge preamorphized wafer undergoes less than half of the junction motion of the boron in the 10 keV Ge case. The effect is even stronger than the figure suggests, because the as-implanted boron profile is already deeper in the 2 keV Ge case, as a result of channelling beyond the amorphous layer depth. The reason can be seen from the sheet resistance data in Fig. 8: a large proportion of the interstitials in the EOR band are trapped in relatively stable boron-interstitial clusters, thus causing initial deactivation in the deeper part of the boron profile.

This conclusion is further confirmed in Fig. 9 which shows data on the time evolution of electrical activity, determined by sheet-Hall measurements [4,5] In the 5 keV and 10 keV cases, activation falls as interstitials from deep EOR defects flow back towards the surface, trapping into boron-interstitial clusters within the boron profile as outlined in Fig. 2c. However, in the 2 keV case, activation increases as the EOR defects, already consisting of boron-interstitial clusters, dissolve during annealing.

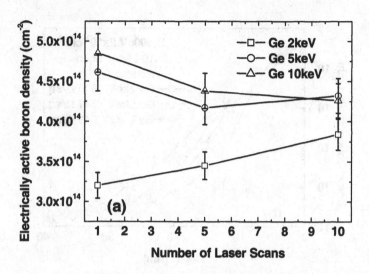

Fig. 9: Evolution of electrically active boron dose as a function of annealing budget (number of laser scans), from Ref. [9]. When EOR defects are located beyond the boron profile (5 keV, 10 keV), interstitial transport towards the surface leads to boron deactivation. When EOR defects are located within the boron profile, initial activation is low because EOR defects are in the form of boron-interstitial clusters (as illustrated in Fig 3c). Subsequent annealing leads to increased electrical activation as boron-interstitial clusters dissolve.

The favourable combination of low diffusion, reasonably high activation, and positioning of sub-microscopic EOR defects within the boron profile, away from depletion regions where junction leakage could occur, makes this configuration highly suitable for ultrashallow junction applications. In particular, it explains the success of heavy molecular boron dopant implants in eliminating defects that are electrically harmful to p-channel devices. Further optimization of the positioning of EOR defects through different choices of implant ion mass, or through co-implantation of other heavy species, may be helpful to further increase sheet doping within the limits set by increased junction leakage.

The advantages shown here for ultrashallow junctions in bulk silicon wafers are equally relevant to thin film devices, where deep amorphisation is not permissible owing to the need to retain a crystalline seed within the depth of the layer. Here again, shallow amorphisation enables highly activated, low-leakage junctions with limited lateral diffusion.

The combination of low diffusion and high activation achievable with octadecaborane leads to exceptionally low $R_s$-$X_j$ values. Fig. 10 shows results obtained with octadecaborane and 2 keV Ge preamorphization, as well as with a range of other methods. These two approaches give almost identical $R_s$-$X_j$ values, close to the best achieved to date with 0.5 keV B energy.

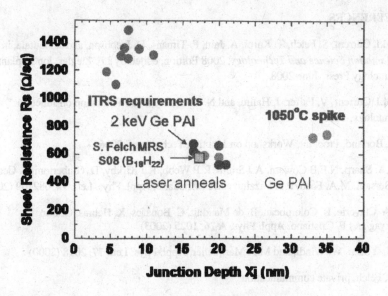

Fig. 10: $R_s$-$X_j$ plot showing similar performances achieved using
(a) octadecaborane (square symbol) [8]; (b) an equivalent energy and
amorphous layer depth obtained using elemental boron and Ge preamorphization
(black symbols).

## CONCLUSIONS

Heavy molecular ion implantation is an efficient method for fabricating ultrashallow p-channel source-drain extensions with high activation, very small (submicroscopic) EOR defects, and low junction leakage. The physics controlling junction depth, activation and defect-assisted junction leakage is closely similar for molecular implantation and preamorphization. Preamorphization offers greater scope for tailoring defect location within the relatively narrow window for good activation and low defect-assisted junction leakage. However, molecular implantation with a suitable (heavy) ion mass provides a simpler method of process control, as it is insensitive to variations in implant energy and other process parameters, thereby ensuring a larger process window.

## ACKNOWLEDGEMENTS

Part of this work was supported by the ATOMICS and PULLNANO European projects. We thank Michael Current and Bartolomej Pawlak for valuable discussions.

# REFERENCES

1. M.I. Current, S. Felch, T. Kuroi, A. Jain, P. Timans, D. Jacobson, and J. Hautala, in *Ion Implantation Science and Technology*, 2008 Edition, edited by J.F. Ziegler, Ion Implantation Technology Press, June 2008

2. M.I. Current, V. Faifer, J. Halim, and N. Ohno, Proc. Int. Workshop on Junction Technology, 2007

3. J. Borland, Proc. Int. Workshop on Junction Technology, 2006

4. J.A. Sharp, N.E.B Cowern, A.J Smith, R.P Webb, K.J Kirkby, D. Giubertoni, S. Gennaro, M. Bersani, M.A. Foad, P.F. Fazzini, and F. Cristiano, Appl. Phys. Lett. **92**, 082109 (2008)

5. A. Claverie, B. Colombeau, B. de Mauduit, C. Bonafos, X. Hebras, G. Ben Assayag and F. Cristiano, Appl. Phys. A **76**, 1025 (2003)

7. X.-Y. Liu, W. Windl, and M.P. Masquelier, Appl. Phys. Lett. **77**, 2018 (2000)

8. S. Felch, private communication

Mater. Res. Soc. Symp. Proc. Vol. 1070 © 2008 Materials Research Society

# Efficacy of Damage Annealing in Advanced Ultra-Shallow Junction Processing

Paul Timans[1], Yao Zhi Hu[1], Jeff Gelpey[2], Steve McCoy[2], Wilfried Lerch[3], Silke Paul[3], Detlef Bolze[4], and Hamid Kheyrandish[5]

[1]Mattson Technology, Inc., Fremont, CA, 94538
[2]Mattson Technology Canada, Inc., Vancouver, V6P 6T7, Canada
[3]Mattson Thermal Products GmbH, Dornstadt, 89160, Germany
[4]IHP, Frankfurt (Oder), 15236, Germany
[5]CSMA Ltd., Stoke-on-Trent, ST4 7LQ, United Kingdom

## ABSTRACT

Low thermal budget annealing approaches, such as millisecond annealing or solid-phase epitaxy (SPE) of amorphized silicon, electrically activate implanted dopants while minimizing diffusion. However, it is also important to anneal damage to the crystal lattice in order to minimize junction leakage. Annealing experiments were performed on low-energy B implants into both crystalline silicon and into wafers pre-amorphized by Ge implantation. Some wafers also received As implants for halo-style doping, and in some cases the halo implants were pre-annealed at 1050°C before the B-doping. The B-implants were annealed by either SPE at 650°C, spike annealing at 1050°C, or by millisecond annealing with flash-assisted RTP™ (fRTP™) at temperatures between 1250°C and 1350°C. Residual damage was characterized by photoluminescence and non-contact junction leakage current measurements, which permit rapid assessment of damage removal efficacy. Damage from the heavy ions used for the halo and pre-amorphization implants dominates the defect annealing behaviour. The halo doping is the critical factor in determining junction leakage current. Spike annealing removes the damage very effectively but causes excessive dopant diffusion, whereas fRTP at temperatures >1250°C minimizes diffusion while also providing better damage annealing than is possible through SPE.

## INTRODUCTION

Advanced CMOS devices require ultra-shallow junctions (USJ) with abrupt doping profiles and very high concentrations of electrically active dopants. USJ are usually formed by ion implantation followed by thermal annealing that electrically activates the implanted dopants and anneals damage. Minimizing residual defects in the depletion regions of p-n junctions is essential because they can greatly increase junction leakage current [1]. CMOS scaling typically increases the channel doping in order to control short-channel effects, which reduces the width of depletion regions, raising the electric field and band-to-band tunneling (BTBT) leakage. Residual defects can exacerbate the problem, for example through trap-assisted tunneling. Conventional annealing cannot electrically activate implanted dopants without simultaneously causing excessive dopant diffusion. Millisecond annealing at temperatures just below the melting point of silicon can overcome this challenge and also provide some defect annealing, but several studies have shown that conventional spike annealing is more effective in removing defects [2-5]. Another path to USJ is through low-temperature (< 650°C) SPE, where dopants are incorporated on lattice sites during recrystallization of an amorphized silicon layer [6].

However, SPE does not remove end-of-range (EOR) defects from the crystalline silicon just below the original position of the amorphous/crystalline interface.

These challenges from defects make it essential for advanced low-thermal budget annealing approaches to be evaluated for their defect annealing capability. Electrical activation is often characterized through 4-point probing (4PP) of sheet resistance, $R_S$, but this is difficult if junctions leak because of residual defects or BTBT effects [7]. Defect annealing can be studied by transmission electron microscopy, but this is time consuming and expensive. Junction leakage in test diodes is very sensitive to damage, but requires device fabrication. These limitations have stimulated interest in alternatives, especially non-invasive methods such as measurement of junction photovoltages [7] and photoluminescence [8], which can provide very rapid characterization of the residual damage.

## EXPERIMENT

Experiments were performed to probe the relative efficacy of spike annealing, millisecond annealing and SPE for activating dopants and removing damage. 200 mm diameter, n-type (100) prime silicon wafers with 10-15 Ωcm resistivity were cleaned in SC1 and SC2 solutions followed by an HF dip before ion implantation. The wafers were doped by implants of $10^{15}$ B/cm$^2$ at 500 eV into crystalline (c-Si) or amorphous silicon (a-Si) surfaces. The latter were formed by preamorphization implants (PAI) of $10^{15}$ Ge/cm$^2$ at 30 keV, which creates an a-Si layer ~50 nm thick [3]. Before the B or PAI implants, some wafers were implanted with $4 \times 10^{13}$ As/cm$^2$ at 40 keV to simulate a strong halo doping condition [9]. Some halo-doped samples were annealed for 10 s at 1050°C before the PAI or B implants. Spike anneals at 1050°C were performed in a Mattson 3000 RTP system, as were SPE anneals of 5 s at 650°C. Millisecond annealing was performed using the fRTP approach, where the wafer was ramped at 150 K/s to an intermediate temperature of either 700°C or 750°C and then its front surface was heated by a pulse of energy from a bank of flash-lamps [2,3]. Peak temperatures were in the range from 1250°C to 1350°C. An ultra-fast pyrometer measured the temperature of the top surface of the wafer and a second pyrometer monitored the bottom surface.

Doping profiles were characterized by secondary ion mass spectrometry (SIMS), using a Cameca IMS 7F instrument with 500 eV O$_2$ primary ions under non-roughening conditions. Fig. 1 shows typical results for the B profile after fRTP at 1300°C or spike annealing, together with a prediction of the As halo profile obtained from a simulation using the SRIM program [10]. For an As diffusivity of ~$10^{-14}$ cm$^2$/s at 1050°C, the 10 s pre-anneals at 1050°C would be expected to introduce ~ 6 nm of diffusion, if transient-enhanced diffusion (TED) effects are negligible, so this anneal should only have a minor effect on the As doping profiles [11]. With halo doping, Fig. 1 shows that the crossover in doping occurs at high concentrations, > $6 \times 10^{18}$ cm$^{-3}$, and we expect the junctions to be prone to leakage issues. The halo As implant dose is close to the threshold for amorphization and introduces heavy damage [12]. Fig. 2 shows reflection spectra of several unannealed samples, as measured with a KLA-Tencor UV1250SE system. The peaks near 3.4 and 4.6 eV are very sensitive to long-range crystalline order. A boron implant reduces these peaks and surface amorphization eliminates them. Interference effects in the amorphous layer cause the reflectivity oscillations seen at lower photon energies [13]. The PAI greatly alters the spectrum, consistent with formation of a ~50 nm-thick amorphous layer, while the halo implant also produces a very heavily damaged surface layer. Fig. 1 shows the approximate location of the EOR damage from the PAI implant.

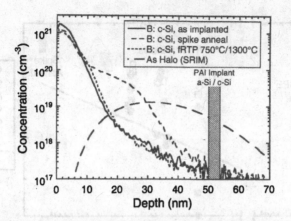

Fig. 1. Doping profiles for the B doping and for the halo-style As implant. End-of-range defects from the PAI are expected to form near the original location of the a-Si / c-Si interface.

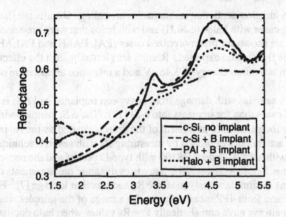

Fig. 2. Reflectance spectra for samples that had not been annealed.

The SIMS profiles were used to determine junction depths, $X_J$, at a concentration of $7 \times 10^{18}/cm^3$ and the amount of profile movement during annealing, $\Delta X_J$. The thermal budget for each process recipe was assessed with an effective process time, $t_{eff}$, which is the time a sample would have to spend at a constant reference temperature, $T_{ref}$, to produce a process effect equivalent to that induced by the recipe [14,15]. $t_{eff}$ was calculated for simple boron diffusion, which is assumed to have an activation energy, $E_A$, of 3.46 eV and choosing $T_{ref} = 1050°C$ [15]. Fig. 3 shows how $\Delta X_J$ varied with $t_{eff}$ for the various samples studied, as well as the diffusion length deduced from $[4D_B t_{eff}]^{1/2}$, where $D_B$ is the intrinsic diffusion coefficient for B at 1050°C [16]. For low $t_{eff}$, the implants with PAI show notably more diffusion than those into c-Si, partly because diffusion already occurs during the SPE process, as seen for the lowest $t_{eff}$ points. TED

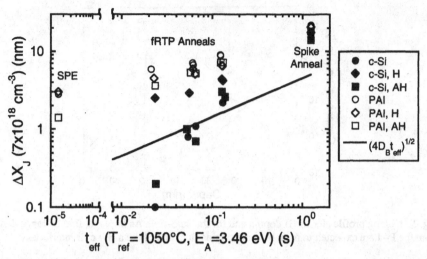

Fig. 3. The shift in junction depth that accompanies annealing. Results are shown for implants into c-Si, including cases with halos (c-Si,H) and with halos that were pre-annealed (c-Si,AH). Analogous results are shown for preamorphized cases (PAI, PAI,H and PAI,AH). The solid line is an estimate of the B diffusion expected. Results are plotted against the effective annealing time, defined for an activation energy of 3.46 eV and a reference temperature of 1050°C.

of B often occurs in samples with damage from heavy-ion implants, and $\Delta X_J$ is expected to be greater for the PAI cases than for implants into c-Si [17]. The c-Si samples with unannealed halo implants also show increased $\Delta X_J$, as a result of the defects created by the As implant.

The electrical activation was probed by measuring $R_S$ with several techniques, including conventional 4PP with a KLA-Tencor RS100 with type-D probes and the non-contact RsL™ method from Frontier Semiconductor. The latter uses dynamic measurements of photo-induced voltages in junctions to simultaneously measure $R_S$ and junction leakage [7]. Fig. 4 compares $R_S$ obtained from RsL and from 4PP measurements, for a range of the samples studied here. The 4PP measurements always gave unrealistically low $R_S$ values when halo doping was present, whereas the two measurements generally showed good agreement on lightly-doped substrates. In a lightly-doped substrate a thick depletion region electrically isolates the doped surface layer from the substrate, and the junction depth is likely to be beyond the EOR damage, so junction leakage effects are minimized. In contrast, the halo doping leads to a very narrow depletion region in a region that may contain residual defects, and hence junction leakage is much higher. RsL provided realistic $R_S$ values even with halo doping, although for extremely high junction leakage the very small photovoltages precluded accurate measurements. Fig. 5 shows $R_S$ vs. $X_J$ results from non-halo cases, with the $R_S$ taken from 4PP measurements, since leakage was not a significant issue for the lightly-doped substrates. fRTP provided superior dopant activation with reduced dopant diffusion as compared to the spike anneals. In this comparison, no benefit is seen from the use of PAI.

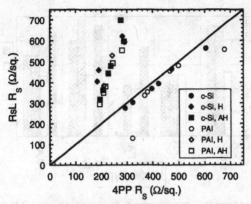

Fig. 4. Comparison of sheet resistances measured by conventional 4PP and by the non-contact RsL method. For halo doped samples, the 4PP measurements consistently showed large errors.

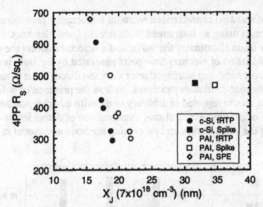

Fig. 5. Sheet resistance versus junction depth for samples without halo doping.

RsL also allows non-contact junction leakage current density, $J_L$, measurements. Fig. 6 summarizes results for a variety of implant combinations and annealing methods. B implants into crystalline silicon gave $J_L$ below the measurement limit of 0.1 μA/cm$^2$. Halo doping increased leakage, as would be expected from the greatly reduced depletion region width. Indeed, increasing the substrate doping makes the RsL $J_L$ measurement extremely sensitive to residual damage. Pre-annealing of the halo implant greatly reduced $J_L$ for implants into c-Si, but had less effect when the PAI was used. This should be expected, since much of the halo implant damage should be located in the region subsequently amorphized by the PAI. SPE produces very leaky junctions, while spike anneals consistently give low leakage, regardless of the implant scheme adopted. PAI cases still showed significant leakage after millisecond anneals at 1250°C, but 1300°C anneals were more effective. The results show that optimized millisecond annealing significantly improves damage annealing relative to low temperature SPE anneals.

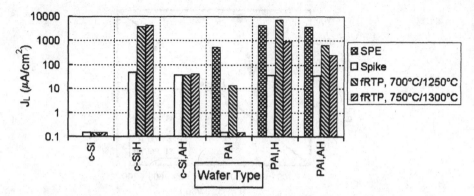

Fig. 6. Leakage current from RsL measurements for samples annealed by various methods. Results are shown for implants into c-Si, including cases with halos (c-Si,H) and with halos that were pre-annealed (c-Si,AH). Analogous results are shown for PAI cases (PAI,H and PAI,AH).

Residual damage was also characterized by room temperature photoluminescence (PL) measurements performed using an instrument from Accent (now Nanometrics). In these measurements a laser beam illuminates the wafer and a detector collects the optical emission from radiative recombination of electron-hole pairs generated by the laser beam. This signal is quite weak in an indirect band-gap semiconductor such as silicon, but it is very strongly affected by factors that alter the recombination processes, such as the presence of defects. The signal is interpreted as a defect metric reported in arbitrary units, with a higher number corresponding to a greater concentration of defects. Fig. 7 shows a comparison of defect levels deduced from the PL measurements for the same wafers and processing methods as covered in Fig. 6.

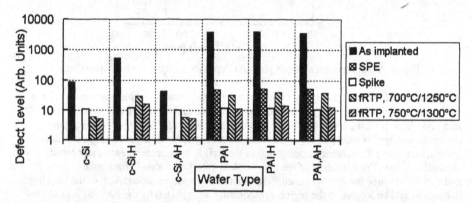

Fig. 7. The defect level as reported by photoluminescence measurements for samples annealed by various methods. Results are shown for the same implants and anneals as in Fig. 6.

The as-implanted samples show high defect levels, which reduce with annealing. Implants into c-Si without the halo implant or with the pre-annealed halo implant show very similar trends with annealing. For c-Si cases where the halo doping implant is not annealed the level is higher, which is consistent with the damage expected. In contrast, halo doping has little effect for the PAI cases, because the PAI amorphizes most of the region affected by halo doping. For the same reason, pre-annealing the halo has little effect in the PAI cases. The trends in the defect levels show some interesting similarities to those observed for $J_L$ in Fig. 6. For example, the spike anneal produces similar, low values in all cases. The SPE anneals lead to rather high defect levels, while the 1250°C fRTP anneals slightly reduce them, and 1300°C anneals reduce them further.

## CONCLUSIONS

High temperature millisecond anneals were shown to provide optimized dopant activation with minimal diffusion, while significantly reducing the effect of defects relative to what can be achieved by SPE alone. The selection of the halo implant conditions was shown to be critical, and the damage introduced by this implant must be considered in the optimization. In this study the halo dose was rather high, leading to very thin depletion regions and also introducing significant damage. Preannealing the damage from the halo reduced the junction leakage. Both PL measurements and the RsL leakage current measurement were shown to be very useful non-invasive schemes for rapid assessment of residual damage and the effects of annealing.

## ACKNOWLEDGMENTS

The authors would like to acknowledge the help of H. Phan, G. Stuart and D. Camm of Mattson Technology, M. Current, T. Nguyen and V. Faifer of Frontier Semiconductor for the RsL measurements and A. Buczkowski, Z. Li and T. Walker of Accent (now Nanometrics) for the PL measurements. This work has been partially supported by the European Union (IST project 027152 ATOMICS) program.

## REFERENCES

1. R. Duffy, A. Heringa, J. Loo, E. Augendre, S. Severi and G. Curtatola, ECS Trans. **3(2)** (2006) 19.
2. P. Timans, J. Gelpey, S. McCoy, W. Lerch and S. Paul, Mater. Res. Soc. Symp. Proc. **912**, (2006) 3.
3. W. Lerch, S. Paul, J. Niess, S. McCoy, T. Selinger, J. Gelpey, F. Cristiano, F. Severac, M. Gavelle, S. Boninelli, P. Pichler and D. Bolze, Mater. Sci. Eng. B **124–125**, (2005) 24.
4. R. A. Camillo-Castillo, M. E. Law, K. S. Jones, L. Radic, R. Lindsay and S. McCoy, Appl. Phys. Lett. **88**, (2006) 232104.
5. T. Noda, S. Felch, V. Parihar, C. Vrancken, T. Janssens, H. Bender and W. Vandervorst, Mater. Res. Soc. Symp. Proc. **912**, (2006) 191.
6. W. Lerch, S. Paul, J. Niess, F. Cristiano, Y. Lamrani, P. Calvo, N. Cherkashin, D. F. Downey and E. A. Arevalo, J. Electrochem. Soc. **152**, (2005) G787.

7. V. N. Faifer, D. K. Schroder, M. I. Current, T. Clarysse, P. J. Timans, T. Zangerle, W. Vandervorst, T. M. H. Wong, A. Moussa, S. McCoy, J. Gelpey, W. Lerch, S. Paul, D. Bolze and J. Halim, J. Vac. Sci. Technol. B **25**, (2007) 1588.

8. A. Buczkowski, B. Orschel, S. Kim, S. Rouvimov, B. Snegirev, M. Fletcher and F. Kirscht, J. Electrochem. Soc. **150**, (2003) G436.

9. S. Severi, E. Augendre, S. Thirupapuliyur, K. Ahmed, S. Felch, V. Parihar , F. Nouri, T. Hoffman, T. Noda, B. O'Sullivan, J. Ramos, E.San Andrés, L. Pantisano, A. De Keersgieter, R. Schreutelkamp, D. Jennings, S. Mahapatra, V.Moroz, K. De Meyer, P. Absil, M. Jurczak and S. Biesemans, in *Technical Digest of the International Electron Devices Meeting 2006* (IEEE, Piscataway, 2006) p. 859.

10. *The Stopping and Range of Ions in Matter*, http://www.srim.org.

11. S. T. Dunham and C. D. Wu, J. Appl. Phys. **78**, (1995) 2362.

12. S. Prussin, D. I. Margolese and R. N. Tauber, J. Appl. Phys. **57**, (1985) 180.

13. P. J. Timans, N. Acharya and I. Amarilio, ECS Proceedings **PV 2000-9**, (2000) 375.

14. A. T. Fiory and K. K. Bourdelle, Appl. Phys. Lett. **74**, (1999) 2658.

15. A. Mokhberi, P. B. Griffin, J. D. Plummer, E. Paton, S. McCoy and K. Elliott, IEEE Trans. Electron. Dev. **49**, (2002) 1183.

16. S. Banerjee, in *Handbook of Semiconductor Manufacturing Technology*, edited by R. Doering and Y. Nishi, (CRC Press, Boca Raton, FL, 2008) p. 8-1.

17. C. Bonafos, M. Omri, B. de Mauduit, G. BenAssayag, A. Claverie, D. Alquier, A. Martinez and D. Mathiot, J. Appl. Phys. **82**, (1997) 2855.

# Optimization of Flash Annealing Parameters to Achieve Ultra-Shallow Junctions for sub-45nm CMOS

Pankaj Kalra[1,2], Prashant Majhi[1], Hsing-Huang Tseng[1], Raj Jammy[1], and Tsu-Jae King Liu[2]
[1]SEMATECH, Austin, TX, 78741
[2]EECS, University of California, Berkeley, Berkeley, CA, 94720

## ABSTRACT

The use of millisecond annealing to meet ultra-shallow junction requirements for sub-45nm CMOS technologies is imperative. In this study, the effect of flash anneal parameters is presented. Reduced dopant diffusion and lower sheet resistance $R_s$ is achieved for intermediate temperature $T_{int}$ = 700°C (vs. 800°C). Significantly lower $R_s$ is achieved with peak temperature $T_{peak}$ = 1300°C (vs. 1250°C). Multiple shots provide for lower $R_s$, albeit at the expense of increased dopant diffusion. Based on a simple quantitative model, an optimal flash anneal can achieve 82% dopant activation efficiency.

## INTRODUCTION

One of the major challenges for MOSFET scaling in the sub-45nm regime is the formation of ultra-shallow junctions (USJs). The *International Technology Roadmap for Semiconductors* (ITRS) indicates that future generations of CMOS technology should have source/drain-extension junctions that are ~12-15nm deep, with sheet resistance $R_s$~1000 Ω/☐, in order to keep pace with historical improvements in high-performance logic devices [1]. The USJ requirement stems from the need to suppress short channel effects (SCE), and dictates a very limited thermal annealing budget to limit dopant diffusion; the low sheet resistance requirement is necessary to ensure low parasitic resistance, and dictates a high annealing temperature to maximize dopant activation. These requirements are especially difficult to meet for $p+/n$ junctions, due to boron transient enhanced diffusion (TED). Flash annealing is a potentially attractive alternative to conventional spike annealing for source/drain dopant activation, because it provides for shorter annealing time (~ms) and higher peak temperature (~1300°C).

## EXPERIMENTAL DETAILS

Blanket USJ formation studies were performed using Si (100) wafer substrates. 2-nm-thick thermal oxide was grown before ion implantation, followed by either a shallow boron implant or a shallow arsenic implant. A pre-amorphization Ge implant (20keV, $10^{15}$ cm$^{-2}$) was done prior to boron implantation to eliminate ion channeling. Nitrogen and fluorine were co-implanted with boron. These co-implanted species interact with excess Si interstitials generated during implantation and reduce the impact of boron TED. A spike anneal (1070°C) or flash anneal was used to activate the implanted dopants. Figure 1 highlights the features of the flash annealing process used in this work [2]. In a flash annealing process, the wafer is first heated up to an intermediate temperature ($T_{int}$), and then a millisecond flash is applied to heat the device side of the wafer to the peak temperature ($T_{peak}$). Various combinations of flash anneal parameters ($T_{int}$, $T_{peak}$, and $n$, where $n$ is the number of flash anneals) were included in the study to elucidate interdependencies. Note that the time between flash anneals was relatively long, for the samples which received multiple flash anneals, as illustrated in Figure 1.

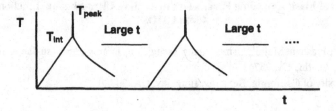

**Figure 1.** Temperature-*vs.*-time profile for multiple-shot flash annealing. For each shot, the wafer temperature is first ramped up to an intermediate temperature ($T_{int}$), and then the device side of the wafer is heated to the peak temperature ($T_{peak}$).

## RESULTS AND DISCUSSION

### SIMS Analyses

Secondary Ion Mass Spectroscopy (SIMS) analysis results for boron- and arsenic-implanted samples before and after activation anneal are shown in Figure 2. Spike annealing results in junctions that are too deep for sub-45nm nodes. (The junction depth $X_j$ is taken to be the depth at which the dopant concentration falls to $5 \times 10^{18}$ cm$^{-3}$.) Co-implantation helps to reduce $X_j$ but not enough to meet ITRS specifications for highly scaled CMOS technologies. Minimal diffusion is seen for the boron and arsenic profiles after flash annealing. Clearly, flash annealing reduces dopant diffusion as compared with spike annealing. These results are consistent with those of other studies [3, 4].

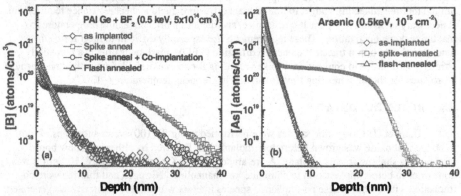

**Figure 2.** SIMS analyses of samples which received low-energy implants of (a) boron and (b) arsenic. Negligible diffusion is observed for the flash annealed samples ($T_{int} = 700°C$, $T_{peak} = 1300°C$, 1 shot) compared to the spike annealed samples. Co-implant specie was F (11keV, $10^{15}$cm$^{-2}$) for this experimental split.

## Sheet Resistance Measurements

Accurate sheet resistance measurement is critical for assessing USJ formation processes. Measurements obtained with the conventional contact method and a non-contact method [5] are compared in Figure 3. The conventional method yields erroneously low values of measured $R_s$ for ultra-shallow junctions, due to probe-tip penetration to a depth lower than $X_j$. The error increases exponentially as $X_j$ decreases, exceeding 1000 $\Omega/\square$ for $X_j < 15$nm. Therefore, non-contact measurement is needed for accurate USJ characterization.

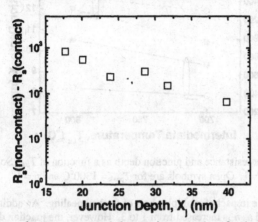

**Figure 3.** Difference between measured sheet resistance values for the non-contact method [5] and the conventional contact method. The conventional method yields erroneously low values of measured $R_s$ as compared to the non-contact method. The error is severe for $X_j$ <30nm. Starting substrate resistivity was 4-10 $\Omega$-cm and implants studied were B (0.2keV, $5\times10^{14}$cm$^{-2}$) and BF$_2$ (0.5-1.0keV, $5\times10^{14}$cm$^{-2}$).

## Impact of $T_{int}$, $T_{peak}$ and $n$

Figures 4-6 show the effects of intermediate temperature ($T_{int}$), peak temperature ($T_{peak}$), and number of flash shots ($n$) on sheet resistance ($R_s$) and junction depth ($X_j$), for samples which received BF$_2$ implants (0.8 keV, $10^{15}$ cm$^{-2}$). $X_j$ was determined by SIMS, and $R_s$ was determined by the non-contact method. Figure 4 shows the impact of $T_{int}$ for two different values of $T_{peak}$, for single-shot flash annealing. For both 1250°C and 1300°C peak temperature, $T_{int} = 700$°C results in shallower $X_j$ and lower $R_s$. Thus, 700°C is preferred over 800°C for the intermediate temperature.

Figure 5 shows the impact of $T_{peak}$. Significant reduction in $R_s$ (26% for $T_{int} = 700$°C and 20% for $T_{int} = 800$°C) is observed when $T_{peak}$ is increased from 1250°C to 1300°C. This reduction can be attributed to higher solid solubility limit of boron at 1300°C compared to 1250°C. However, the reduction in $R_s$ comes at the expense of a small increase in $X_j$ (<1nm).

The combination of $T_{int} = 700°C$ and $T_{peak} = 1300°C$ appears to be optimal for achieving shallow $X_j$ with low $R_s$.

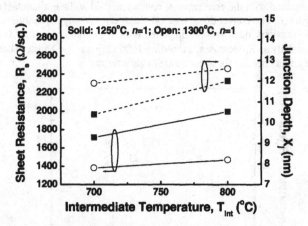

**Figure 4.** Measured sheet resistance and junction depth as a function of $T_{int}$. Solid symbols are for $T_{peak} = 1250°C$ and $n = 1$. Open symbols are for $T_{peak} = 1300°C$ and $n = 1$.

Figure 6 shows the impact of multiple shots of flash annealing. An additional 20% decrease in $R_s$ is achieved as $n$ is increased from 1 to 3. However, the junction depth is also increased, by ~1.5nm. Therefore, multiple shots of flash annealing can be used to further reduce $R_s$ with a trade-off in $X_j$.

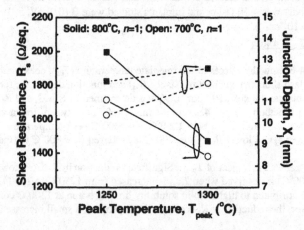

**Figure 5.** Measured sheet resistance and junction depth as a function of $T_{peak}$. Solid symbols are for $T_{int} = 800°C$ and $n = 1$. Open symbols are for $T_{int} = 700°C$ and $n = 1$.

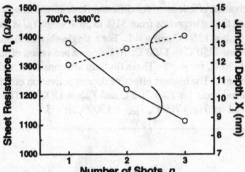

**Figure 6.** Measured sheet resistance and junction depth as a function of $n$ for $T_{int} = 700°C$ and $T_{peak} = 1300°C$.

## Dopant Activation Efficiency

A simplified model is used here to compare the dopant activation efficiencies for different flash anneal conditions. The effective dopant activation efficiency, $\eta$, is defined as

$$\eta \equiv \frac{R_{SIMS}}{R_{meas}} \tag{1}$$

where $R_{SIMS}$ is the sheet resistance calculated from the SIMS profile, and $R_{meas}$ is the measured sheet resistance of the same junction. $R_{SIMS}$ is calculated by dividing the doped semiconductor into multiple parallel slices, as shown in Figure 7.

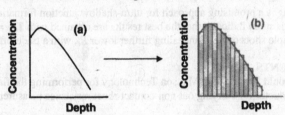

**Figure 7.** (a) Measured SIMS profile, (b) SIMS profile divided into multiple parallel slices.

Each slice of width $\Delta x$ is assumed to be uniformly doped to a concentration $N(x)$, so that it has a sheet resistance

$$R_{s,j} = \frac{(qN(x)\mu)^{-1}}{\Delta x} \tag{2}$$

The sheet resistance of the entire junction is then obtained using the following equation:

$$\frac{1}{R_{SIMS}} = \sum_j \frac{1}{R_{s,j}} \tag{3}$$

The effective dopant activation efficiencies for different flash anneal conditions are summarized in Figure 8. It is observed that $\eta$ improves from 51% to 58% when $T_{int}$ is reduced from 800°C to 700°C, respectively, for $T_{peak} = 1250$°C and $n = 1$. For a single-shot flash anneal, $\eta$ is improved when $T_{peak}$ is increased from 1250°C to 1300°C: 70% dopant activation efficiency is achieved for $T_{int} = 700$°C and $T_{peak} = 1250$°C, for $n = 1$. These findings are consistent with conclusions drawn from $R_s$ and $X_j$ measurements. The highest effective dopant activation efficiency ($\eta \sim 82\%$) is achieved with 3 flash anneal shots for $T_{int} = 700$°C and $T_{peak} = 1300$°C. The $R_s \times X_j$ product was also found to be minimized for $T_{int} = 700$°C, $T_{peak} = 1300$°C, $n = 3$.

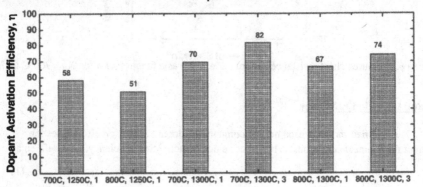

**Figure 8.** Effective dopant activation efficiencies for various flash anneal conditions.

## CONCLUSIONS

Flash annealing is a promising approach for ultra-shallow junction formation for sub-45nm MOSFETs. This study indicates that the best results are obtained with $T_{int} = 700$°C and $T_{peak} = 1300$°C. Multiple shots of flash annealing further lower $R_s$, with a trade-off in $X_j$.

## ACKNOWLEDGMENTS
The authors would like to thank Mattson Technology for performing the flash annealing and Frontier Semiconductor for carrying out non-contact sheet resistance measurements.

## REFERENCES
1. International Technology Roadmap for Semiconductors, 2006 Update.
2. W. Lerch, S. Paul, J. Niess, S. McCoy, T. Selinger, J. Gelpey, F. Cristiano, F. Severac, M. Gavelle, S. Boninelli, P. Pichler and D. Bolze, Mat. Sci. Eng. B **124–125**, (2005) 24.
3. T. Ito, K. Suguro, T. Itani, K. Nishinohara, K. Matsuo, and T. Saito, *Symp. on VLSI Technology Tech. Dig.*, (2003) 53.
4. K. Adachi, K. Ohuchi, N. Aoki, H. Tsujii, T. Ito, H. Itokawa, K. Matsuo, K. Suguro, Y. Honguh, N. Tamaoki, K. Ishimaru, and H. Ishiuchi, *Symp. on VLSI Technology Tech. Dig.*, (2005) 142.
5. V. N. Faifer, M. I. Current, T. Nguyen, T. M. H. Wong, and V. V. Souchkov, *Ext. Abs. the Fifth International Workshop on Junction Technology*, (2005) 45.

Mater. Res. Soc. Symp. Proc. Vol. 1070 © 2008 Materials Research Society          1070-E04-06

## Doping of Sub-50nm SOI Layers

Bartek J. Pawlak[1], Ray Duffy[1], Mark van Dal[1], Frans Voogt[2], Robbert Weemaes[3], Fred Roozeboom[4], Peer Zalm[3], Nick Bennett[5], and Nick Cowern[5]

[1]NXP-TSMC Research Center, Kapeldreef 75, Leuven, 3001, Belgium
[2]NXP Semiconductors, Nijmegen, Netherlands
[3]Philips, Eindhoven, Netherlands
[4]NXP Semiconductors, Eindhoven, Netherlands
[5]University of Newcastle, Newcastle, United Kingdom

## ABSTRACT

Doping of thin body Si becomes very essential topic due to increasing interest of forming source/drain regions in fully depleted planar silicon-on-isolator (SOI) devices or vertical Fin field-effect-transistors (FinFETs). To diminish the role of the short-channel-effect (SCE) control, the Si layers thicknesses target the 10 nm range. In this paper many aspects of thin Si body doping are discussed: dopant retention, implantation-related amorphization, thin body recrystallization, sheet resistance ($R_s$) and carrier mobility in crystalline or amorphized material, impact of the annealing ambient on $R_s$ for various SOI thicknesses. The complexity of 3D geometry for vertical Fin and the vicinity of the extended surface have an impact on doping strategies that are significantly different than for planar bulk devices.

## INTRODUCTION

Understanding of dopant behavior in thin body Si is crucial for Fin source and drain formation.

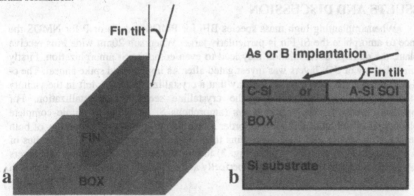

**Figure 1.** Schematic drawings of implants into Fin (a) when source and drain is formed and of the experiment on implants into crystalline or amorphous SOI blanket wafer. (b) The tilt angle as indicated is consistent with Fin doping geometry. The a-Si layer mimics the full Fin amorphizing implant condition.

The vicinity from both sides of a large Si/SiO$_2$ interface introduces differences from regular bulk Si doping with respect to dopant incorporation, material crystallinity, defect generation/annihilation, dopant activation, and diffusion. The study of dopant implantation in thin body Si can be conducted on SOI wafers. The sidewall implantation and diffusion in a vertical Fin can be mimicked by the tilt angle adjustment (Fig. 1) [1]. In that case the electrical and chemical analysis is simple to conduct on planar structures rather than on free standing vertical structures. The implant angles specified in the text and figure captions relate to what the process engineer would specify. The tilts specified here refer to what the sidewall would receive, thus 90-(tilt angle). e.g. 20° discussed in the figures corresponds to 90°-20°=70° received by the planar sample. – see Fig. 1b.

## EXPERIMENTAL DETAILS

Blanket 200 mm SOI wafers were used in the experiment. The thickness of crystalline SOI layer was controlled by etch-based thinning. Some wafers were prepared in order to mimic the fully amorphizing implant. The amorphous SOI layers were obtained by Si deposition on SiO$_2$. The thickness of Si layers was confirmed by cross-section transmission electron microscopy (XTEM) and Rutherford backscattering spectroscopy (RBS-c) to be 10nm, 24nm, 38nm and 65nm for c-SOI (obtained by thinning) and 10nm, 20nm, 30nm and 65nm for a-SOI (obtained by Si deposition). Implantations of B and As were carried out at various tilt angles to the wafer plane at 83°, 30°, 20°, and 10°, in that way the results are meaningful for side-wall doping of the Fin (Fig. 1). The simulation of as-implanted profiles has been performed by means of Monte Carlo approach, SRIM 2006 [2]. The amount of retained dopants was estimated by Rutherford Backscattering Spectroscopy in the channeling mode (RBS-c) for As and from simulation for B. Activation anneal has been done in the rapid thermal annealing system by spike activation anneal at 1050°C in various ambient gases containing either pure N$_2$ or mixture of 10% H$_2$ or O$_2$. R$_s$ values have been taken in the 4 point probe system, carrier mobilities were extracted from Hall measurement.

## RESULTS AND DISCUSSION

When implanting high mass species BF$_2$ for PMOS and As or P for NMOS the chance to amorphize the Si Fin is particularly large. When sub-20nm wide Fins receive implant, the damage accumulation may lead to even complete Si amorphization. Firstly the morphology of SOI layers was investigated after As implant and spike anneal. The c-SOI wafers were implanted in such a way that a crystalline layer was left in the vicinity of the backside interface offering the crystalline seed for recrystallization. For comparison, we have used a-SOI layers (amorphous Si on SiO$_2$) to mimic complete implantation-related amorphization. In order to see the morphological properties of both materials with similar and typical for Fins thicknesses we have done XTEM studies of SOI after spike RTA at 1050 °C (see Fig. 3) in pure N$_2$ ambient. Crystalline SOI layers that were partially amorphized become perfectly and uniformly crystalline after regrowth

(Fig. 2a) with a good surface roughness. Amorphous material forms after spike anneal poly-Si grains of random orientation stuck next to each other (Fig. 2b). There is no clear evidence that any spots of the material remain amorphous. The poly-Si grains fit firmly next to each other having only small impact on the surface roughness. The morphological non-homogeneity of the thin layer has an impact on carrier conducting properties that are discussed in the next section.

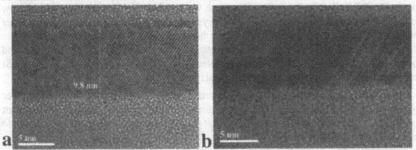

**Figure 2.** XTEM analysis of crystalline (a) and amorphous (b) SOI layer subjected to shallow As implant and spike RTA at 1050 °C. The c-SOI remains smooth and perfectly defect-free material, the a-SOI turns into poly-Si structure with grains of various crystallographic orientations.

Further experiments were focused to investigate the relation between the resulting crystalline quality of SOI layers and doping efficiency for both PMOS and NMOS. C-SOI and a-SOI wafers were implanted with B and As and their electrical properties have been measured, see Fig. 3.

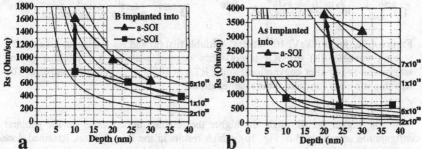

**Figure 3.** $R_s$ analysis of B (2keV, $10^{15}$ at./cm$^2$, 83° tilt) (a) and As-implanted (2keV, $10^{15}$ at./cm$^2$, 83° tilt) (b) SOI layers when original material is crystalline (■) or amorphous (▲) Si after conventional spike RTA at 1050 °C in 100% N$_2$ ambient. The $R_s$ increment from crystalline to amorphous material has been observed to be around 50% - 100% for B and approximately 1 order of magnitude for As. The trend lines are activation curves for a given active concentration taking into account standard carrier mobilities and box-like profiles.

The thinner the SOI layers the higher the $R_s$. The $R_s$ is higher for B and As in wafers that were originally amorphous and formed poly-Si structure after spike anneal. The strongest impact of poor crystallinity is observed for As doped layers, where the ~10 times $R_s$ increment is attributed to electron mobility degradation. The electron Hall mobility for 24 nm crystalline SOI layer is 75 cm$^2$/Vs, which is consistent with previously published data [3], and for the 20 nm amorphous SOI (which becomes poly-Si after spike RTA) is only 8 cm$^2$/Vs. Therefore we conclude that in order to maintain low resistance of doped thin body Si a good crystalline material has to be obtained after implant and activation anneal. However the main impact on low resistance of the thin body Si has material quality, we investigated also the role of the tilt angle and corresponding dopant retention. Some examples of recrystallization issues of thin body Si vertical Fins have been reported before by Duffy *et al.* [4].

Dopant incorporation plays also important role in forming good conducting layer that is set by the species implant energy, dose and tilt angle [1]. Setting the implant energy at 2 keV and the dose at $10^{15}$ at./cm$^2$ we investigate the role of tilt angle selectively for B and As, see Fig. 4.

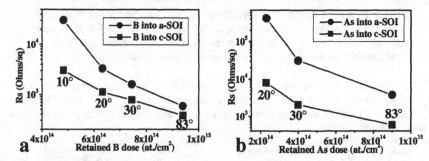

**Figure 4.** $R_s$ for B (2keV, $10^{15}$ at./cm$^2$, variable tilt) (a) and As-implanted (2keV, $10^{15}$ at./cm$^2$, variable tilt) (b) SOI layers when original material is crystalline (■) or amorphous (●) Si after conventional spike RTA at 1050 °C in 100% $N_2$ ambient. The $R_s$ value is a clear function of the retained dose.

The higher the tilt angle the higher the retained dopant dose (in the geometry configuration as indicated in Fig. 1b), which results in the $R_s$ reduction. Retained doses versus implant tilt angle for B have been simulated, whereas for As they have been obtained from RBS-c analysis. The impact of the tilt angle on dopant retention seems to be a similar problem for B and As as well as for crystalline and amorphous SOI material. For low tilt angles the dose retention is the main problem. Low tilt implantation is necessary in order to form low resistive thin body Si for dense Fin structures.

The behavior of B in thin body Si has been previously investigated with respect to its thermal stability [5] by amorphizing the SOI wafer close the backside interface. Also the diffusion properties of B seem to be affected by the close vicinity of interfaces from both sides [6]. Both phenomena are related to sinking of Si interstitials to the interface between Si/SiO$_2$. Therefore in the coming paragraph the impact of Si interstitials injection by varying the ambient gas [7] is investigated with respect to conduction properties of the layer.

Apart from controlling good crystalline quality of the sample and maximizing the dopant retention, the type of ambient gas used during RTA also affects thin Si body resistivity by enhancing or reducing the dopant activation. The regular $N_2$ gas can be mixed with $O_2$ (for interstitial injection) [8] or $H_2$ (for B dose loss) [9] to alternate dopant activation within SOI layer. Fig. 6 shows the effect of the ambient type during RTA annealing for B when dopants are positioned deeper (a) during 83° tilt implant or shallower (b) during 20° tilt implant for various SOI thicknesses, see Fig. 5.

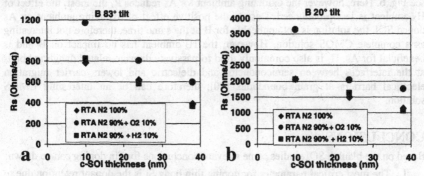

Figure 5. $R_s$ of crystalline SOI layers implanted with B 2keV, $10^{15}$ at./cm$^2$ at 83° tilt (a) and at 20° tilt (b) and subsequently annealed at various gas ambients. The oxidizing ambient results in $R_s$ increase especially when B is close to the Si/SiO$_2$ interface (thin SOI or shallow implant – 20° tilt).

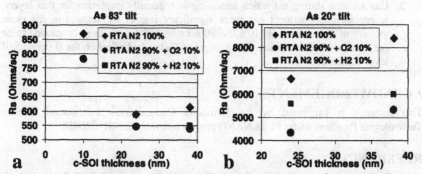

Figure 6. $R_s$ of crystalline SOI layers implanted with As 2keV, $10^{15}$ at./cm$^2$ at 83° tilt (a) and at 20° tilt (b) and subsequently annealed at various gas ambients. The oxidizing and hydrogenating ambient results in $R_s$ decrease with respect to regular $N_2$ 100% ambient.

The conventional $N_2$ 100% ambient offers equally with $N_2$ 90% + $H_2$ 10% best $R_s$ values. In case of B doping a strong $R_s$ increment is observed when an $N_2$ 90% + $O_2$ 10% is used. The $O_2$ ambient anneal induced strong B deactivation, especially when the dopant is positioned close to the interface e.g. for 10 nm wide SOI thickness. Oxidizing ambient injects Si interstitials resulting in higher $R_s$ for B doped SOI layers (that have confined volume), which is in a direct contrast with bulk Si properties, where injection of Si interstitials promotes B diffusion and lowers $R_s$. Therefore the oxidizing ambient should be avoided, also when the source of the oxygen is in the material used for processing like new spacer or liner. Identical studies have been performed for As doped crystalline SOI, see Fig. 6. Here, however the oxidizing ambient for As reduces $R_s$ the most, the effect of $H_2$ ambient is less pronounced. Despite the positive effect of oxidizing ambient on As doped SOI, the solution is not applicable for B at the same time, therefore not interesting as a complete CMOS solution. However, the $H_2$ ambient has no impact on B and is beneficial for As. $H_2$ is also commonly used to passivate damage-related dangling bonds at the interfaces between semiconductor and dielectric and lowers carrier migration electrical barriers at grain boundaries [10], therefore can be an interesting CMOS solution.

## CONCLUSIONS

Based on the blanket SOI studies some relevant conclusions for Fin doping can be drawn:

1. The most critical parameter for doping thin body Si is the dopant retention due to high tilt angles used for sidewall of vertical Fins.

2. Some heavy elements like As or P may amorphize the thin body Si resulting in formation of complete amorphous layer. In that case the $R_s$ is high, mainly due to poor carrier mobility.

3. Gas ambient during activation annealing is especially important for thin layers, where surface treatment has more significant impact on dopants in the close vicinity of the surface. $H_2$ 10%, $N_2$ 90% or $N_2$ 100% have been concluded to be the best for CMOS solution. The mixture of $O_2$ has bad effect on the B doped SOI layer conduction.

## ACKNOWLEDGEMENTS

This research is supported by the European Commission's Information Society Technologies Program, under PULLNANO project contract No. IST-026828.

## REFERENCES

[1] R. Duffy, G. Curatola, B. J. Pawlak, G. Doornbos, K. van der Tak, P. Breimer, J. G. M. van Berkum, and F. Roozeboom, J. Vac. Sci. Technol. B **26**, 402 (2008).

[2] J.F. Ziegler, M.D. Ziegler, J.P. Biersack, SRIM 2006.

[3] S. Matsumoto, T. Niimi, J. Murota, E. Arai, J. Elect. Soc. **127**,1650 (1980).

[4] R. Duffy, M. J. H. Van Dal, B. J. Pawlak, M. Kaiser, R. G. R. Weemaes, B. Degroote, E. Kunnen, and E. Altamirano, Appl. Phys. Lett. **90**, 241912 (2007).

[5] J. J. Hamilton, K. J. Kirkby, N. E. B. Cowern, E. J. H. Collart, M. Bersani, D. Giubertoni, S. Gennaro, and A. Parisini, Appl. Phys. Lett. **91**, 092122 (2007).

[6] S. M. Kluth, D. Álvarez, St. Trellenkamp, J. Moers, and S. Mantl, J. Kretz, and W. Vandervorst, J. Vac. Sci. Technol. B **23**, 76 (2005).

[7] S. Mizuo, T. Kusaka, A. Shintani, M. Nanba, and H. Higuchi, J. Appl. Phys. **54**, 3860 (1983).

[8] N. Guillemot, D. Tsoukalas, C. Tsamls, J. Margail, and M. Paon, J. Appl. Phys. **71**, 1713 (1992).

[9] L. Zhong, Y. Kirino, Y. Matsushita, Y. Aiba, K. Hayashi, R. Takeda, H. Shirai, H. Saito, J. Matsushita, and J. Yoshikawa, Appl. Phys. Lett. **68**, 1229 (1996).

[10] T. Yamazaki, Y. Matsumura, Y. Yraoka, and T. Fuyuki, Jpn. J. Appl. Phys., **45**, 342 (2006).

Mater. Res. Soc. Symp. Proc. Vol. 1070 © 2008 Materials Research Society 1070-E04-08

# Optimization of Stressor Layers Created by ClusterCarbon™ Implantation

Karuppanan Sekar[1], Wade A Krull[1], and Thomas N Horsky[2]

[1]Process Technology, SemEquip Inc, 34 Sullivan Rd, North Billerica, MA, 01862
[2]Technology Group, SemEquip Inc, 34 Sullivan Rd, North Billerica, MA, 01862

## ABSTRACT

Si:C layers are interesting candidates as stressor layers for NMOS transistors. Growth of such a Si:C layer has been realized by an expensive epitaxial growth process for devices to produce tensile strain in the channel, leading to enhanced mobility and device performance. Use of a monomer carbon ion implant in conjunction with a Ge pre-amorphizing implant (Ge-PAI) in Si is an alternative, lower cost approach to obtaining such SiC layers. This approach has not yielded desired device performance owing to low carbon substitutionality $[C]_{sub}$, and also the presence of end-of-range (EOR) defects and large leakage currents due to the Ge-PAI implant. In this study we will show the formation of a Si:C layer using a ClusterCarbon approach that creates self-amorphization in Si thus avoiding an extra Ge-PAI implant step. We show that more than 2% substitutional carbon can be realized by using solid-phase-epitaxial-regrowth (SPER) and millisecond anneal. Si:C layers are characterized by using High Resolution X-ray Diffraction (HRXRD) and SIMS techniques.

## INTRODUCTION

Carbon ion implantation into Si acts as a barrier to boron diffusion. Also, C implantation in Si followed by annealing is used to synthesize Si:C layers. These SiC layers are interesting candidates for stressor layers in NMOS device transistors. In such devices, Si:C layers are created in the source and drain regions of the device and act as stressors, producing lateral tensile strain in the channel to enhance electron mobility. Formation of such Si:C layer is constrained by the limitation of carbon atom incorporation into substitutional sites in Si. It has been shown that the probability of placing a carbon atom into substitutional site decreases rapidly when the total carbon concentration exceeds 2% [1]. There are two primary means of incorporating carbon: implantation and epitaxial growth. Growing Si:C films by epitaxy is a slow and an expensive process. The second means is to form a Si:C layer through carbon ion implantation and annealing. Several studies of carbon implantation have been reported. Lower fluence implants produce only partially damaged layers and there is no clear interface is observed [2]. Damage within a disordered region surrounded by a single crystal matrix is difficult to completely anneal. Reordering of the damaged region is absent in such layers [3]. Si:C layer formation with monomer carbon implant has been shown [4], including slight improvement of NMOS transistor characteristics.

Such studies have identified that the necessary process to achieve high substituionality is the formation of a carbon doped amorphous silicon layer, followed by the recrystallization of the layer. SPER is a good dopant activation technique that does not involve high processing temperatures [5]. The requirements for stressor layer formation align well with the properties of ClusterCarbon implantation. Cluster implantation has been shown to produce amorphous layers without PAI; this is called self-amorphization. By suitably optimizing the implant parameters (dose and energy) with multiple implants one can tailor the concentration profile and the thickness of the amorphous region. After an SPER anneal at 750°C for various anneal times we obtained a percentage of substitutional carbon of roughly 1.5%. This number is consistent with the studies of Kramer and Thompson that beyond 1.4% atomic carbon, the carbon does not go substitutional but forms SiC precipitates [6]. Their absorption studies on carbon implanted Si showed a decrease in IR absorption beyond 1.4% atomic carbon. A higher absorbance indicates a higher level of substitutional carbon [7]. With an additional millisecond anneal we achieved an enhanced level of carbon activation close to 2% substitutional carbon.

In this study we show the results for a Si:C layer formed by cluster ion implantation with low temperature SPER anneals and high temperature millisecond (ms) anneals. We will present HRXRD data showing the stress, strain and the % of substitutional carbon due to these implants after anneal. We will show that self-amorphizing property of clustercarbon [8] plays an important role in such enhanced carbon substitutionality in Si.

## EXPERIMENTAL

The starting substrates used in this study were 200mm, n-type Si(100) silicon substrates. The wafers were implanted with ClusterCarbon at different energies and doses using $C_7H_7$ (from $C_{14}H_{14}$ material) from a ClusterIon® source. We selected $C_7H_7$ part of the mass spectrum that provided a cluster implant with 7 carbon atoms. SPE anneals in this work were performed with an RTP 200mm SUMMIT XT tool, while flash RTP (fRTP) anneals were done with Mattson Technology Canada's Millios flash anneal system. Secondary Ion Mass Spectrometry (SIMS) measurements were carried out using commercially available facilities, as were the HRXRD profiles.

## RESULTS AND DISCUSSION

Fig. 1 shows the mass spectrum for $C_{14}H_{14}$ material. There are multiple mass peaks corresponding to various ClusterCarbon species $C_xH_x$. The peak corresponding to 91amu and 182 amu are $C_7H_7$ and $C_{14}H_{14}$, respectively.

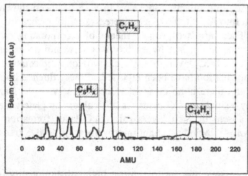

Fig. 1. Mass spectrum of $C_{14}H_{14}$ feed material. All the implanted ions used in this work are from the peak around 92 amu ($C_7H_x$).

The $C_7H_7$ specie was used for all our implants. The reason for choosing $C_7H_7$ is that we can go to higher implant energies (with lower mass) and still preserve the self-amorphizing property of the cluster ions.

The XTEM images of Fig. 2 show the formation of the self-amorphized layer for the implantation of both $C_7H_7$ and $C_{16}H_{10}$. In the case $C_{16}H_{10}$ the amorphous layer thickness (~313Å) is only slightly larger than in the case of $C_7H_7$ (~307Å). This indicates that at this energy of 7 keV per carbon atom, the amorphous layer thickness is dose limited rather than being limited by the mass of the cluster. Another feature one can observe in the XTEM pictures is the amorphous-crystalline (a-c) interface roughness. The a-c interface roughness is smoother and lower in the case of the heavier $C_{16}H_{10}$ ClusterCarbon implant when compared to $C_7H_7$ implant. In the $C_{16}H_{10}$ case the a-c interface roughness is (25±3)Å and for $C_7H_7$ it is (38±3)Å. By extrapolating these two data we can say that in the case of a monomer carbon implant, the interface roughness will be so high as to lead to a partial amorphization. As discussed earlier, such partial amorphization will lead to incomplete re-crystallization. This illustrates that a PAI implant is necessary to use with a monomer implant to obtain a re-crystallized layer.

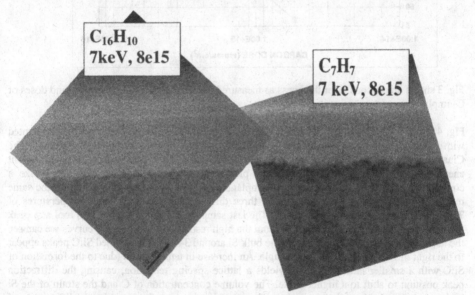

Fig. 2 shows XTEM images of $C_{16}H_{10}$ (Fig. 2(a)) and $C_7H_7$ (Fig. 2(b)) at 7keV per atom at an equivalent carbon dose of 8e15 atoms/cm$^2$.

Fig. 3 shows the amorphous layer thickness for $C_7H_7$ and $C_{16}H_{10}$ implants at two different energies. From the chart it is clear that the amorphous layer thickness is larger at higher energy. At a given energy for two different cluster masses ($C_{16}$ & $C_7$), the mass effect dominates at lower doses. The amorphous layer thickness at 7keV, 2e15 atoms/cm$^2$ for $C_{16}$ is (253±5)Å and for $C_7$ it is (207±5)Å. At higher doses at a given energy, the mass does not play a major role (as shown in Fig. 2). The energy factor plays a dominant role in determining the amorphous layer thickness at higher doses. It has been shown the size of the carbon cluster and hence the amorphous layer thickness is important in obtaining enhanced carbon substitutionality [2].

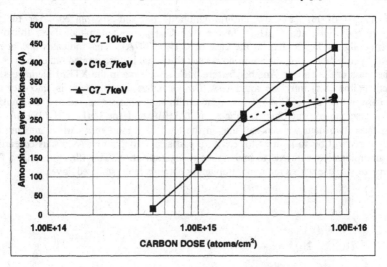

Fig. 3 shows amorphous layer thickness as measured by XTEM for various energies and doses of $C_7$ implants.

Fig. 4 shows high resolution diffraction curves of the Si(400) reflections for samples implanted with multiple energy ClusterCarbon implants annealed at various temperatures. The ClusterCarbon implant energies are 2keV, 5keV and 8keV. The implant dose for each implant energy is 3e15 atoms/cm$^2$. With accurate placement of carbon dopants one can realize a controlled profile with multiple sequence implants [6]. Three samples implanted with the same multiple implant recipe were annealed at three different temperatures, SPE temperatures of 750°C, 800°C and 850°C for 5 seconds. The last sample is annealed with a flash tool at a peak temperature of 1060 °C ( Ti = 750°C ). From the high resolution x-ray rocking curves we can see the large diffraction peak coming from the bulk Si around 34.57°. The strained Si:C peaks appear to the right of Si bulk peak at a larger angle. An increase in tensile strain (due to the formation of Si:C with a smaller lattice constant) yields a lattice spacing reduction, causing the diffraction peak position to shift to a higher angle. The volume concentration of C and the strain of the Si lattice varies in depth according to the carbon concentration profile, resulting in different local conditions for nucleation and growth of Si-C complexes and SiC particles. Hence one obtains a strained gradient in depth from the region with highest concentration.

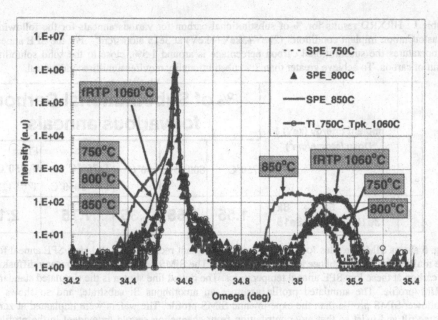

Fig. 4 shows HRXRD spectrum of a multiple $C_7$ implant sequence :
(2keV+5keV+8keV) of 3e15 atoms/cm$^2$ each.

The shoulders in the left hand side of the bulk Si peak is due to the presence of interstitial carbon. Looking at the profiles, the strained peak slowly shifts towards higher angle with increasing temperature except at 850°C. The substitutional carbon percentage as determined using Vegard's law [9] is estimated to be ~1.55% at 750°C, ~1.66% at 800°C and ~1.75% using fRTP at 1060°C. However, at 850°C it drops to 1.35% indicating diffusion of carbon from substitutional sites to interstitial sites. This is also evident from the broadening of the profile for 850°C to the left of the bulk Si peak. Beyond 850°C formation of SiC precipitates hinder the stability of carbon atoms in substitutional sites [10]. To trap the carbon atoms in substitutional sites one requires a high temperature millisecond anneal, where the atoms could be trapped in their positions in a non-equilibrium condition. All these above results are summarized in Table I. From Table I, it is seen that at a peak flash anneal temperature of 1275°C, a carbon substitutionality of greater than 2% can be achieved. The high temperature millisecond anneal not only provides higher substitutional carbon but also helps in re-crystallization of the damaged Si lattice.

Table I : HRXRD results for % of substitutional carbon for various anneals for the following ClusterCarbon implant condition: (2keV+5keV+8keV) at 3e15 atoms/cm². For the SPE anneal temperatures the substitutional carbon percentage is around 1.5%, close to the solid solubility limit of carbon. To achieve greater than 2% substitutionality requires a millisecond anneal.

| # | Implant Energy (keV) & Dose (atoms/cm²) | % of Substitutional Carbon for various anneals | | | | |
|---|---|---|---|---|---|---|
| | | 750°C | 800°C | 850°C | Flash $T_i$ 750°C $T^{peak}$ 1060°C | Flash $T_i$ 750°C $T^{peak}$ 1275°C |
| 1 | 2k + 5k + 8k 3e15 + 3e15 + 3e15 | 1.55 | 1.66 | 1.35 | 1.75 | 2.19 |

Fig. 5 shows SIMS profiles for a multiple carbon implant recipe before and after SPE anneal for the multiple implant sequence as discussed earlier. The SIMS profiles do not show any diffusion of carbon at these low SPE anneal temperatures. The dotted line shown is the simulated standard TRIM profile. The simulated profile assumes an amorphous Si substrate, and so shows a shallower SIMS profile than the experimental SIMS profile. The wafers were implanted at zero degree tilt and twist so that any contribution from channeling should be evident in the profile. The simulated profile shown reproduces fairly accurately the peak concentration and width of the profile.

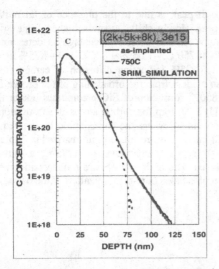

Fig. 5 Simulated and measured SIMS profiles for sequential 3e15 ClusterCarbon implants of (2keV+5keV+8keV) .

# ACKNOWLEDGEMENTS

We would like to thank our SemEquip GSD team Brian Haslam and Dennis Klesel for their support and would like to extend our thanks to Jeff Gelpey, Steve McCoy and Jason Chan of Mattson Technology, Canada, for performing flash anneals.

# CONCLUSION

A functional ClusterCarbon implant approach as an alternative to epitaxial Si:C process has been presented. It provides a perfect implant solution when compared to a monomer sequence solution. The monomer approach suffers from residual defects after annealing due to a Ge-PAI implant that degrades leakage current performance. The self-amorphizing property of ClusterCarbon implants plays an ideal role in the incorporation of substitutional carbon that is essential to providing a stressor layer in devices. With accurate placement of carbon dopants one can realize a controlled profile with a sequence of multiple implants. Using the ClusterCarbon approach, we have shown that one can achieve greater than 2% substitutional carbon that is otherwise difficult to achieve with a Si:C epi process.

# References:

1) King-Jien Chui, Kah-Wee Ang, Narayanan Balasubramanian, Ming-Fu Li, Ganesh S. Samudra and Yee-Chia Leo, IEEE Transactions of Electron Devices, 54, 249 (2007)
2) A. Li-Fatou et al, Proceeding of the Electro-chemical Society Meeting, 212, 1305 (2007)
3) J. Gyulai and P. Revesz, Conf. Serv.-Inst. Phys., 46, 128 (1979)
4) Y. Liu et al, *Symposium on VLSI Technology Digest of Technical Papers, 44 (2007)*
5) S. T. Picraux, Ann. Rev. Mater. Sci. 14, 335 (1984)
6) K. M. Kramer and M. O. Thompson, J. Appl. Phys. 79, 4118 (1996)
7) Melendez-Lira et al, J. App. Phys. 82, 4246 (1997)
8) K. Sekar, et al, Proc. International Workshop on INSIGHT in Semiconductors Device Fabrication, Metrology and Modeling, 141 (2007)
9) J. Hornstra and W. J. Bartels, J. Cryst. Growth, 44, 513 (1978) ; M. Berti et al , J. Appl. Phys. 72, xx (1998)
10) M. S. Goorsky et al , Appl. Phys. Lett. 60, 2758 (1992) ; J. W. Stane et al, J. Appl. Phys. 76, 3656 (1994)

Mater. Res. Soc. Symp. Proc. Vol. 1070 © 2008 Materials Research Society    1070-E04-09

# Optimization of Si:C Source and Drain Formed by Post-Epi Implant and Activation Anneal: Experimental and Theoretical Analysis of Dopant Diffusion and C Evolution in High-C Si:C Epi Layers

Yonah Cho[1], Victor Moroz[2], Nikolas Zographos[3], Sunderraj Thirupapuliyur[1], Lucien Date[1], and Robert Schreutelkamp[1]

[1]Applied Materials, Inc., 974 E. Arques Ave., Sunnyvale, CA, 94085
[2]Synopsys, Inc., 700 East Middlefield Road, Mountain View, CA, 94043
[3]Synopsys Switzerland LLC, Zurich, Switzerland

## ABSTRACT

Experimental and simulated P and As dopant diffusion profiles in Si:C epi films containing high C (>1 atomic %) are presented. A new set of physical effects were incorporated to accurately model P or As diffusion in the presence of high level of C. Evolution of substitutional C ($C_{sub}$) profile in the Si:C epi film through dopant implant and activation anneal was characterized by high-resolution x-ray diffraction (HRXRD) technique. Three-layer analysis was utilized to obtain non-uniform $C_{sub}$ profile. Dependency of $C_{sub}$ retention on anneal thermal budget is studied. It is shown the initial $C_{sub}$ in the epi layer is lost during dopant implantation and conventional spike anneal sequence. Use of advanced millisecond (ms) laser anneal resulted in near 100% $C_{sub}$ retention in P-implanted Si:C epi film without compromising junction depth. Measured $C_{sub}$ (by HRXRD) and total C (by SIMS) profiles are compared with the ones predicted by the newly developed compact modeling in this study.

## INTRODUCTION

nFET device performance improvement using recessed source and drain (S/D) Si:C approach has been demonstrated [1-2] and considered for added strain solution for nFET performance for 32nm technology and beyond. High substitutional C ($C_{sub}$) is required for high in-film and channel strains. Figure 1 shows the calculated longitudinal ($\sigma_{xx}$) and vertical ($\sigma_{zz}$) stress distribution in channel between Si:C S/D of 1.8% $C_{sub}$ and expected channel mobility improvement ($\Delta\mu_e$) as a function of $C_{sub}$ in the recessed S/D.

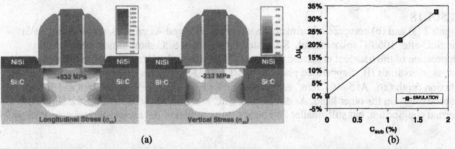

(a)                                                                                (b)

**Figure 1.** (a) Longitudinal ($\sigma_{xx}$) and vertical ($\sigma_{zz}$) stress distribution in the channel between Si:C stressors of 1.8% $C_{sub}$ and 80nm thick in a 60nm recess after NiSi formation. (b) Expected mobility improvement is plotted as a function of $C_{sub}$% based on known piezo-resistance values.

The key challenges are to obtain high $C_{sub}$ Si:C during the epitaxy and to retain it during post-epitaxy integration sequence. As a parallel approach to in-situ doped Si:C epi process, post-epi implant and activation anneal of undoped Si:C epi was explore for S/D junction formation. Proper junction formation and retention of $C_{sub}$ in the Si:C S/D for maximum channel strain are critical in fully realizing performance benefit, therefore, accurate prediction of n-dopant and C profiles is necessary. In this paper, a physical model is developed to analyze experimentally measured P and As diffusion and $C_{sub}$ evolution in high C-containing Si:C epi films. Using the model, the impacts of both the conventional spike anneal and the advanced millisecond laser anneal on dopant diffusion and $C_{sub}$ retention are explained.

## EXPERIMENTS

Undoped Si:C epi layers > 1.35 % $C_{sub}$ were prepared by low temperature RP CVD. Post-epi implant was performed with conditions emulating deep nMOS S/D formation, followed by conventional spike anneal and millisecond (ms) laser anneal. Total C ($C_{tot}$) and dopant profiles were measured by SIMS whereas $C_{sub}$ was characterized by high-resolution x-ray diffractometer (HRXRD) with multilayer analysis capability. Comparison of $C_{tot}$ and $C_{sub}$ revealed near 100% of the total C atoms in are at substitutional sites.

## MODELING

P and B diffusion models were previously developed in C-implanted Si with moderate level of C concentrations ($4 \times 10^{20}$/cm$^3$) [3]. However, these previous models did not reconcile the experimental dopant profiles in Si:C epi layer containing > 1% or $5 \times 10^{20}$/cm$^3$ C. Starting with basic carbon cluster reactions, $C_S + CI \Leftrightarrow C_2I; C_2I + CI \Leftrightarrow C_3I_2; C_3I_2 \Leftrightarrow C_3I_3 + V$, a new set of effects were introduced to explain P and As behavior in epitaxial Si:C films with C > 1%. Formation of carbon-rich $C_2$ clusters, full cascade Monte Carlo implant model instead of a plus factor, and impact of carbon-induced stress on $E_g$, I*, V*, and pair diffusion were incorporated. Most prominent effect for a high C environment is formation of mobile C-I pair whenever as-implanted I is lower than C, resulting in unusual vacancy super-saturation (VSS) effect that strongly affects As but only marginally affects P. Initial conditions for diffusion and clustering after As implantation leading to VSS phenomenon is shown elsewhere [4]. Modeling of P, As, and C implantation, diffusion and clustering is performed with Sentaurus Process[2] in this study. This modified model is general enough to accurately describe both the P, As, and C behavior in this work as well as the B, C, Sb, and P diffusion in Si with lower carbon content [3,5-6]

## RESULTS

Figure 2 (a) and (b) compare experimental and simulated P and As profiles, respectively, in Si and Si:C after 1050°C spike anneal. Simulated P profiles in Si:C show almost complete suppression of the transient enhanced diffusion (TED) due to the C trapping of implant-induced excess interstitials (I). Suppressed phosphorous TED in Si:C resulted in significantly smaller junction depth ($x_j$). At $5 \times 10^{18}$ /cm$^3$, $\Delta x_j$ formed by P 6keV and 12keV were 23nm and 50nm, respectively. On the other hand, As diffusion in Si:C is significantly higher than the zero-TED thermal equilibrium, but still smaller than in Si. Introduction of the excess vacancies or VSS was necessary to accurately describe the As behavior in Si:C. Without incorporation of VSS effect, modeled As diffusion profile appeared shallower than experimental SIMS profile. $\Delta x_j$ formed by As 12keV and 25keV at $5\times10^{18}$ /cm$^3$ are 2nm and 5nm, respectively. Since shallower junction is formed Si:C compared to Si, implant energy needs to be increased in Si:C to meet the same junction depth as Si.

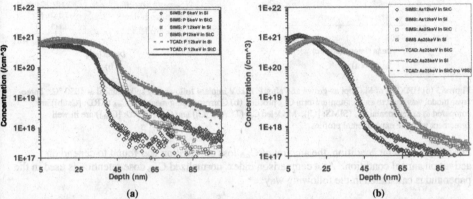

**Figure 2.** Measured and simulated dopant profiles in Si and Si:C after (a) P implant between 6 and 12 keV followed by 1050 °C spike anneal and (b) after As implant between 12 and 25keV.

Step-by-step analysis of $C_{sub}$ profile evolution by HRXRD indicates the top part Si:C epi layer is amorphized after dopant implantation and recrystallized after activation anneal. Fig.3(a) compares the HRXRD of a Si:C epi layer as-grown and after P12keV implant and spike anneal sequence. Good crystallinity and uniform $C_{sub}$ are manifested in well-defined film peak and fringes in the as-grown Si:C film. After implant and anneal, asymmetric and broadened film peak and fringes are observed, indicating non-uniform distribution of $C_{sub}$ across the epi thickness. 3-layer XRD model was fitted to the measured spectrum to accurately characterize non-uniform $C_{sub}$ profile after implant-anneal. Measured $C_{sub}$ profile is then compared to total C ($C_{tot}$) profiles measured by SIMS in Fig.3(b). HRXRD and HRTEM (shown in later section in this paper) indicate the top 40nm was amorphized by the implantation and re-crystallized by activation anneal. The level of $C_{sub}$ was much lower than $C_{tot}$ in the top amorphized layer indicating that as much as 60% of C atoms initially at substitutional sites are lost to interstitial sites after re-crystallization through the implant-anneal sequence. On the other hand, layer below the amorphized layer showed comparable $C_{sub}$ to $C_{tot}$. Simulated $C_{tot}$ and $C_{sub}$ profiles using the compact modeling developed in this study agree well with the measured profiles, confirming self-consistency of the model developed in this study.

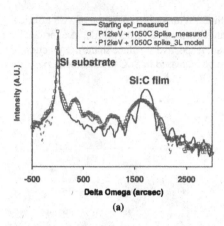 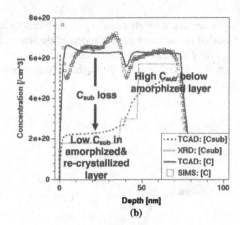

(a)

(b)

**Figure 3 (a)** HRXRD of Si:C epi as-grown and after P 12keV implant followed by spike anneal at 1050 °C. Three-layer model was used to extract non-uniform $C_{sub}$ profile. (b) Corresponding measured $C_{sub}$ (XRD: [Csub]) are compared to experimental $C_{tot}$ (SIMS: [C]). Modeled $C_{tot}$ (TCAD: [C]) and $C_{sub}$ (TCAD: [$C_{sub}$]) are in well agreement with the experimental profiles.

For a given implant condition, the amount of $C_{sub}$ loss or retention was found to depend on activation anneal condition. As a comparison index, normalized $C_{sub}$ dose retention is used in the paper and is calculated in the following way:

$$C_{sub} \text{ dose retention (\%)} = \frac{\int_{0}^{t_{epi}} C_{sub}(x)dx}{C_{sub,o} \cdot t_{epi}} \cdot 100\%$$

$C_{sub}(x)$ is the final substitutional C concentration profile after post-epi processes, $C_{sub,o}$ is the uniform substitutional C concentration in the starting epi layer, and $t_{epi}$ is the thickness of Si:C epi layer. Fig. 4 shows % retention of $C_{sub}$ dose after P 12 keV implant followed by spike anneal between 950 and 1050 °C or ms laser anneal. A trend of higher $C_{sub}$ retention for lower thermal budget anneal condition is evident. Of all conditions, diffusion-less ms laser anneal with higher temperature but shortest time resulted in near 100% retention of $C_{sub}$. $C_{sub}$ profile after P implant and ms laser anneal was observed to be almost intact without much of the loss observed in the case of spike anneal. Figure 5 (a) shows the HRXRD spectrum of Si:C epi layer after P12 keV implant and ms laser anneal compared to the starting epi. In contrast to spike anneal shown in Fig. 3(a), Si:C film peak and fringes are recovered in the case of ms anneal indicating, recovery of $C_{sub}$ similar to the initial profile. Corresponding $C_{sub}$ is compared to $C_{tot}$ in Fig. 5(b) and shows no loss of $C_{sub}$ in the top 40nm layer after amorphization and re-crystallization. Both experimental and simulated $C_{tot}$ and $C_{sub}$ agree well for the ms anneal case. HRTEM images in Fig. 6 shows amorphization by dopant implantation and complete re-crystallization of the amorphized layer after ms laser anneal without obvious signs of defects. The thickness of amorphized layer formed by P12keV implant in the image is shown to be about 40nm and is consistent with amorphized layer thickness determined by HRXRD.

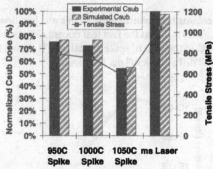

**Figure 4** Experimental and simulated values of normalized $C_{sub}$ dose (with respect to starting epi film) retention and corresponding tensile stress as a function of anneal condition.

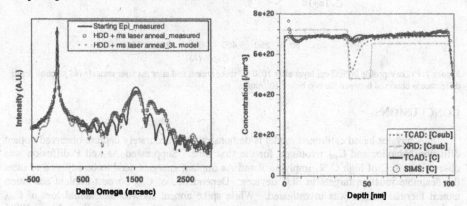

**Figure 5** (a) HRXRD of Si:C epi as-grown and after P 12keV implant followed by ms laser anneal. Three-layer model was used to extract non-uniform $C_{sub}$ profile. (b) Corresponding measured $C_{sub}$ (XRD: [Csub]) are compared to experimental $C_{tot}$ (SIMS: [C]). Modeled $C_{tot}$ (TCAD: [C]) and $C_{sub}$ (TCAD: [$C_{sub}$]) are also outlined and are in good agreement with experimental profiles.

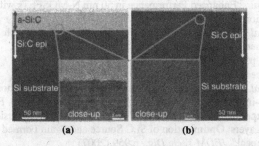

**Figure 6.** HRTEM images of Si:C epi layer after (a) P 12keV implant and after (b) P 12keV implant and ms laser anneal. The top implant-induced amorphized layer (a-Si:C) after implant is re-crystallized by ms annealing.

Compared to spike anneal, ms laser anneal resulted in same junction depth in Si:C below $10^{19}$ $cm^3$ as shown in Fig. 7. This is due to the fact that dopant diffusion is already suppressed (no TED) in the presence of C in the case of spike anneal, and it cannot be further suppressed in the case of what is known as "diffusion-less" ms laser anneal

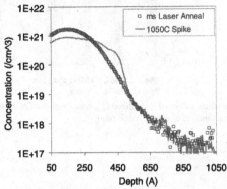

**Figure 7.** P12keV profile in Si:C epi layer after 1050 °C spike anneal and after ms laser anneal. No junction depth difference is observed between the two below $10^{19}/cm^3$.

## CONCLUSIONS

A physics based calibrated model is demonstrated to accurately capture observed dopant diffusion behavior and $C_{sub}$ evolution for the first time. Suppressed As and P diffusion was observed in Si:C of high C %, implying P and As implant energies need to be adjusted to meet the desirable junction targets in Si:C devices. Dependence of $C_{sub}$ on post-implant activation anneal thermal budget was investigated. While spike anneal showed substantial loss of $C_{sub}$ during the post-epi implant-anneal doping sequence, use of ms laser anneal allowed near 100% $C_{sub}$ retention without further compromising junction depth.

## REFERENCES

[1] K.-W. Ang et al., "Enhanced Performance in 50 nm N-MOSFETs with Silicon-Carbon Source/Drain Regions," *IEDM Tech. Dig.* p.1069 (2004).
[2] K.-W. Ang et al., "Thin Body Silicon-on-Insulator N-MOSFET with Silicon-Carbon Source/Drain Regions for Performance Enhancement," *IEDM Tech. Dig.*, p. 20.3.1 (2005)
[2] Sentaurus Process User's Manual, v. 2007.12, Synopsys Inc., 2007
[3] C. Zechner et al., "Modeling Ultra Shallow Junctions Formed by Phosphorus-Carbon and Boron-Carbon Co-implantation," *Mat. Res. Soc. Symp. Proc.*, v. 994, p. F11 (2007)
[4] Cho et al., "Experimental and Theoretical Analysis of Dopant Diffusion and C Evolution in High-C Si:C Epi Layers: Optimization of Si:C Source and Drain Formed by Post-Epi Implant and Activation Anneal," *IEDM Tech. Dig.*, p.959 (2007)
[5] H. Ruecker et al., "The impact of supersaturated carbon on transient enhanced diffusion," *Appl. Phys. Lett.* 73, 1682 (1998)
[6] P. Laveant, *PhD thesis*, MPI, Germany (2002)

# Solid Phase Epitaxial Regrowth

Mater. Res. Soc. Symp. Proc. Vol. 1070 © 2008 Materials Research Society 1070-E05-01

# Indirect Diffusion Mechanism of Boron Atoms in Crystalline and Amorphous Silicon

Salvo Mirabella[1], Davide De Salvador[2,3], Enrico Napolitani[2,3], Elena Bruno[1], Giuliana Impellizzeri[1], Gabriele Bisognin[2,3], Emanuele Francesco Pecora[1,4], Alberto Carnera[2,3], and Francesco Priolo[1,4]

[1]MATIS, CNR-INFM, Via Santa Sofia, 64, Catania, I-95123, Italy
[2]Physics Department, University of Padova, Via F. Marzolo, 8, Padova, I-35131, Italy
[3]MATIS, CNR-INFM, Via F. Marzolo, 8, Padova, I-35131, Italy
[4]Physics and Astronomy Department, University of Catania, Via Santa Sofia, 64, Catania, I-95123, Italy

## ABSTRACT

The diffusion of B atoms in crystalline and amorphous Si has been experimentally investigated and modeled, evidencing the indirect mechanism of these mass transport phenomena. The migration of B occurs after interaction with self-interstitials in crystalline Si (c-Si) or with dangling bonds in amorphous Si (a-Si). In the first case, an accurate experimental design and a proper modeling allowed to determine the microscopic diffusion parameters as the B-defect interaction rate, the reaction paths leading to the diffusing species and its migration length. Moreover, by changing the Fermi level position, B atoms are shown to interact preferentially with neutral or doubly positively charged self-interstitials. As far as the amorphous case is concerned, B diffusion is revealed to have a marked transient character and to depend on the B concentration itself. In particular, boron atoms can move after the interaction with dangling bonds whose density is transiently increased after ion implantation or permanently enhanced by the presence of boron atoms themselves. Unexpectedly, B diffusivity in a-Si is seen to be orders of magnitude above than in c-Si and to depend on the thermal history, i.e. the relaxation status of the amorphous phase. These data are presented and their implications discussed.

## INTRODUCTION

Atomic diffusion in semiconductors is an intriguing, elementary process relevant for many applications in microelectronic device fabrication. Boron is the main $p$-type dopant in crystalline silicon (c-Si), and its migration in Si matrices represents a crucial issue for the continuous scaling-down of microelectronic devices. Today, B-doped c-Si can be prepared with unprecedented purity and accuracy, representing a proper ideal system for fundamental diffusion studies. In fact, B diffusion in c-Si has been widely investigated in the last 30 years, evidencing the indirect feature of B migration which can occur after interaction with a Si self-interstitial (I) [1-5]. A single diffusion event needs the mediation of an I which, through a kick-out reaction with substitutional B ($B_S$), generates a BI complex (at a rate $g$) moving throughout the crystal for a free mean path ($\lambda$), until a kick-in reaction re-gives the immobile $B_S$ [2-5]. In such a picture, the macroscopic B diffusivity ($D$) comes out as $D=g\lambda^2$, and if the overall migration process involves many single events (i.e., $gt$ is well higher than 1, $t$ being the diffusion time), a standard description by the Fick equations is adequate [2]. Inversely, whenever $gt$ is around or below 1 (the average broadening is comparable to $\lambda$), the overall diffusion significantly deviates from a Gaussian process and an atomistic description is needed accounting for the BI formation and migration [2].

In the last condition, proper experimental design and data analysis can measure the microscopic diffusion parameters ($D$, $g$ and $\lambda$) and strengthen new insights on the migration process.

The B diffusion process in c-Si is known to be strongly dependent on the free charge availability, as B diffusivity linearly increases with $p$-type doping over a wide temperature range (870-1250 °C) [6]. This evidence suggests that the charge status of all the atomistic species involved in the diffusion (B, I and BI) is crucial, since the population of charged species depends on the Fermi level position. The linear trend of $D$ over hole concentration ($p$) suggested the interaction of substitutional B with a single positively charged I, whose population linearly increases with $p$-type doping [7]. On the contrary, a theoretical work proposed $I^{++}$ defects [4] as the interstitial charge state that promotes the BI complex formation, while a modeling paper [8], based on recent Si self-diffusion experiments [9-10], considered $I^-$, $I^0$ and $I^+$.

Our recent results, based on $g$ and $\lambda$ investigation versus the Fermi level, have shown that the interactions with $I^0$ and $I^{++}$ are the main channels for B diffusion. The $BI^-$ and $BI^+$ complexes formed by these interactions switch to $BI^0$ before diffusing. The linear trend of diffusivity versus the hole concentration is due to the fact that the overall charge change of a B atom in order to diffuse is +1 [11]. These results have been recently confirmed by Bracht and coworkers by an independent methodology that simultaneously investigates the Si self diffusion, by isotope tracing, and B diffusion [12].

On the other hand, B diffusion in amorphous Si (a-Si) represents a fascinating process within a network with properties very different from the c-Si and continuously changing upon annealing. In a-Si 4-fold coordinated Si atoms form an ideal continuous random network where short range order is maintained while long range order is lost. Point defects in a-Si are represented by dangling bonds (db), i.e. 3-fold coordinated Si atoms, and floating bonds (fb), i.e. 5-fold coordinated atoms [13-15]. The point defect populations depend on the thermal history of the sample, a-Si being rich of defects just after ion implantation (unrelaxed state) and poorer and poorer upon annealing through defect annihilation and bond rearrangement (relaxed state) [16]. In such a matrix, B diffusion investigation is fascinating as well as technologically relevant since B is often introduced in a-Si prior to epitaxy and many phenomena occurring in a-Si heavily determine the B features in re-crystallized Si [17]. Despite the advanced comprehension of B diffusion mechanism in c-Si, the counterpart in a-Si has been totally neglected until only very recently when initial evidences of B diffusion [18-19] and clustering [20,21] in the amorphous phase have been provided.

In this work, some experimental evidences of the microscopic mechanism leading to B diffusion in both the crystalline and amorphous phases of Si will be given, together with appropriate and advanced modeling. In c-Si, we will review our results regarding the B migration via a neutral BI as a function of the Fermi level. We will show how, in $p$-type Si, two diffusion channels arise since B interacts with neutral or with double positive I, by increasing hole concentration. The detailed reaction paths will be explained. On the other hand, in a-Si boron atoms are shown to have indirect diffusion too, since they migrate after interaction with dangling bonds. B diffusivity in a-Si is measured to be orders of magnitude above that in c-Si and to depend on the thermal history, i.e. the relaxation status, of the amorphous phase and on the B concentration itself.

## EXPERIMENT

In order to study the B diffusion in c- and a-Si, very sharp B profiles were obtained by molecular beam epitaxy (MBE) growth and then measured by secondary ion mass spectrometry (SIMS). As far as the B diffusion in c-Si is concerned, the following experiment plan was carried out (see figure 1).

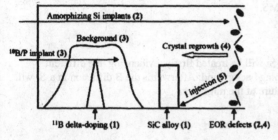

**Figure 1.** B diffusion in c-Si: experiment plan. (1) Epitaxial growth of silicon with delta-doped B and a $Si_{1-x}C_x$ alloy barrier. (2) Amorphizing Si implants. (3) Multiple $^{10}$B (or $^{31}$P) implants to insert a B (or P) background around delta-doped B. (4) Solid phase epitaxial re-growth. (5) Diffusion annealing.

The basic idea is to analyze the diffusion of a sharp $^{11}$B delta doping, determining the microscopic diffusion parameters ($D$, $g$ and $\lambda$) while varying the background doping around the delta [21]. The samples were grown by inserting a B spike at the depth of 180 nm, using a natural isotopic abundance solid B source, and a $Si_{1-x}C_x$ alloy layer (x=0.3 at.%) between 390 and 440 nm. The structure was amorphized down to a depth of 500 nm by implanting Si ions and afterwards, multiple $^{10}$B (or $^{31}$P) implants were performed to produce the background doped box in the range between 110 and 250 nm. Background doping varied in the range $5\times10^{18}$ - $7.5\times10^{19}$ cm$^{-3}$ in the case of $^{10}$B, while it was $3\times10^{18}$ cm$^{-3}$ in the case of $^{31}$P. Subsequent rapid thermal annealing (RTA) at 700 °C, in inert $N_2$ atmosphere, was performed to induce the solid phase epitaxy (SPE), thus producing the starting sample for diffusion studies. Further annealing at the same temperature in the time interval from 100 s to 40 h were performed to induce the diffusion. The I injection from the residual damage below the amorphization depth (end of range defects) is screened by the $Si_{1-x}C_x$ alloy layer [22]. Finally, to increase the SIMS sensitivity and accuracy, we performed chemical profiling with the sample frozen at -70 °C, also flooding the samples with a jet of $O_2$ gas as described in Ref. [23].

As far as the B diffusion in a-Si is concerned, the following experimental plan was employed (see figure 2), still based on MBE growth and SIMS measurements.

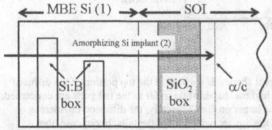

**Figure 2.** B diffusion in a-Si: experiment plan. (1) Epitaxial growth of silicon with two, narrow B doped region upon a SOI substrate. (2) Amorphizing Si implants. (3) Diffusion annealing.

To better investigate the phenomenon, we need B doped region to remain in the amorphous state as much as possible and thus we used the Silicon On Insulator (SOI) approach [19] in order

195

to block the SPE process, combined with the MBE growth. Upon a SOI substrate we grew by MBE a 350 nm- thick Si layer with embedded two 20 nm- thick B-doped regions at concentrations of $7x10^{19}$ and $8x10^{20}$ B/cm$^3$ (hereafter named as low- or high- B box, respectively) [24]. The structure was amorphized, by implanting Si ions, down to about 1 μm, well below the oxide layer which stops any SPE re-growth starting from the original a/c interface. Thermal annealing in the range 450 – 650 °C were performed to induce the B diffusion in a-Si for times shorter than the onset of the random nucleation and growth phenomenon [24]. Finally, B profiles were measured by SIMS.

## DISCUSSION

In the following the B diffusion in c-Si will be treated firstly, evidencing the different migration paths while modifying the doping background. Afterwards the B diffusion in a-Si will be presented underlining the indirect feature of the migration.

### B diffusion in crystalline Si

In order to evidence the different migration feature for B in c-Si with different Fermi level position, we compared two diffusion processes performed on the same $^{11}$B delta but surrounded by a n-type or p-type background doping. In figure 3, two 700 °C diffused profiles of the $^{11}$B delta are plotted for the $^{10}$B doping ($6x10^{18}$ cm$^{-3}$, triangles) or the P doping ($3x10^{18}$ cm$^{-3}$, circles) case, both starting from the initial profile (squares). Different annealing times were used to achieve a comparable overall diffusion (1 or 7.5 hours for the p or n-type case, respectively).

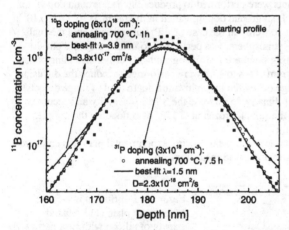

**Figure 3.** SIMS measurements (symbols) of concentration profiles of $^{11}$B spike before (squares) and after diffusion annealing at 700 °C for the samples with $^{10}$B doping (triangles) or with $^{31}$P doping (circles). The continuous lines are the best fits to the data (the so extracted $\lambda$ and $D$ values are indicated).

The two profiles present quite different shapes. If we look at the top portion of the diffused profiles, the B doping case presents the highest shape, while as far as the tail parts are concerned, the same sample shows unexpectedly the larger profile. Typically, the diffusion broadening proceeds via a concomitant lowering of the top and widening of the tails. In our case, the experimental data point out two diffusion processes with quite different migration lengths, the longer in the B doping case. The lines are the best fits to the diffused data obtained by numerically "diffusing" the as-regrown profile according to the equations of Ref. [2]. The fits allow to quantify the two different mean free path values (together with $g$ and $D$) and confirm the above qualitative description. The different $D$ values are in good agreement with a linear function of hole concentration, as reported in Ref. [6].

Afterwards, we varied the hole concentration by two orders of magnitude and extracted the values of $D$, $g$ and $\lambda$ by means of fitting procedure with the same accuracy as illustrated above. Those values are plotted in figure 4 as a function of the hole concentration $p$ normalized by the intrinsic hole concentration in an undoped Si ($n_i$, equal to $0.92 \times 10^{18}$ cm$^{-3}$ at the diffusion temperature of 700 °C [7]). $p$ is calculated starting from the background B concentration according to Boltzmann statistics and considering a full ionization of substitutional B. The whole set of $D$, $g$ and $\lambda$ furnishes very detailed information about B diffusion mechanism under different Fermi level position.

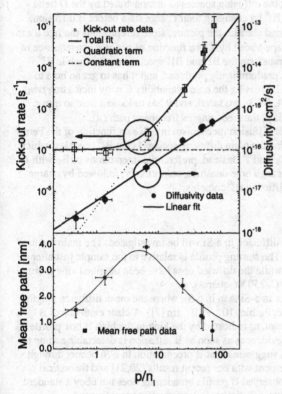

**Figure 4.** Experimental data on diffusivity (closed circles), kick-out rate (open squares) and mean free path (closed squares) as a function of the hole excess introduced by the doping background ($p/n_i$). The continuous lines are the best fits to the data, while dotted (dashed) line is the quadratic (constant) dependence of the kick-out rate over the doping background.

The trend of $D$ is very well fitted by a linear function in the whole range [$D = D_0(p/n_i)$], extending the trend proposed in the literature [6] to a lower temperature. Such a trend is univocally connected to the charge state of the mobile BI complex since it says that the whole diffusion process proceeds by the net exchange of a single positive charge (whose density increases linearly with $p/n_i$). A $B_S$ (which is singly negative charged before diffusion starts) has to interact with one I and get one hole to form the $BI^0$ complex able to diffuse.

The $g$ and $\lambda$ trends explain into much deeper details the migration paths involved at different Fermi level positions. The $g$ trend has two components, as a function of the hole density: a constant one and a quadratic one. This can be thought as if the interaction with I, that promotes the formation of the BI complex, is driven by neutral $I^0$ or by a doubly positive $I^{++}$. The populations of such species have a constant and a quadratic trend with $p/n_i$ [7], thus introducing a constant $g_0$ and a quadratic $g_{++}(p/n_i)^2$ term in the kickout frequency trend versus $p/n_i$ : $g = g_0 + g_{++}(p/n_i)^2$. As a consequence of this, at low (high) $p/n_i$ values a BI complex in negative (positive) charge states is produced, as follows:

$$(\text{low } p/n_i) \ B_S^- + I^0 \to BI^-, \qquad \text{or} \qquad (\text{high } p/n_i) \ B_S^- + I^{++} \to BI^+. \tag{1}$$

Both these two BI complexes are not the diffusing species as demonstrated by the $D$ trend analysis, but they have to transform into $BI^0$ by changing their charge state before B diffusion, through a free carrier exchange. The $\lambda$ trend clarifies the picture, since it represents the free mean path of the diffusing species. The bell shape traced by $\lambda$ as a function of $p/n_i$ is a consequence of the charge exchange that must occur to transform the $BI^-$ and $BI^+$ species into the mobile $BI^0$ species. When $p/n_i$ is lower than 4, $BI^-$ is predominantly produced, and it has to get an hole to move, so BI mean free path increases by increasing the hole availability (i.e. by increasing $p/n_i$). On the contrary when $p/n_i$ is larger than 4, $BI^+$ is produced, and it has to loose a hole to move. This is less probable to occur the higher is $p$, thus reducing its free mean path ($\lambda$).

The data so far presented clarify the B diffusion mechanism in c-Si as a function of the Fermi level position. It has been consolidated that the main diffusing species is the $BI^0$ complex, which does not follow a simple interaction of $B_S^-$ and $I^+$. Instead, preferential interactions of $B_S^-$ with $I^0$ or with $I^{++}$ have been evidenced at low or high hole densities, respectively, followed by charge exchange towards the formation of the diffusing $BI^0$ complex.

<u>**B diffusion in amorphous Si**</u>

In the following the mechanism of B diffusion in a-Si will be investigated. The main features of such a diffusion are given in figure 5. The starting profile is relative to the sample just after the amorphization implant (dotted line), while the diffused ones have been obtained after thermal annealing at 500°C, 8 h (circles) or 650 °C, 250 s (squares).

B diffusion appears to be much higher in a-Si than in c-Si, where the mean diffusion length after similar thermal budgets would be negligible ($10^{-2} \div 10^{-3}$ nm [7]). A clear evidence of B clustering above ~ $1\text{-}2\text{x}10^{20}$ B/cm$^3$ is shown, as indicated by the kink in the high B box profile (see the arrow). The clustering has been evidences as soon as B diffusion is detectable and up to the longest time used at all temperatures, suggesting that B precipitation in a-Si occurs through quick formation of a B complex, in agreement with our recent results [20,21] and theoretical predictions [25]. On the other hand, the observed B profile broadening does not obey a standard Fick's law with constant and homogeneous diffusivity ($D_B$), since after annealing the concentration gradient significantly increases. Both B boxes assume a "box-like" shape with wide shoulders and narrow tails, suggesting a B diffusivity univocally increasing with B concentration ($n_B$). Still, a significant difference in the broadening of the two boxes is recorded, the high B box showing a larger diffusion than the low B box for a fixed B concentration. In such a manner, B diffusivity seems to feel the global amount of B present in the surroundings, being higher in the high B box. Solid lines in figure 5 represent the simulated diffusion profiles based upon the model proposed below.

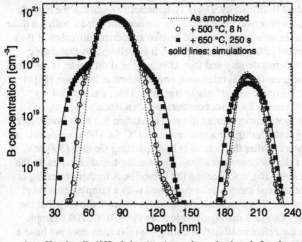

**Figure 5.** B diffusion in amorphous Si. SIMS measurements of B concentration profiles in the as-amorphized sample (dotted line) and after annealing at 500 °C, 8 h (circles) or at 650 °C, 250 s (squares). Best fit results (solid lines) are also shown.

An effective B diffusivity ($D_{eff}$) can be calculated, for the two boxes, as the difference between the squared variance ($\sigma^2$) of the diffused B distribution with respect to the as-amorphized case ($\sigma^2_0$), divided by twice the annealing time ($t$): $D_{eff} = (\sigma^2 - \sigma^2_0)/2t$. The result of such an exercise is reported in figure 6 for the high (squares) and low (circles) B boxes, for all the annealing processes.

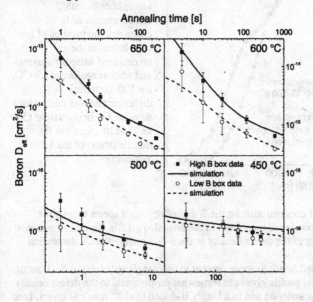

**Figure 6.** Time evolution of the effective B diffusivity in amorphous Si extracted from the high (squares) and the low (circles) B boxes for all the annealing processes. Simulations of $D_{eff}$ (lines) are also shown.

With such an approach, the clustered fraction (assumed constant) gives no contribution, being cancelled out in the variance difference. Under a standard Fick's law diffusion process, the so calculated effective B diffusivity should be constant and homogeneous, while our data show a transient and concentration dependent B diffusivity. $D_{eff}$ is systematically higher in the high B box, and an evident decrease of $D_{eff}$ is observed in all the cases, for both the boxes, within a quite wide time scale. In order to explain these features we could invoke an indirect diffusion of B in a-Si, mediated by some defects present in the amorphous phase. It is well-known that during annealing, a-Si relaxes reducing the defect density and thus affecting the B diffusion, if mediated by such defects. Still, at all temperatures, the a-Si relaxation time reported in literature [16] is well shorter (10 to 100 times) than the observed diffusivity transient. Thus, a not trivial hypothesis has to be found, also accounting for the concentration dependence.

The effect of the defect density modification during thermal relaxation has been investigated into deeper details by the following test. During the annealing at 650 °C for 100 s, we de-relaxed the sample (breaking the thermal process after the first 50 s) by implanting Ge ions (300 keV, $6\times10^{13}$ /cm$^2$, corresponding to a 0.1 displacement per atom). This Ge implantation raises up the defect density as in the un-relaxed state [16], not affecting the B profile. A further annealing at 650 °C, 50 s was done on such a de-relaxed sample to be compared with a sample annealed at 650 °C, 100 s without any de-relaxation. The results are presented in figure 7 showing that the de-relaxed sample gives a diffusion (stars) not distinguishable from the 650 °C, 100 s sample (circles). Thus, we could infer that the defect annihilation during the relaxation does not have a significant role on the overall B diffusion.

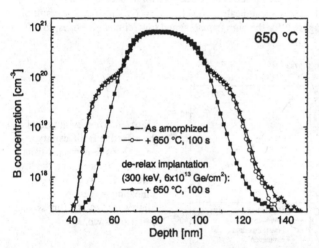

Figure 7. B diffusion in amorphous Si. SIMS measurements of B concentration profiles of the high B box in the as-amorphized sample (squares) and after annealing at 650 °C, for 100 s (circles). A diffusion test was done with a de-relax Ge implantation (300 keV, $6\times10^{13}$ /cm$^2$) in the middle (stars) of the 100 s annealing.

On the other hand, the effect of B concentration on the B diffusivity itself seems to be quite important, giving out the well pronounced box-shape in the annealed profile. To investigate this aspect, we tried to give a profiling of the defect density in the a-Si by using the Cu decoration technique (see figure 8) [26].

Cu atoms bind with unsaturated bonds as those present in db (dangling bonds) or fb (floating bonds) defects, thus a Cu chemical profile gives out somewhat proportional to the defect density. We compared this Cu signal in an undoped and in a highly B-doped ($8\times10^{20}$ /cm$^3$) Si layers, both in the unrelaxed and in the relaxed state (i.e. after annealing at 500 °C, 1h [16]).

We prepared these four samples and then implanted them with Cu (15 keV, $1 \times 10^{16}$ /cm$^2$). Afterwards, we induced Cu diffusion and defect decoration by annealing the samples at 200 °C for 1h (such an annealing do not have significant effect on the relax status of a-Si). Figure 8 presents the Cu profiles in those four samples. We observed that B doped a-Si layers are richer in defects than undoped matrices, both in the unrelaxed and in the relaxed state. Thus, a significant effect of B in steadily increasing the defect density has to be assumed. Similarly, Muller *et al.* [27] have observed that in thermally relaxed a-Si, impurities are usually incorporated within a 3-fold coordination site and a large kinetic barrier exists against the conversion to the tetrahedral site. Thus, it is very reasonable that B atom, 4-fold coordinated in c-Si, upon amorphization is accommodated into a 3-fold coordinated site, breaking one Si-B bond and generating an excess of db in the network.

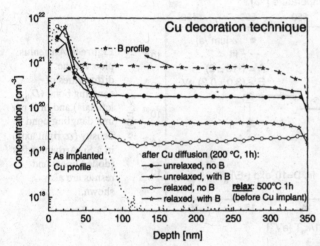

**Figure 8.** SIMS measurements of Cu concentration profiles in amorphous Si after implantation (dotted line) and after Cu diffusion annealing (200°C, 1h) in unrelaxed (closed symbols) or relaxed (500 °C, 1h annealing, open symbols) Si matrices. Cu concentration in undoped Si (circles) is lower than in B doped matrices (stars), regardless of the relax status. B concentration in doped Si is also plotted.

In the following, such a db excess is exploited to model the collected data by assuming that B migration in a-Si occurs via an indirect mechanism. Boron jumps between adjacent 3-fold coordination sites through the temporary restoring of a metastable, 4-fold coordinated B, by the capture and release of one db, as follows:

$$B_3 + db \leftrightarrow B_4. \tag{2}$$

The time evolution of the B profile has to be modelled in conjunction with the time evolution of the db population, whose density is transiently increased just after ion implantation or permanently enhanced by the presence of boron atoms themselves. This model, whose details and rate equations are presented in Ref. [24], accounts for the transient and concentration dependence of B diffusion. The continuous lines in figures 5-7 are best fits to the data following this model.

The non Fick-like diffusion is explained considering that the higher the B density, the more db are present, promoting a faster B diffusion. The excess of db is higher in the higher doped regions, accounting for the wider tail broadening in the high B box. The transient diffusion is related both to the db–fb annihilation (for the very early stages of annealing) and to the progressive reduction of db density from the B doped region due to db diffusion. Both db and B diffusivities can be extracted within this model, spanning over six orders of magnitude and show activation energies of 2.6±0.4 and 3.0±0.2 eV, respectively (figure 9). B diffusivity ($\alpha$) is intended per unit dangling bond density. These activation energies are very similar, meaning that B migration barrier is approximately fixed by db diffusion barrier.

These data and modelling explain the B diffusion mechanism in a-Si as an indirect migration mediated by db defect. B diffusivity in a-Si is some orders of magnitude higher than in c-Si, and presents a transient character which cannot be ascribed to the well-known a-Si relaxation. Moreover, B migration is increased with B concentration up to $2 \times 10^{20}$ B/cm$^3$, above which clustering occurs.

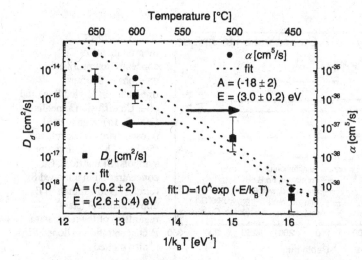

**Figure 9.** Arrhenius plots for the diffusivities of dangling bond ($D_d$, left axis) and of B per unit dangling bond density ($\alpha$, right axis) in a-Si. Arrhenius fits (dotted lines) and results are also shown.

To attempt some parallelism between B diffusion in c- and in a-Si, we can consider that, more generally, the indirect diffusion depends on the transport capacity (concentration multiplied by the diffusivity) of the diffusion-mediator species and on the mean free path of the diffusing species ($\lambda$). In the a-Si case, the last parameter is very small, for all the investigated temperatures, since simulation attempts (not shown) confirmed that the migrating species have a migration length well below 2 nm. Indeed, the similarity of the activation energies for diffusion of B and db suggests that the interaction between them produces a metastable state, as indicated in eq. (2), which lasts for a while, leading to B diffusion with very short $\lambda$. In c-Si, $\lambda$ has been measured to be 1-4 nm at 700 °C (see figure 4), slightly increasing with decreasing the diffusion temperature [2]. The small difference in $\lambda$ in the two cases does not account for the great variation in the overall B diffusivity. As far as the transport capacity of the mediators is concerned, in a-Si this term could be well higher than in c-Si, explaining the higher B diffusion. On one hand, in a-Si, the diffusivity of db has been shown to have an activation energy of more than 2 eV (see figure 9) while for I diffusion the energy barrier is less than 1 eV in c-Si [28]. On the other hand, in a-Si the db density has been shown to be very high and substantially supported by the B itself, while in c-Si the I concentration has a well high energetic barrier around 3-4 eV [28]. Therefore, very reasonably, the higher B diffusivity in a-Si is due to the higher density of the diffusion-mediator species, which are furnished by the presence of B itself, regardless of the temperature, with respect to the case of c-Si in which a relatively high I density requires a very high thermal budget to be created.

## CONCLUSIONS

The microscopic mechanism of B diffusion in crystalline and in amorphous Si has been experimentally fixed and justified by appropriate modeling. In both the Si matrices, B migrates by the mediation of point defects in different and peculiar fashion.

B migrates within c-Si mainly via a neutral BI complex, whose formation depends on the Fermi level. For moderate $p$-type doping, $B_S$ interacts preferentially with $I^0$, while at high $p$-type doping the same impurity couples with $I^{++}$. In both cases, a free charge exchange between the BI complex and the hosting matrix is needed to have B diffusion, leading to the formation of the diffusing $BI^0$ complex.

B diffusion in a-Si has been also investigated, showing that it is an indirect event too. The diffusion of B in a-Si is shown to have a transient and concentration dependent behaviour. Moreover, it has been shown that the presence of B itself enhances the density of dangling bonds in the matrix, since boron is incorporated in 3-fold coordinated sites. Finally, boron atoms are modelled to diffuse after the interaction with dangling bonds promoting the formation of a metastable, 4-fold coordinated B.

## ACKNOWLEDGMENTS

The authors wish to thank C. Percolla, S. Tatì (CNR-INFM MATIS), A. Marino (CNR-IMM), R. Storti (University of Padova) for technical contribution, M. Mastromatteo, and M. Pesce (University of Padova) for useful discussions and experimental contribution.

## REFERENCES

1. U. Goesele and T.Y. Tan in *Defects in Semiconductors II*, edited by J.W. Corbett and S. Mahayan (North-Holland, New York, 1983), p. 45.
2. N. E. B. Cowern, K. T. F. Janssen, G. F. A. vandeWalle, D. J. Gravesteijn, *Phys. Rev. Lett.* 65, 2434 (1990); N. E. B. Cowern, G. F. A. vandeWalle, D. J. Gravesteijn, C. J. Vriezema, *Phys. Rev. Lett.* 67, 212 (1991).
3. J. Zhu et al., *Phys. Rev. B* 54, 4741 (1996).
4. B. Sadigh et al., *Phys. Rev. Lett.* 83, 4341 (1999).
5. W. Windl, M.M. Bunea, R. Stumpf, S.T. Dunham, M.P. Masquelier, *Phys. Rev. Lett.* 83, 4345 (1999).
6. B.R. Fair, P.N. Pappas, *J. Electrochem. Soc.* 122 1241 (1975).
7. P.M. Fahey, P.B. Griffin, J.D. Plummer, *Rev. Mod. Phys.* 61 289 (1989).
8. I. Martin-Bragado et al., *Phys. Rev. B* 72 35202 (2005).
9. H.H. Silvestri, *Mater. Res. Soc. Proc.* 719 F13.10 (2002).
10. H.A. Bracht, H.H. Silvestri, E.E. Hallerb, *Solid State Commun.* 133 727 (2005).
11. D. De Salvador *et al.*, *Phys. Rev. Lett.* 97, 255902 (2006).
12. H. Bracht, H. H. Silvestri, I. D. Sharp, and E. E. Haller, *Phys Rev. B* 75, 035211 (2007)

13. S.T. Pantelides, *Phys. Rev. Lett.* **57**, 2979 (1986) ; P. C. Kelires and J. Tersoff, *Phys. Rev. Lett.* **61**, 562 (1988).
14. N. Bernstein, J. L. Feldman and M. Fornari, *Phys. Rev. B* **74**, 205202 (2006).
15. S. Roorda, S. Doorn, W.C. Sinke, P.M.L.O. Scholte, E. vanLoenen, *Phys. Rev. Lett.* **62**, 1880 (1989); S. Roorda *et al.*, *Appl. Phys. Lett.* **56**, 2097 (1990).
16. P. A. Stolk *et al.*, *J. Appl. Phys.* **75**, 7266 (1994) and references therein.
17. B.J. Pawlak *et al.*, *Appl. Phys. Lett.* **86**, 101913 (2005).
18. R. Duffy *et al.*, *Appl. Phys. Lett.* **84**, 4283 (2004).
19. V.C. Venezia *et al.*, *Mat. Sci. Eng. B* **124-125**, 245 (2005).
20. D. De Salvador *et al.*, *Appl. Phys. Lett.* **89**, 241901 (2006).
21. D. De Salvador *et al.*, *J. Vac. Sci. Tech. B* **26**, 382 (2008)
22. S. Mirabella *et al.*, *Phys. Rev. B* **65**, 045209 (2002).
23. E. Napolitani et al., *J. Vac. Sci. Technol. B* **24** 394 (2006); E. Napolitani et al., *Phys. Rev. Lett.* **93** 055901 (2004).
24. S. Mirabella *et al.*, *Phys. Rev. Lett.* (2008) in press.
25. A. Mattoni and L. Colombo, *Phys. Rev. B* **69**, 45204 (2004).
26. S. M. Myers and D. M. Follstaedt, *J. Appl. Phys.* **79**, 1337 (1996).
27. G. Muller *et al.*, *Phyl. Mag. B* **73**, 245 (1996); G. Muller, *Curr. Opin. Sol. State Mater. Sci.* **3** (1998) 364.
28. P. Pichler, *Intrinsic point defects, impurities, and their diffusion in silicon,* edit by S. Selberherr, (Springer, Wien-NewYork, 2004).

Mater. Res. Soc. Symp. Proc. Vol. 1070 © 2008 Materials Research Society          1070-E05-02

# Effect of Elevated Implant Temperature on Amorphization and Activation in As-implanted Silicon-on-insulator Layers

Katherine L. Saenger[1], Stephen W. Bedell[1], Matthew Copel[1], Amlan Majumdar[1], John A. Ott[1], Joel P. de Souza[1], Steven J. Koester[1], Donald R. Wall[2], and Devendra K. Sadana[1]

[1]IBM Semiconductor Research and Development Center, T.J. Watson Research Center, Yorktown Heights, NY, 10598

[2]IBM Microelectronics Division, Hopewell Junction, NY, 12533

## ABSTRACT

The ion implantation steps used in fabricating field effect transistors in ultrathin (6 to 30 nm) silicon-on-insulator (UTSOI) substrates present many challenges. Deep source/drain (S/D) implants in UTSOI layers are a particular concern, since it can be difficult to implant the desired dose without amorphizing the entire SOI thickness. In a first study, we investigated the effect of implant temperature (20 to 300 °C) on the sheet resistance (Rs) of 28 nm thick SOI layers implanted with As$^+$ at an energy of 50 keV and a dose of 3 x 10$^{15}$ /cm$^2$, and found Rs values after activation sharply lower for samples implanted at the highest temperature. In a second study, on 8 nm thick SOI layers implanted with As$^+$ at an energy of 0.75 keV and doses in the range 0.5 to 2 x 10$^{15}$ /cm$^2$, the benefits of the elevated implantation temperature were less clear. Explanations for these effects, supported by microscopy, medium energy ion scattering (MEIS), and optical reflectance data, will be discussed.

## INTRODUCTION

Historically, most performance improvements in semiconductor field-effect transistors (FETs) have been achieved by scaling down the relative dimensions of the device. Implementation of this approach for silicon-on-insulator (SOI) technology entails continued reduction of the SOI layer thickness. However, ion implantation doping of ultrathin (6 to 30 nm) silicon-on-insulator (UTSOI) layers presents some unique challenges that are not present with thicker SOI layers and bulk substrates. In particular, it can be difficult to perform deep source/drain (S/D) implants with high doses of heavy ions such as As$^+$ without amorphizing the entire SOI thickness. In the absence of some crystalline Si to act as a seed or template, the amorphized Si (a-Si) will be unable to recrystallize by solid phase epitaxy (SPE), instead recrystallizing with a polycrystalline Si (poly-Si) morphology by a random nucleation and growth process. In addition, the implanted dopants in UTSOI layers are in close proximity to both the top SOI interface with the overlying screen oxide and the bottom SOI interface with the underlying buried oxide (box) layer. Such interfaces are a known route to dopant loss [1], but the bottom SOI interface in thicker SOI layers is typically too far from the implanted dopant to be much of a concern.

It has long been known that the threshold dose for amorphization increases with increasing substrate temperature [2]. This effect is due to the fact that the crystalline-to-amorphous transition depends on a competition between the temperature-independent rate at which damage and disorder are generated by the implanted ions and the temperature-dependent rate at which the implant damage self-repairs. For example, data from Ref. 2 indicate that the threshold dose required for amorphization of (100)-oriented single crystal Si with 200 keV B increases by a factor of 40 (from ~2 x $10^{15}$ /cm$^2$ to ~8 x $10^{16}$ /cm$^2$) as the Si substrate temperature is increased from 200 K to 300 K.

The present work was motivated by the thought that heated implants might provide useful benefits for UTSOI. In this paper we look at the effect of implant temperature [room temperature (RT) vs. 300-400 °C] on the sheet resistance (Rs) two sets of samples, UTSOI (28 nm) wafers implanted with 50 keV As$^+$ and UTSOI (8 nm) wafers implanted with 0.75 keV As$^+$, and attempt to explain the results in terms of sample morphology and As concentration profiles.

## EXPERIMENT

UTSOI substrates with SOI layers 8 and 28 nm in thickness were prepared from bonded Si/SiO$_2$(145 nm)/Si(55 nm) substrates by one or more cycles of thermal oxidation and surface oxide removal by wet etching in aqueous dilute HF. The 28 nm UTSOI substrates were implanted through a native surface oxide at three substrate temperatures (RT, 150, and 300 °C) with 50 keV As$^+$ at a dose of 3 x $10^{15}$ /cm$^2$ and a tilt angle of 20° using the ion implantation facilities at the State University of New York at Albany. The 8 nm UTSOI substrates were implanted through a 2-3 nm screen oxide at RT and at 300-400 °C with 0.75 keV As$^+$ at doses ranging from 0.5 to 2 x $10^{15}$ /cm$^2$ and a tilt angle of 7°. SRIM [3] calculations were used to estimate projected range (Rp) values and knock-on O concentrations forward scattered from the surface oxide.

After implantation, sample pieces were annealed in N$_2$ by rapid thermal annealing at 500 or 900 °C for 1 min or 1000 °C for 5 s. Rs was measured by 4-point probe. Selected samples were further analyzed by cross-sectional transmission electron microscopy (TEM) to assess Si morphology and defect structure, by optical reflectance (n&k$^{TM}$) to assess the presence of surface amorphous layers, by Bragg-Brentano θ-2θ x-ray diffraction (XRD) with Cu Ka (λ = 0.15418 nm) to assess the presence or absence of poly-Si, and by medium energy ion scattering (MEIS) with 100 keV protons in both channeled and (random) non-channeled geometry to assess the As depth distribution and implant damage recovery. Details on the MEIS apparatus are described elsewhere [4].

## RESULTS AND DISCUSSION

### 28 nm SOI (50 keV As$^+$, 3 x $10^{15}$ /cm$^2$)

As$^+$ implanted into Si at 50 keV has an Rp value of about 38 nm and would ordinarily completely amorphize a 28 nm thick SOI layer at a RT dose of 3 x $10^{15}$ /cm$^2$. Table I shows the Rs values for these samples as-implanted and after annealing at 500 or 900 °C for 1 min, or 1000 °C for 5 s. There is a clear benefit for the heated implants: after annealing in N$_2$ at 900° C for 1 min, the 26 and 150°C samples have Rs in the range of 8-11 kΩ/sq, consistent with recrystallization of the amorphous SOI layer to polycrystalline Si; in contrast, the 300°C sample has a Rs of 790 Ω/sq, consistent with doped single-crystal Si. In addition, the implanted As$^+$ is quite substantially activated even as-implanted (Rs ~ 15.5 kΩ/sq) and is more activated (with an Rs of 4.4 kΩ/sq) after very mild annealing (500°C/1 min) than the 26 and 150°C samples are after 900 °C/1 min.

**Table I.** Rs ($\Omega$/sq) for 28 nm SOI samples implanted with $3 \times 10^{15}$ /cm$^2$ 50 keV As$^+$.

| Anneal/Implant Temperature | RT | 150 °C | 300 °C |
|---|---|---|---|
| as-implanted | >1000 k | >1000 k | 15.5 k |
| 500 °C/1 min | >1000 k | >1000 k | 4.4 k |
| 900 °C/1 min | 10.7 k | 8.2 k | 790 |
| 1000 °C/5 s | 5.2 k | 6.2 k | 670 |

The optical reflectance vs. wavelength data of Fig. 1 corroborates this interpretation: the peaks at 270 and 395 nm associated with crystalline Si are present in both the non-implanted control sample (trace A) and the as-implanted sample for the 300 °C implant (trace D). However, these peaks are absent from traces B and C for the SOI layers implanted at RT and 150 °C, clearly indicating at least a surface amorphization. XRD of these samples after 900 °C annealing shows that poly-Si peaks are present in the samples implanted at RT and 150 °C but absent from the samples implanted at 300 °C, thus confirming the poly-Si explanation for the higher Rs.

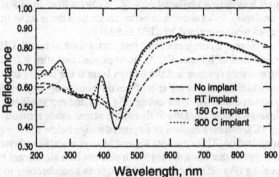

**Figure 1.** Reflectance vs. wavelength for 28 nm thick UTSOI layers implanted with $3 \times 10^{15}$ /cm$^2$ of 50 keV As$^+$ at three different temperatures and a control sample that did not receive an implant.

## 8 nm SOI (0.75 keV As$^+$, 0.5 to 2.0 $\times 10^{15}$ /cm$^2$)

SRIM simulations show As$^+$ implanted into Si at 0.75 keV to have a Rp value of about 3.1 nm, with the entirety of the implanted As atoms falling within the upper 6 nm of the 10 nm bilayer comprising the 8 nm SOI and 2-3 nm screen oxide overlayer. One thus expects that such implants will not amorphize the entire thickness of the SOI layer. Table II shows the Rs values for RT and 400 °C implanted samples, as-implanted and after annealing at 900 °C for 1 min, where the values shown are representative best values, since there was some scatter in the data due to unintentional variations in annealing temperature and SOI thickness. As with the earlier 28 nm SOI results, the heated implant dramatically lowers the as-implanted Rs values, in this case reducing them from about 30-50 k$\Omega$/sq to about 4-6 k$\Omega$/sq. As expected, there is a reduction in Rs after the 900 °C/1 min activation anneals. However, the initial benefit of the heated implants has completely disappeared: the Rs values of the 400 °C-implanted samples are now about the same or higher than the Rs values of the RT-implanted samples. The Rs values for the annealed RT implants also scale better with dose, i.e., the same factor of 4 increase in concentration lowers Rs by a factor of 2.9 for the RT implant vs. a factor of only 1.2 for the heated implant.

Several mechanisms may be playing a role in the observed Rs behavior. For example, As concentrations (peak values of $2 \times 10^{21}$ /cm$^3$ for the lowest dose implant) are substantially above the expected solid solubility limit of 1-3 $\times 10^{20}$ /cm$^3$ [5], and As clustering is non-negligible [6].

**Table II.** Rs ($\Omega$/sq) for 8 nm SOI samples implanted with various doses of 0.75 keV As$^+$.

| Treatment | as-implanted | | 900 $^\circ$C/1 min | |
| --- | --- | --- | --- | --- |
| Dose/Implant Temperature | RT | 400 $^\circ$C | RT | 400 $^\circ$C |
| 0.5 x 10$^{15}$ /cm$^2$ | 36-45 k | 6.2 k | 2.3 k | 2.1 k |
| 1 x 10$^{15}$ /cm$^2$ | 32 k | 4.8 k | 1.3 k | 1.8 k |
| 2 x 10$^{15}$ /cm$^2$ | 52 k | 4.0 k | 800 | 1.7 k |

Other complicating factors include high concentrations of O from screen oxide recoil mixing and dopant loss during implantation (e.g., via volatilization of AsOx, a potential concern for the 400 $^\circ$C implants). In an attempt to better understand the relative importance of these factors, the samples were examined by TEM and MEIS.

Figure 2 compares TEM images for a set of four 1 x 10$^{15}$ /cm$^2$ samples: implant temperatures of RT and 400 $^\circ$C, as-implanted and after 900 $^\circ$C/1 min annealing. In each image, the top layer is carbon and the bottom layer is the box. In the as-implanted RT sample (Fig. 2A), three additional layers are seen; from the bottom up they are SOI, a continuous a-Si layer, and a top screen oxide. After annealing (Fig. 2B), the recrystallized a-Si layer becomes indistinguishable from the SOI, and the screen oxide remains. In the as-implanted 400 $^\circ$C sample (Fig. 2C), one sees regions of implant damage below the screen oxide, but no clear signs of any amorphization. This contrasts the continuous amorphous layer seen in the RT-implanted sample, and suggests that epitaxial regrowth will not be a significant mechanism for As activation. After annealing (Fig. 2D), the damaged Si regions again become indistinguishable from the SOI.

MEIS has previously been used to investigate bulk Si implanted with 1 x 10$^{15}$ /cm$^2$ of 1 keV As$^+$ [7] and bulk Si and 200 nm-thick SOI layers implanted with 0.5 x 10$^{15}$ /cm$^2$ of 20 keV As$^+$ [8]. The data of Fig. 3 for the 2 x 10$^{15}$ /cm$^2$ RT samples before and and after annealing at 900 $^\circ$C for 1 min show that this technique is well suited for thinner SOI samples as well. Simulation of the non-channeled data showed reasonable agreement with the implanted dose for as-implanted sample and no As loss after annealing. Comparison of the as-implanted and annealed samples indicated a substantial redistribution of the As: the As layer is initially confined to the top half of the SOI, but after annealing about 40% of the As is in the bottom half of the SOI layer and about 10-15% (3-4 x 10$^{14}$ /cm$^2$, approximately a third of a monolayer) is at the box/SOI interface. Comparison to the channeling data indicates that this diffused As is on lattice sites, and that implant-induced damage to the bottom half of the SOI layer is repaired by the anneal. In contrast, the As in the top half of the SOI layer shows no sensitivity to

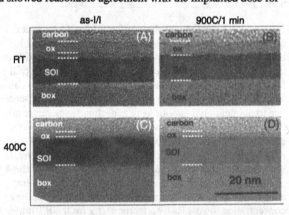

**Figure 2.** TEM images of 8 nm thick UTSOI layers implanted with 1 x 10$^{15}$ /cm$^2$ of 0.75 keV As$^+$: RT-implanted, as-implanted (A) or after 900 $^\circ$C/1 min annealing (B); 400 $^\circ$C-implanted, as-implanted (C) or after 900 $^\circ$C/1m annealing (D).

channeling, indicating that the As is located within 0.5-1 nm of the SOI/screen oxide interface.

Figure 4 compares MEIS data for as-implanted samples implanted with $2 \times 10^{15}$ /cm$^2$ As$^+$ at RT or 400 °C, conditions for which the heated sample was at ~400 °C for about 9 min. The channeled data indicate that the 400 °C implant produces about 50% less Si lattice damage than the implant at RT, consistent with the morphology shown by the TEM images of Figs. 2A and 2B for the 1 x 10$^{15}$ /cm$^2$ samples. As expected, the 400 °C sample shows no sign of the As pileup at the SOI/box interface seen in the RT sample after 900 °C annealing. Integrated As intensities indicate that only about 60% of the implanted As dose is retained in the 400 °C sample, a significant loss. However, this loss cannot entirely account for the higher Rs of this 400 °C samples after 900 °C/1 min annealing, since the 1 x 10$^{15}$ /cm$^2$ RT sample has both a lower Rs than the 2 x 10$^{15}$ /cm$^2$ 400 °C sample (1.3 kΩ/sq vs. 1.8 kΩ/sq) and a lower concentration of retained As (1.0 x 10$^{15}$ /cm$^2$ vs. 60% x 2.0 x 10$^{15}$ /cm$^2$).

SRIM calculations of O concentrations from recoil mixing of the screen oxide suggest that large amounts of O (tens of atomic %) will be present in the top half of the SOI layer for the 1 x 10$^{15}$ /cm$^2$ As$^+$ doses

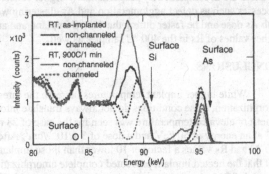

**Figure 3.** MEIS data for UTSOI samples (8 nm SOI with a 2-3 nm screen oxide overlayer) implanted with $2 \times 10^{15}$ /cm$^2$ of 0.75 keV As$^+$ at RT before (black) and after (gray) annealing at 900 °C for 1 min. The solid (dotted) lines are for the non-channeling (channeling) condition.

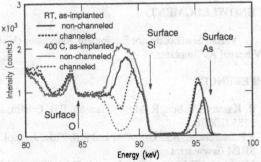

**Figure 4.** MEIS data for UTSOI samples (8 nm SOI with a 2 nm screen oxide overlayer) implanted with 2 x 10$^{15}$ /cm$^2$ of 0.75 keV As$^+$ at RT (black) or 400 °C (gray). The solid (dotted) lines are for the non-channeling (channeling) condition.

used in these experiments. The details of how this O interacts with the Si lattice and the implanted As are not entirely clear, but it seems likely that the O will interfere with dopant activation and that an electrically inactive "dead" layer will form in regions where the O concentration is high. We speculate that the thickness of this dead layer will grow if the O atoms are allowed to diffuse, and that dynamic (during implantation) annealing makes this O diffusion significantly faster for the heated implants. Heated implant samples would thus have thicker dead layers and higher Rs. This explanation is consistent with the increasing differences between the post-activation Rs values for the RT and 400 °C implants as As dose is increased, since higher

As doses are accompanied by proportionately higher concentrations of recoiled O. However, processes such as defect agglomeration and As clustering would also be expected to increase with As dose and be faster during the 400 °C implant, and are likely to be contributing to the higher values of Rs in the 400 °C samples as well.

## CONCLUSIONS

While higher implant temperatures increase the threshold $As^+$ implant dose for Si amorphization, these conditions do not always translate into lower Rs values in UTSOI. A clear benefit of elevated temperature was seen for the case of 28 nm thick SOI layers implanted with $As^+$ at an energy of 50 keV and a dose of 3 x $10^{15}$ /cm$^2$. After activation, samples implanted at 300 °C had Rs values a factor of 10 lower than those implanted RT, a finding attributed to the fact that the heated implant prevented complete amorphization of the SOI layer. For the case of 8 nm thick SOI layers implanted with $As^+$ at an energy of 0.75 keV, amorphization appeared to be eliminated with the 400 °C implants. The effects of elevated implantation temperature on Rs were negligible at low doses, but clearly unfavorable (by about a factor of 2) at higher doses. This behavior in the 400 °C samples, while not entirely understood, is believed to arise from a combination of factors including As loss during implantation, the absence of epitaxial regrowth as a mechanism for activation, and undesirable diffusion-driven processes whose rates are increased by dynamic annealing at elevated temperatures.

## ACKNOWLEDGMENT

Varian Semiconductor Equipment Associates (VSEA) is thanked for performing the 0.75 keV heated $As^+$ implants.

## REFERENCES

1. R. Kasnavi, Y. Sun, R. Mo, P. Pianetta, P.B. Griffin, and J.D. Plummer, J. Appl. Phys. **87**, 2255 (2000).
2. F.F. Morehead, B.L. Crowder, and R.S. Title, J. Appl. Phys. **43**, 1112 (1972).
3. SRIM (www.srim.org).
4. M. Copel, IBM J. Res. Develop. **44**, 571 (2000).
5. Lietoila, J.F. Gibbons, and T.W. Sigmon, Appl. Phys. Lett. **36**, 765 (1980).
6. Krishnamoorthy, K. Moller, K.S. Jones, D. Venables, J. Jackson, and L. Rubin, J. Appl. Phys. **84**, 5997 (1998).
7. S. Whelan, V. Privitera, G. Mannino, M. Italia, C. Bongiorno, A. La Magna, and E. Napolitani, J. Appl. Phys. **90**, 3873 (2001).
8. M. Dalponte, H. Boudinov, L.V. Goncharova, D. Starodub, E. Garfunkel, and T. Gustafsson, J. Appl. Phys. **96**, 7388 (2004).

Mater. Res. Soc. Symp. Proc. Vol. 1070 © 2008 Materials Research Society          1070-E05-05

# Boron Enhanced H Diffusion in Amorphous Si Formed by Ion Implantation

Brett C. Johnson[1], Armand J. Atanacio[2], Kathryn E. Prince[2], and Jeffrey C. McCallum[1]

[1]School of Physics, University of Melbourne, Melbourne, 3010, Australia

[2]Australian Nuclear Science & Technology Organisation, PMB 1, Menai, NSW, 2234, Australia

## ABSTRACT

Boron enhanced H diffusion in amorphous Si (a-Si) layers formed by ion implantation is observed using secondary ion mass spectroscopy (SIMS). Constant concentrations of B were achieved using multiple energy B implantations into thick a-Si layers. The evolution of single H implanted profiles centered on the uniformly B-implanted regions was studied for partial anneals at temperatures in the range 380 - 640 °C. Boron enhanced diffusion is observed and the enhanced diffusion coefficient shows trends with temperature typically associated with a Fermi level shifting dependence. A modified form of the generalized Fermi level shifting model is considered in light of these results.

## INTRODUCTION

The presence of H in amorphous Si (a-Si) plays an important role in determining the electronic and structural properties of the material.[1] However, the behavior of H can be quite complex and is not well understood. In a-Si formed by ion implantation, H diffusion at low concentrations (< 0.06 at.%) can be described by an Arrhenius-type equation of the form, $D = D_o \exp(-E_a / kT)$ where the activation energy is $E_a = 2.7$ eV and the pre-exponential factor is $D_o = 2.2 \times 10^4$ cm$^2$/s.[2] In deposited hydrogenated a-Si (a-Si:H), where the H concentration is much larger (8-40 at.%), thermal limits to H stability are observed and diffusion also becomes more complex, having H concentration and time dependencies.[3] The activation energy for H diffusion in good electronic quality deposited a-Si:H films is 1.4 - 1.5 eV which is considerably lower than that in a-Si formed by ion implantation.[1,4-7] This suggests that the rate limiting step of the H diffusion mechanism is different in the two materials.

In a-Si with a low implanted H concentration, it has been argued that H diffusion is governed by a dangling bond mediated mechanism and closely related to the solid phase epitaxial (SPE) crystallization process in a-Si.[2] Indeed, the H concentration was similar to the concentration of dangling bond type defects and no H concentration dependence of the diffusion was observed. At higher H concentrations of 1-2 at.%, Acco et al. have reported an activation energy of 2.26 eV in a-Si formed by ion implantation.[8,9] For higher H concentrations the activation energy was further reduced. However, for implanted H concentrations above 20 at.% a non-diffusing component of H was also observed. This component was associated with the formation of H$_2$ bubbles. The diffusion coefficient was observed to have only a slight H concentration dependence. At present, these are the only two studies performed in a-Si formed by ion implantation in the literature, neither of which consider doping effects.

In deposited a-Si:H, however, there have been several studies of dopant-enhanced H diffusion and several models have been proposed to account for the observed dependence. [1,4-7,10] Beyer suggests that shifts in the Fermi level affect the Si-H bond rupture energy.[10] This assumes that the H diffusion process is limited by Si-bonds acting as traps. Once the Si-H bond is broken, H becomes an interstitial where it can move freely until trapped by another Si-bond.

211

Alternatively, Street *et al.* consider the movement of H from one Si bond directly to another without moving via interstitial sites.[6] In this model the rate of H diffusion is then proportional to the rate of diffusion of dangling bonds, the concentration of which is sensitive to the position of the Fermi level.

In the present paper, we present the first studies of H diffusion in the presence of dopants in *a*-Si formed by ion implantation. The diffusion coefficients are determined for temperatures in the range 380 - 640 °C. As a possible means of identifying the H diffusion mechanism and the origin of the doping dependence, we have applied a modified form of the generalized Fermi level shifting model to gain insight into the H diffusion mechanism. This model was originally developed to explain the observed dopant-dependence of SPE crystallization,[11,12] but its applicability to H diffusion in *a*-Si naturally follows if the proposition of Roth *et al.* that H diffusion and SPE are mediated by the same defect turns out to be correct. [2] This is discussed further below.

## EXPERIMENT

Boron enhanced H diffusion was studied using thick surface *a*-Si layers formed by self-ion-implantation in *n*-type, 5-10 Ω.cm, Si(100) Czochralski grown wafers. Three samples were produced with surface amorphous layers 4.1 μm thick formed by multiple-energy Si implants. Two of these samples were then implanted with B ions at multiple energies resulting in approximately constant concentration profiles between 1.4 and 2.4 μm. Peak concentrations of 1 x $10^{20}$ and 1.5 x $10^{20}$ B/cm$^3$ were formed, respectively. Finally, single H implants at an energy of 205 keV were performed into all three samples. This H implant has a projected range of 1.92 μm and an as-implanted peak concentration of 3 x $10^{20}$ H/cm$^3$. This concentration is less than the H concentration required to form the H$_2$ bubbles observed by Acco. The implanted H and B profiles were at a sufficient depth below the surface so that in-diffusion of H from the surface did not affect the measurements.

Anneals were performed in air over a temperature range of 380 - 640 °C while the samples were held on a resistively heated vacuum chuck. The temperature of the samples during the anneals was calibrated by comparing the reading of a type-K thermocouple embedded in the sample stage with the melting points of various suitably encapsulated metal films evaporated onto Si wafers. The error associated with the temperature reading was found to be ±1 °C.

The depth profiles of the B and H were determined with a Cameca IMS-5F secondary ion mass spectrometer equipped with a Cs ion source. A primary ion beam of Cs$^+$ with a net impact energy of 14.5 keV was applied for the depth profiling. The Cs$^+$ beam was rastered over an area of 250 μm$^2$. Negative secondary ions were collected from the central part of the rastered region (55 μm in diameter) in order to avoid crater sidewall effects. The SIMS detection limits were 2 x $10^{17}$ and 5 x $10^{15}$ for H and B, respectively. The depth scale was quantified by a stylus profilometer with an accuracy of within 2 %.

## RESULTS AND DISCUSSION

Figure 1 shows the B and H concentration depth profiles obtained by SIMS in the sample containing 1.5 x $10^{20}$ B/cm$^3$. The dashed curves show the profiles predicted by Stopping and Range of Ions in Matter (SRIM) simulations.[13] The SIMS and SRIM data show general agreement. The B profile was produced with six multi-energy implants to achieve a constant B concentration profile. The peaks of five of the implants are apparent in the SIMS data with a sixth implant contributing to a shoulder at 1.7 μm. The B concentration varies between 1.2 – 1.7 $\times 10^{20}$ B/cm$^3$ over the depth range of 1.4 – 2.4 μm.

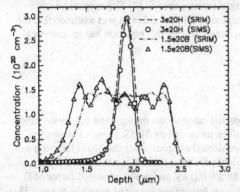

**Figure 1.** B (Δ) and H (O) concentration profiles obtained by SIMS and SRIM simulations.[13]

Figure 2 shows the H concentration profile in the intrinsic and B implanted samples after an anneal at 460 °C for 5 minutes. Under these conditions the crystalline-amorphous interface is expected to move only 4 Å and therefore does not affect the region of interest. The initial H profile is shown as a dashed line for comparison. As can be seen, the anneal causes the H profile to broaden and decrease in peak concentration. In B implanted samples the broadening is much more pronounced. The H concentration gradient becomes steep at the boundary of the B implanted region in the higher B concentration sample at 1.3 and 2.5 μm. For samples annealed for longer than 5 minutes at 460 °C, this effect becomes more apparent and eventually a box-like H concentration profile is formed. This is due to the relatively slow transport of H in the undoped region.

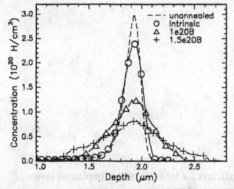

**Figure 2.** The H concentration profile of an as-implanted sample (dashed line) as well as an intrinsic sample (O) and samples implanted with B to concentration of 1 x 10$^{20}$ B/cm$^3$ (Δ) and 1.5 x 10$^{20}$ B/cm$^3$ (+) after an anneal at 460 °C for 5 minutes.

Annealing times at each temperature were limited to a range where the evolution of the H-profile was not strongly affected by the boundary of the uniform doped region. Within these time scales, no time dependence of the diffusion coefficient was observed. Indeed, no time dependence was observed for H diffusion in intrinsic amorphous Si.[2] The crystalline amorphous interface does not interact with the region of interest either. It was assumed that the diffusion obeyed a simple Fick's law. For an implanted profile, this equation has an analytical solution given by:[14]

$$C(x,t) = \frac{Q_I}{\sqrt{\pi}(\Delta R_p^2 + 4Dt)^{1/2}} \exp\left(\frac{-(x - R_p)^2}{2\Delta R_p^2 + 4Dt}\right),$$ (1)

where $Q_I$ is the fluence in H/cm³, $R_p$ is the projected range of the implant and $\Delta R_p$ is the range straggling. This equation was fitted to the H profiles measured by SIMS. Figure 3 shows the results of these fits for the intrinsic and dopant implanted samples. Over the broad temperature range of 380 - 640 °C, an Arrhenius-type dependence for each data set is observed. For the intrinsic sample an activation energy of $E_a = (1.88 \pm 0.02)$ eV and pre-exponential factor of $D_o = (0.4 \pm 0.2)$ cm²/s was found. This activation energy is lower than expected possibly due to a H concentration dependence as discussed by Beyer and Zastrow.[3] For the boron implanted samples a larger diffusion coefficient was measured. In both the 1 x10²⁰ B/cm³ and 1.5 x 10²⁰ B/cm³ implanted samples the activation energy for H diffusion was found to be $E_a = (1.58 \pm 0.03)$ eV. It should be noted that an apparent change in the activation energy with dopant concentration is not necessarily indicative of a change in the diffusion mechanism. Temperature dependent changes in the pre-exponential factor, $D_o$, will also affect the slope of the Arrhenius plot. For example, if we consider the dangling bond model proposed by Street, the pre-exponential factor is given by $D_o = a^2 (N_D / 6 N_H)v_o$ where $a$ is the hopping distance, $v_o$ is the activation prefactor, $N_D$ is the dangling bond concentration and $N_H$ is the H concentration.[6] A number of these terms may have a temperature dependence.

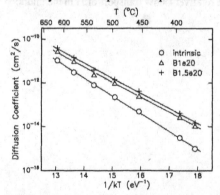

**Figure 3.** Arrhenius plots of the H diffusion coefficient for intrinsic, and B implanted layers. The solid lines are least-squares fits of the data using an Arrhenius-type equation.

The generalized Fermi level shifting (GFLS) model also considers temperature dependent changes in the pre-exponential term. We have previously applied this model to dopant-enhanced

SPE crystallization in *a*-Si and *a*-Ge to some success.[11,12] In adapting the model to the case of dopant-enhanced H diffusion, we consider the motion of an H atom to depend on the concentration of some defect, $X^0$, and its positively or negatively charged counterpart, $X^\pm$, that mediates H migration. This defect could be the dangling bond as *per* Street.

The type of defect is not specified by the model, however, it is assumed that $X^0$ and $X^\pm$ are in thermal and electronic equilibrium with the latter being determined by the band structure and density of states of the amorphous material. The diffusion coefficient is expected to be proportional to the total concentration of $X^0$ and $X^\pm$ defects. This model predicts the diffusion coefficient will be enhanced by the dopants and that this enhancement will be greatest for the lowest temperatures where the Fermi level shifts the most. The derivation of the model follows similar arguments to those presented in reference [11] but full details of modifications of the model for application to H diffusion will be presented elsewhere.

Figure 4 shows the boron enhanced diffusion coefficient for the samples implanted with 1 x $10^{20}$ B/cm$^3$ and 1.5 x $10^{20}$ B/cm$^3$. The errors were estimated by considering the ±1 °C temperature variation and from the fits of equation 1 to the SIMS data. For clarity, the error is indicated under the legend and not on the data. It can be seen that the enhancement is greatest for the lowest temperatures. These trends are very similar to the doping enhancement of SPE rates found in B doped *a*-Si and are indicative of a Fermi level dependent process. For the lowest temperatures the Fermi level is expected to be closer to the valence band and hence increase the population of $X^+$ relative to the intrinsic sample. In turn, the diffusion coefficient in the doped sample increases as shown in Fig. 4.

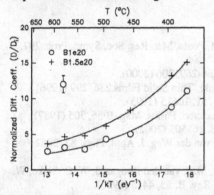

**Figure 4.** The B-enhanced H diffusion coefficient normalized to the corresponding intrinsic diffusion coefficient as a function of temperature for samples containing 1 x $10^{20}$ B/cm$^3$ and 1.5 x $10^{20}$ B/cm$^3$. The associated errors are indicated under the legend.

The solid lines in figure 4 show the expected trends in the GFLS model. As yet we do not have sufficient knowledge of the electronic properties of *a*-Si in this doping regime to be able to extract meaningful *g* and $E^+$ values of $X^+$ and are not presented at this time. These findings add further weight to the possibility that H diffusion and SPE are mediated by the same defect as postulated by Roth *et al.*[2] Alternatively, it has also been suggested that the dangling bond concentration is increased by the incorporation of B onto three-fold coordinate sites in the amorphous lattice.[17] The full implications of this latter model in the context of H diffusion have not yet been elucidated and will therefore be presented elsewhere. Also, experiments to determine whether a similar dependence is observed for n-type doping and whether compensation doping yields similar values to intrinsic a-Si are underway.

## CONCLUSIONS

The diffusion coefficients of intrinsic and B implanted a-Si formed by ion implantation were determined over the temperature range 380 - 640 °C. B-enhanced H diffusion was observed and found to have the typical trends associated with a Fermi level shifting dependence. The enhancements were found to be quite similar to B-enhanced SPE. A modified form of the generalized Fermi level shifting model provided excellent fits to the data. However, the model is yet to incorporate the semiconductor statistics of doped a-Si.

## ACKNOWLEDGMENTS

The Department of Electronic Materials Engineering at the Australian National University is acknowledged for their support by providing access to ion implanting facilities. This work was supported by a grant from the Australian Research Council. The authors would like to thank the Australian Institute of Nuclear Science and Engineering for providing financial assistance (Award No AINGRA08035) to enable work on the SIMS to be conducted.

## REFERENCES

1. R. A. Street, Physica B 170, 69 (1991).
2. J. A. Roth, G. L. Olson, D. C. Jacobson, and J. M. Poate, Mat. Res. Soc. Symp. Proc. 297, 291 (1993).
3. W. Beyer and U. Zastrow, J. Non-Crys. Solids. 266-269, 206 (2000).
4. K. Tsukamoto, S. Iwasaki, T. Sadoh, and Y. Kuroki, Thin Solid Films 286, 299 (1996).
5. C. G. V. de Walle and R. A. Street, Phys. Rev. B. 51, 10615 (1995).
6. R. A. Street, C. C. Tsai, J. Kakalios, and W. B. Jackson, Philos. Mag. B 56, 305 (1987).
7. Y.-S. Su and S. T. Pantelides, Phys. Rev. Lett. 88, 165503 (2002).
8. S. Acco, W. Beyer, E. E. van Faassen, and W. F. van der Weg, J. Appl. Phys. 82, 2862 (1997).
9. S. Acco, D. L. Williams, P. A. Stold, F. W. Saris, M. J. van den Boogaard, W. C. Sinke, W. F. van der Weg, S. Rooda, and P. C. Zalm, Phys. Rev. B. 53, 4415 (1996).
10. W. Beyer, Physica B 170, 105 (1991).
11. B. C. Johnson and J. C. McCallum, Phys. Rev. B. 76, 045216 (2007).
12. B. C. Johnson and J. C. McCallum (2008), submitted to Physical Review B.
13. J. P. Biersack and L. G. Haggmark, Nucl. Inst. Meth. 174, 257 (1980).
14. J. W. Mayer and S. S. Lau, Electronic Materials Science For Integrated Circuits in Si and GaAs (Macmillan Publishing Company, 1990).
15. J. Bourgoin and M. Lannoo, Point defects in semiconductors II (Springer, Berlin, 1983).
16. C. N. Waddell, W. G. Spitzer, J. E. Fredrickson, G. K. Hubler, and T. A. Kennedy, J. Appl. Phys. 55, 4361 (1984).
17. G. Muller, Curr. Opin. Solid State Mater. Sci. 3, 364 (1998).

Mater. Res. Soc. Symp. Proc. Vol. 1070 © 2008 Materials Research Society     1070-E05-06

# Intrinsic and Dopant-Enhanced Solid Phase Epitaxy in Amorphous Germanium

Brett C. Johnson[1], Paul Gortmaker[2], and Jeffrey C. McCallum[1]

[1]School of Physics, University of Melbourne, Swanston Street, Melbourne, Australia
[2]Department of Electronic Materials Engineering, Australian National University, Canberra, Australia

## ABSTRACT

The kinetics of intrinsic and dopant-enhanced solid phase epitaxy (SPE) are studied in thick amorphous germanium ($a$-Ge) layers formed by ion implantation on <100> Ge substrates. The SPE rates for H-free Ge were measured with a time-resolved reflectivity (TRR) system in the temperature range 300 – 540 °C and found to have an activation energy of $(2.15 \pm 0.04)$ eV. Dopant enhanced SPE was measured in $a$-Ge layers containing a uniform concentration profile of implanted As spanning the concentration regime 1 - 10 x $10^{19}$ cm$^3$. The generalized Fermi level shifting model shows excellent fits to the data.

## INTRODUCTION

Crystallization of ion implanted materials via solid phase epitaxy (SPE) is a common processing step during device fabrication for its ability to achieve high dopant activation with a low thermal budget.[1] Ge is an ideal alternative to Si in which to gain unique insight into the SPE process.[2] Ge is also regaining importance due to recent developments in nano-scale electronics and opto-electrical devices.[3] However, unlike Si, only a few researchers have reported on SPE measurements in Ge.[2,4-9] SPE is a thermally activated process and the velocity of the crystalline-amorphous ($c$-$a$) interface through the amorphous phase can be described by an Arrhenius-type equation of the form, $v = v_o \exp(-E_a/kT)$ where $v_o$ and $E_a$ are the velocity pre-exponential factor and activation energy of SPE, respectively. The activation energies reported in previous studies range between 2.0 and 2.3 eV with a velocity prefactor lying between 6.1 x $10^6$ and 7 x$10^7$ m/s. These factors are not yet known to an accuracy comparable to the corresponding Si values which are accepted to be 2.7 eV and 4.64 x $10^7$ m/s for H-free SPE.[10] Furthermore, the thickest $a$-Ge layers used to date in SPE measurements were 0.8 µm thick, while the majority of the measurements involved layers 0.5 µm or less in thickness. Roth *et al.* have demonstrated that hydrogen in-diffusion can effect the Si SPE rate for interface depths up to 2 µm.[10,11] Before the current study it was unclear whether the existing thin layer Ge SPE measurements were afflicted by the same problem.

The very mechanism by which atoms rearrange during SPE is still an area of considerable debate. For Si at least, it has been shown that the SPE rate is sensitive to shifts in the Fermi level caused by the presence of dopants and that both neutral and charged defects may be responsible for the SPE process.[3,12-14] There have been very few studies on the SPE rates in dopant implanted $a$-Ge.[15] However, such information could be an important key to understanding the atomic rearrangement processes responsible for SPE.

In this paper, we present comprehensive SPE measurements for intrinsic $a$-Ge layers over a temperature range of 300 - 540 °C, which is substantially greater than that used in other SPE measurements in $a$-Ge. We also present new As-enhanced SPE data and show that a similar

dependence on concentration and temperature is observed to that for dopant-enhanced SPE in a-Si layers.[12,13] The generalized Fermi level shifting model is applied and shows excellent
agreement with measurement, giving increased confidence in the effectiveness of this model in describing the doping dependence of SPE.

## EXPERIMENT

Ge wafers from different suppliers, and with different background doping levels were used in an effort to determine if the origin of the material had any effect on the SPE rate. Amorphous layers 1.5 μm thick were formed with a National Electrostatics Corp. 1.7 MV tandem accelerator. During implantation, substrates were affixed with Ag paste to the implanter stage, which was held at 77 K. The samples were tilted 7° off the incident beam axis to avoid channeling and also rotated about the surface normal by a similar amount to prevent any remaining possibility of planar channeling.[16] One set of samples was created with the same sequence of ion energies but with only 20 % of the fluence to investigate any possible dependence of the SPE rate on the amorphization fluence. Uniform As concentration profiles over the depth range 0.25 - 0.55 μm were also formed with multiple energy implants. Fluences were chosen to result in peak concentrations of 1, 5 and 10 x $10^{20}$ /cm$^3$.

The SPE rates were determined in air using a time resolved reflectivity (TRR) system equipped with a laser collecting data at $\lambda = 1.152$ μm. The samples were held on a resistively heated vacuum chuck and annealed at temperatures in the range 300 - 540 °C. The temperature of the samples during the anneals was calibrated by comparing the reading of a type-K thermocouple embedded in the sample stage with the melting points of various suitably encapsulated metal films evaporated onto Si wafers. The error associated with the temperature reading was found to be ± 1 °C. Further details on the experimental apparatus are presented elsewhere.[12]

## RESULTS AND DISCUSSION

The a-Ge layers formed by Ge implantation were used to determine the SPE regrowth rates. TRR data were collected in 20 °C intervals from 300 to 540 °C. The determination of the SPE rates from TRR data relies on the refractive index. By comparing Rutherford backscattering spectroscopy measurements with TRR data of partially annealed samples, we have determined the refractive index for a-Ge to be 5.34 ± 0.15. Using the refractive index, SPE rates were measured over the temperature range 300 – 540 °C. In each case, the SPE rate was extracted from TRR measurements for c-a interface depths greater than 0.3 μm below the surface. For depths less than 0.3 μm we observed SPE rate retardation due to H infiltration which was fully investigated using secondary ion mass spectrometry (SIMS). These SIMS measurements are beyond the scope of this paper and will be presented elsewhere.[17]

The SPE rate versus temperature data for intrinsic a-Ge layers are presented in Arrhenius form in Fig. 1 a). The errors were calculated by considering the reproducibility of the data and the RMS noise in the determined velocity curves. The average activation energy and velocity prefactor determined from these measurement sets is $E_a = (2.15 \pm 0.04)$ eV and $v_o = (2.6 \pm 0.5)$ x $10^7$ m/s, respectively.

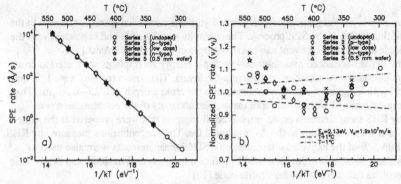

**Figure 1.** The SPE rates for the various a-Ge layers displayed on a) an Arrhenius plot with a fit to the Series 1 data only, giving $E_a = 2.13$ eV and $v_o = 1.9 \times 10^7$ m/s and b) the same data normalized to the Arrhenius fit. The two dashed lines mark the boundaries that are obtained by considering a systematic temperature error of $+1$ and $-1$ °C, respectively. Substrates are: Series 1, undoped (O), Series 2, p-type ($\Diamond$), Series 3, low-fluence amorphization (x), Series 4, $n$-type ($\star$), Series 5, 0.5 mm substrate ($\Delta$). Errors are $\pm$ 15 % for the velocity values and $\pm$ 1.5 °C for the temperature values and are about the size of the symbols in a).

As shown in Fig. 1 b), no statistically significant difference in the SPE behavior was observed for any of the intrinsic a-Ge sample sets containing different background doping, a-Ge layer thickness, or produced under different amorphization conditions. Given that our measurements are not affected by H infiltration, span a temperature range of 300 - 540 °C which is 100 °C greater than any of the other measurements and are based on an independent determination of the refractive index of a-Ge, we believe that the values reported here for the activation energy and prefactor for SPE of intrinsic a-Ge surface layers represent the most reliable data available.

It is interesting to examine the kinetic theory of SPE in light of these improved values for the SPE parameters. The SPE process can be treated as a thermally excited transition from the amorphous to crystalline phase using transition state theory. Lu *et al.* have extended this theory by reconsidering the dangling bond model of Spaepen and Turnbull,[22] resulting in an extended kinetic theory of SPE.[2] Within the context of this model, the velocity prefactor is given by

$$v_o = 2 \sin(\theta) \, v_s \, n_r \exp((\Delta S_f + \Delta S_m)/k) \tag{1}$$

where $\theta = 55°$ for the (100) interface (misorientation from $\{111\}$), $v_s$ is the speed of sound, $n_r$ is the net number of jumps a dangling bond makes before it is annihilated, $\Delta S_f$ is the entropy of formation of a dangling bond pair, and $\Delta S_m$ is the entropy of motion of the dangling bond at the interface. By using the same bounds on the entropy terms as used by Lu *et al.* it is possible to refine their estimate for the number of crystallization events per formation of a dangling bond pair, $N_r = 2 \, r \, n_r$, where $r$ is the ratio of crystallization events to configurational coordinate steps. Substituting in the value $v_o = 2.6 \times 10^7$ m/s from this work results in $65 \leq N_r \leq 2600$. In this case, the additional uncertainty associated with the velocity prefactor value is only ~20 % instead of the previous amount of ~50 x estimated by Lu. The improved bounds on the number of crystallization events per dangling bond pair formed give one somewhat more confidence in the plausibility of this model of the SPE process. These results are plausible and consistent with the idea that a single bond-breaking event can cause multiple crystallization events.[22]

For the dopant-enhanced measurements multiple energy As implants were used to create three different constant concentration profiles in $a$-Ge layers. The concentrations were 1, 5 and 10 x $10^{19}$ /cm$^3$ with a region of constant concentration covering a depth of 0.25 - 0.55 μm. The SPE rates were determined within this depth range. The accuracy of the concentration was determined by RBS measurements on As implanted Si samples that were prepared at the same time and under similar conditions as the As implanted Ge. The concentrations measured by RBS agreed to within 5 % of the expected concentrations. SIMS measurements were also used to verify the As concentration profiles. The SIMS profiles compared well in shape and depth scale to expected profiles calculated with the Profile code.[18]

Figure 2 a) shows the As enhanced SPE rate for crystallization at 340 °C compared to the implanted As concentration profile determined by SIMS. The SPE velocity and the concentration profiles agree well. Also shown is the intrinsic SPE rate. Hydrogen retardation of the SPE rate can again be observed at depths less than 0.3 μm.

Figure 2 b) shows the SPE rates normalized to the intrinsic rates for the H-free depth range at each temperature. The error bars were calculated by considering the reproducibility of the data and the RMS noise in the determined velocity curves. Error bars for the other two concentrations are of similar magnitude and have been omitted for clarity.

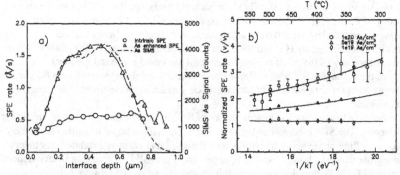

**Figure 2.** a) Comparison of the implanted As profile (dashed line) as determined from a SIMS measurement and the SPE rate enhancement (Δ) due to the implanted As for crystallization at 340 °C. The SPE rate for an intrinsic $a$-Ge layer sample (O) is also shown for comparison. b) The SPE rates of As implanted $a$-Ge normalized against the SPE velocity intrinsic $a$-Ge at each temperature. Three different concentrations are shown: 1, 5 and 10 x $10^{19}$ /cm$^3$. Solid lines are best fits using the GFLS model.

Fits to the data were performed using the generalized Fermi level shifting (GFLS) model and are shown as solid lines in Fig. 2 b). The GFLS model links structural changes related to SPE at the interface to shifts in the Fermi level and has been applied to Si SPE with some success.[13] For the dopant concentrations and temperatures used in this work, the Fermi levels were calculated numerically using degenerate semiconductor statistics. Details of this calculation will be presented elsewhere.[17] Once the Fermi levels are known the normalized SPE velocity data for $n$-type material can be fitted using

$$v/v_i = (1 + g \exp(E_f\text{-}E_k/kT))/(1 + g \exp (E_{fi} - E_k/kT)) \qquad (2)$$

where $E_f$ is the Fermi level and $E_k$ represents the energy level within the band gap of the defect responsible for the SPE process. The degeneracy factor, $g$, associated with $E_k$ is given by $g = Z(D^-)/Z(D^o)$ where $Z(D^-)$ and $Z(D^o)$ are the internal degeneracies of the $D^-$ and $D^o$ defect states, respectively.[19] If a dangling bond defect is responsible for the SPE process then it is expected that $g = 1/2$ if only the spin degeneracy needs to be considered.

The normalized velocity data for the 5 and $10 \times 10^{19}$ /$cm^3$ data sets exhibit a larger temperature variation than the $1 \times 10^{19}$ /$cm^3$ data set. A greater temperature dependence allows more accurate fits to be performed and therefore, only fits to these data will be considered for the purposes of the following discussion.

The values of $E_k$, referenced to the conduction band edge, and $g$ obtained from fitting the $5 \times 10^{19}$ /$cm^3$ and $10 \times 10^{19}$ /$cm^3$ data sets are $(E_c - E_k) = (0.07 \pm 0.03)$ eV and $g = 0.4 \pm 0.2$ and $(E_c - E_k) = (0.06 \pm 0.01)$ eV and $g = 0.6 \pm 0.2$, respectively. The solid lines in figure 2 b) correspond to these values. The error values quoted here are from the fits only. This results in a SPE defect level that is about 0.06 eV below the conduction band edge and a degeneracy value of about a half. If the degeneracy factor is fixed to a value of a half and only $E_k$ is allowed to vary then one obtains $(E_c - E_k) = 0.053$ eV and $(E_c - E_k) = 0.072$ eV for the $5 \times 10^{19}$ /$cm^3$ and $10 \times 10^{19}$ /$cm^3$ data, respectively. The quality of the fits remains fairly reasonable in this case.

The degeneracy value of $g = 1/2$ is consistent with a dangling bond type defect. This value also agrees with similar SPE studies in As implanted Si where a value of a half was obtained.[13] The defect energy level is relatively close to the conduction band edge. As a reference point for this energy level, other defects in $c$-Ge can be found at $(E_c - 0.04)$ eV tentatively assigned to the transition level of a self-interstitial,[20] $(E_c - 0.39)$ eV for a negatively charged divacancy and $(E_c - 0.54)$ eV for a negatively charged vacancy.[21] The defect level found here is entirely consistent with these values. However, to the best of our knowledge the band-gap state for a dangling bond in $c$-Ge is not in the literature.

In $a$-Si layers, fits to As enhanced normalized SPE rates yielded $(E_c\text{-}0.16)$ eV which also compares well to the energy levels of typical defects in $c$-Si.[13] The consistency of the GFLS model for both Si and Ge adds greatly to confidence in the predictive power of the model.

## CONCLUSIONS

The SPE regrowth rate of Ge layers formed by self-ion implantation of various <100> Ge substrates obtained from different suppliers was measured at temperatures between the range 300 - 540 °C. From these measurements, an activation energy of $(2.15 \pm 0.04)$ eV with a velocity prefactor of $(2.6 \pm 0.5) \times 10^7$ m/s was determined. No significant variation was observed between samples with different background doping levels. In applying the kinetic model, our improved data lead to a reduction of the uncertainty, from a factor of 50 to 20 %.

The presence of As to concentrations greater than $1 \times 10^{19}$ /$cm^3$ resulted in enhanced SPE rates similar to those found for Si. The GFLS model yielded excellent fits to the As enhanced SPE data. Degenerate semiconductor statistics were used to fit the model yielding a degeneracy and energy level value of $(E_c - E_k) = (0.06 \pm 0.01)$ eV and $g = 0.5 \pm 0.2$. These results are

221

remarkably consistent with previous studies performed in $a$-Si and may be consistent with a dangling bond mediating the SPE process as suggested by Spaepen and Turnbull. This study has advanced our knowledge of SPE mechanisms by showing that the GFLS model is robust and capable of describing the dopant enhanced SPE process in both Ge and Si.

## ACKNOWLEDGMENTS

The Department of Electronic Materials Engineering at the Australian National University is acknowledged for their support by providing access to SIMS and ion implanting facilities. This work was supported by a grant from the Australian Research Council.

## REFERENCES

1. Inter□national Technology Roadmap for Semiconductors (2003 Edition).
2. G.-Q. Lu, E. Nygren, and M. J. Aziz, J. Appl. Phys. 70, 5323 (1991).
3. B. Depuydt, A. Theuwis, and I. Romandic, Mater. Sci. Semicond. Process. 9, 437 (2006).
4. L. Csepregi, R. P. K ullen, J. W. Mayer, and T. W. S□igmon, Solid State Commun.□ 21, 1019 (1977)
5. E. Donavan, F. Spaepen, D. Turnbull, J. Poate and D. Jacobsen, J. Appl. Phys. 57, 1795 (1985).
6. G. Q. Lu, E. Nygren, M. J. Aziz, D. Turnbull, and C. W. White,□ Appl. Phys. Lett. 56, 137 (1990).
7. P. Kringhøj and R. G. Elliman□, Phys. Rev. Lett. 73, 858 (1994).
8. T. E. Haynes, M. J. Antonell, C. A. Lee, and K. S. Jones, Phys. Rev. B. 51, 7762 (1995).
9. G. Olson and J. Roth, in Handbook of Crystal Growth, edited by D. Hurle (Elsevier Science B. 1994), vol. 3, chap. 7, pp. 255 - 312.
10. J. A. Roth, G. L. Olson, D. C. Jacobson, and J. M. Poate, Appl. Phys. Lett. 57, 1340 (1990).
11. J. A. Roth, G. L. Olson, D. C. Jacobson, and J. M. Poate, Mat. Res. Soc. Symp. Proc. 205, 45 (1992).
12. J. C. McCallum, Appl. Phys. Lett. 69, 925 (1996).
13. B. C. Johnson and J. C. McCallum, Phys. Rev. B. 76, 045216 (2007).
14. J. Williams and R. Elliman, Phys. Rev. Lett. 51, 1069 (1983).
15. I. Suni, G. Göltz, M. Nicolet, and S. Lau, Thin Solid Films 93, 171 (1982).
16. A. Armigliato, R. Nipoti, G. Bentini, M. Mazzone, M. Bianconi, A. N. Larsen, and A. Gasparotto, Mat. Sci. Eng. B2, 63 (1989).
17. B. C. Johnson, P. Gortmaker, and J. C. McCallum, submitted to PHYSICAL REVIEW B. (2008)
18. Profile, Ion Beam Profile Code version 3.20, Implant Sciences Corp. 107 Audubon Rd., No. 5, Wakefield, MA 01880.
19. J. C. Bourgoin and P. Germain, Phys. Lett. 54A, 444 (1975). 30 Y. Jeon, M. Becker, and R. Walser, Mat. Res. Soc. 157, 745 (1990).
20. H. Haesslein, R. Sielemann, and C. Zistl, Phys. Rev. Lett. 80, 2626 (1998).
21. J. Coutinho, V. J. B. Torres, A. Carvalho, R. Jones, S. Olberg, and P. R. Briddon, Mater. Sci. Semicond. Process. 9, 477 (2006).
22. F. Spaepen and D. Turnbull, AIP Conf. Proc. 50, 73 (1979).
23. T. Saito and I. Ohdomari, Phil. Mag. B 49, 471 (1984).

Mater. Res. Soc. Symp. Proc. Vol. 1070 © 2008 Materials Research Society     1070-E05-07

# First Principles Study of Boron in Amorphous Silicon

Iván Santos[1,2], Wolfgang Windl[1], Lourdes Pelaz[2], and Luis Alberto Marqués[2]

[1]Dept. Materials Science and Engineering, The Ohio State University, Columbus, OH, 43201

[2]Dpto. Electricidad y Electrónica, Universidad de Valladolid, Valladolid, Spain

## ABSTRACT

We have carried out an *ab initio* simulation study of boron in amorphous silicon. In order to understand the possible structural environments of B atoms, we have studied substitutional-like (replacing one Si atom in the amorphous cell by a B atom) and interstitial-like (adding a B atom into an interstitial space) initial configurations. We have evaluated the Fermi-level dependent formation energy of the neutral and charged ($\pm 1$) configurations and the chemical potential for the neutral ones. For the interstitial-like boron atom, we have find an averaged formation energy of 1.5 eV. For the substitutional case, we have found a dependence of the chemical potential on the distance to Si neighbors, which does not appear for the interstitial ones. From MD simulations, we could observe a diffusion event for an interstitial-like boron atom with a migration barrier of 0.6 eV.

## INTRODUCTION

Future CMOS technology requires ultrashallow junctions with high dopant activation levels and very sharp doping profiles [1]. According to the ITRS [1], solid-phase epitaxial regrowth (SPER) of doped amorphous silicon (a-Si) is a promising technique to minimize channeling [2], transient-enhanced diffusion [3,4] and dopant deactivation [5,6] and it allows in the case of B to obtain high active concentrations of $2\text{-}4 \times 10^{20}$ cm$^{-3}$ [7,8], compared to $2\text{-}9 \times 10^{19}$ cm$^{-3}$ in c-Si [9,10]. Thus, for the fabrication of state-of-the-art electronic devices, preamorphization implants and high dopant implantation fluences are being used. Because of these reasons, there is an increasing interest in studying and modeling dopant processes in a-Si. B has been extensively studied in c-Si and diffusion mechanism [11-14], interaction with Si atoms and segregation processes [15-18] are well known. However, its behavior in a-Si has been studied to a much lesser degree, especially at the atomic level, apart from one early study [19].

The aim of this work is to understand the possible structural environments of B in a-Si and eventually its diffusion. Evidently, there will be a continuous range of configurations, unlike the discrete set obtained in a crystalline environment. The key dynamical features of B in a-Si are naturally formulated as hopping between various metastable configurations in the energy landscape, where of course the a-Si network has a high probability of rearranging and thus changing the energy landscape during annealing. Thus, we believe that this paper is an important step to the larger problem of unraveling the dynamics of B in a-Si.

## SIMULATION DETAILS

We have carried out an *ab initio* study of B in a 64-atom a-Si cell. The most realistic computer models of a-Si are made using the algorithm of Wooten, Weaire and Winer (WWW) [20]. The WWW models are formed using a particular bond-switching algorithm, and extremely simple potentials. The cell used in this work was provided by D. A. Drabold using a model proposed by Barkema and Mousseau (BM) [21], who developed an algorithm significantly improved from the original WWW scheme. The BM scheme guaranteed that there was no 'crystalline remnant' in the WWW. Upon relaxing with *ab initio* interactions, the BM model barely changes at all. To represent the amorphous environment faithfully, it is necessary to work with cells of at least several hundred atoms, and it is desirable to have even more. The difficulty with smaller models is that there are discernible strain effects (most notably in a slightly bloated bond angle distribution relative to experiment). Nevertheless, for local structural properties as we are focused on here, we anticipate that the strain artifacts are small. The cell properly includes the correct local topological disorder of the material, and it is big enough for a high-precision calculation that includes B.

For our *ab initio* calculations, we have used the efficient density-functional theory code VASP [22]. We have performed our calculations with generalized gradient approximation (GGA) ultrasoft pseudopotentials [23], using a Monkhorst-Pack k-point sampling, 4×4×4 for the density of states calculation and 2×2×2 for the ionic relaxations. The energy cutoff chosen was 230 eV. For the substitutional-like B atoms ($B_s$), we replaced one of the Si atoms by B. For the interstitial-like ones ($B_i$), we either placed a B atom in an interstitial site or added an interstitial Si atom next to a $B_s$. Subsequently, we carried out a relaxation of both the atomic positions and cell shape and volume. For our ab-initio MD simulations, we used an energy cutoff of 156 eV and $\Gamma$-point sampling.

## RESULTS AND DISCUSSION

Since we wanted to include charge effects into our study, we first calculated the electronic density of states (DOS) in the amorphous cell to evaluate its band gap. In the case of c-Si, *ab initio* calculations give an incorrected value of 0.63 eV for the gap [24] instead of the experimental value of 1.12 eV (at 300 K). However, *ab initio* calculations can be corrected with scissor, finite-size, and gap-state corrections [24]. For a-Si, the *ab initio* calculation finds a gap of 0.76 eV, 21% larger than c-Si. As a first approximation, we can estimate the corrected gap value to be 21% larger than the experimental value for c-Si. This gives a gap value of 1.36 eV for a-Si, which is in agreement with experiments, where it ranges from 1.2 to 1.6 eV [25, 26].

For charged B atoms in the amorphous cell, the formation energy ($E_f$) for a system with charge $Q$ as a function of the Fermi level ($E_F$) can be calculated from the expression

$$E_f = E_{tot}(Q) - E_{ref} + Q(E_v + E_F), \qquad (1)$$

where $E_{tot}(Q)$ is the total energy of the charged system, $E_{ref}$ is a neutral reference energy ($E_{ref} = E_{tot}(aSi_{63}B) + E_{tot}(aSi_{64})/64$ for $B_i$ in a 64-atom supercell), and $E_v$ is the energy of the top of the valence band. In order to evaluate the variation of energy when adding/removing one electron, we have only considered those cells in which there were not any coordination changes during this process. Then, for the different coordinations of B atoms (fourfold for $B_s$ and three, four and fivefold for $B_i$), we have evaluated the Boltzmann averaged formation energy at 650°C,

$$\langle E_f \rangle = \frac{\sum E_f e^{-E_f / k_B T}}{\sum e^{-E_f / k_B T}}. \qquad (2)$$

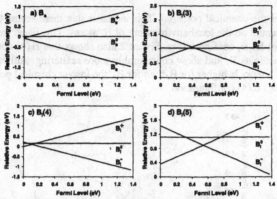

**Figure 1.** Formation energies of $B_s$ and $B_i$ relative to $B_s^0$ as a function of the Fermi level, Boltzmann averaged at 650 °C. In the case of $B_i$, we have distinguished between atoms with b) three ($B_i(3)$), c) four ($B_i(4)$) and d) five ($B_i(5)$) Si neighbor atoms.

In Fig. 1 we plot the formation energy of B atoms relative to $B_s^0$ as a function of the Fermi level. Figure 1(a) shows that $B_s^-$ has a lower energy than both $B_s^0$ and $B_s^+$ for the entire Fermi-level range. Therefore, $B_s$ will be negatively charged for all Fermi level values. In the case of $B_i$, both +1 and −1 charge states can be stable (with a very small stability range for neutral $B_i(3)$). The difference in formation energy between $B_s$ and $B_i(4)$ is small. They also have a similar energy dependence on the Fermi level. This suggests that $B_i(4)$ and $B_s$ are similar (substitutional, but with changed a-Si network). The midgap energies correspond in all cases to the negative charge state. Taking the lowest-energy configuration, $B_s^-$, as reference, they are $E(B_i^-(4)) = 0.30$ eV, $E(B_i^-(3)) = 1.53$ eV, and $E(B_i^-(5)) = 1.55$ eV. Since $B_i(4)$ can be considered as a substitutional-like configuration and $B_i(3)$ and $B_i(5)$ have nearly identical Boltzmann-averaged formation energies, there seems to be a well defined formation energy for interstitial-like B of ~1.54 eV at midgap.

The chemical potential of B atoms, $\mu_B$, can be calculated as

$$\mu_B = E_{tot}(\text{a-Si}_n\text{B}) - n\, E_{tot}(\text{a-Si}_{64})/64 \tag{3}$$

where $n$ indicates the number of Si atoms in the cell, 63 for $B_s$ and 64 for $B_i$ (in our case, $E_{tot}(a\text{Si}_{64})/64 = -5.25$ eV defines the Si chemical potential). Figure 2 shows the B chemical potential as a function of the average bond length between the B atom and its nearest neighbors. 62.5% of $B_s$ configurations are inside the limits marked by the straight dashed lines. Solid line corresponds to a linear fitting of these configurations. Figure 2 shows that there is more scattering in the data for $B_i$ than for the $B_s$ data. Chemical potential values scatter within an interval of ~ 1.4 eV for $B_s$ and ~ 3 eV for $B_i$. This wide distribution of values makes the identification of any particularity in the local structures of B atoms in a-Si tedious. However, it can be seen in Fig. 2 that the $B_s$ configurations within the dashed lines represent a linear trend: the closer the neighbors, the lower the chemical potential. This indicates that there seems to be a dependence of the chemical potential on the local environment of $B_s$ atoms. This linear trend appears to be a lower limit to almost all $B_i$ configurations. Figure 2 also shows that $B_i(4)$ configurations do not display the same trends as $B_s$ and show considerably more scattering. Although the Boltzmann averaged formation energy is higher for $B_i$ than for $B_s$, the lowest chemical potential corresponds to a (substitional-like) $B_i(4)$ configuration.

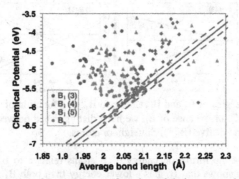

**Figure 2.** Chemical potential of $B_i$ and $B_s$ as a function of their average bond length to nearest Si neighbor atoms for the different coordinations considered. 62.5% of $B_s$ configurations are inside the limits marked by the dashed lines. Solid line corresponds to a linear fitting of these points.

In order to analyze the dynamical features of B in a-Si, we have used several $B_i$ and $B_s$ configurations as the starting point of *ab initio* molecular dynamics (MD) simulations at 1000 K. We found that within the simulated time (around 55 ps) none of the MD simulations of $B_s$ configurations showed any migration of the B atom. However, local rearrangements (bond switching) of Si-Si bonds have been observed in the cell. On the other hand, we observed a diffusion event in one of the $B_i(3)$ configuration. This diffusion event was analyzed using the Nudged-elastic Band Method (NEBM) [27] implemented into VASP in order to obtain the energy barriers along the path. We used configurations obtained with MD as initial, intermediate and final points. Figure 3 shows the nudged-elastic band total energy along the diffusion path together with some snapshots of the B environment at the marked points. The highest forward barrier, which represents the overall migration energy from initial to final configuration, is ~0.6 eV (similar to the value for crystalline Si [14]). Intermediate forward barriers range from ~0.3 to ~0.6 eV and may indicate a possible range for migration barriers. The highest reverse barrier measures ~ 0.9 eV. It can be seen from the snapshots that in some cases the barriers correspond to Si rearrangement. These rearrangements of the amorphous structure at high temperatures could drag B atoms along (or not). Thus, they cannot really be separated from the more traditional B migration (mostly unchanged cell), which makes further studies difficult. However, further studies on diffusion are required to have a statistic result for the migration energy.

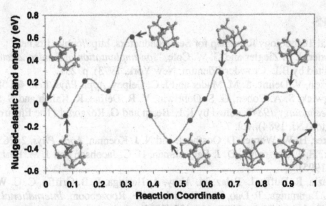

**Figure 3.** Nudged-elastic band total energy along the diffusion path found in the MD simulation. Snapshots show the B environment at the indicated points. B atom is blue, Si atoms are yellow.

## CONCLUSIONS

We have carried out an *ab initio* study of B in a-Si. From the calculation of the DOS in the amorphous reference cell we obtained a band gap higher that in the crystalline case, in agreement with experiments. For B atoms in a-Si, we have started considering initially substitutional-like and interstitial-like configurations. In the analysis of the dependence of their formation energy as a function of the Fermi level, we have found that $B_s$ behaves similarly to the crystalline case. In the case of $B_i$, it is positively charged for highly $p$-doped a-Si for all coordinations and will change to the negative state with increasing value of the Fermi level. At midgap, $B_i$ has a formation energy of ~1.5 eV. From the analysis of the chemical potential, we found that most of the $B_s$ configurations display a linear trend: the closer the Si neighbors, the lower the chemical potential. This trend can be used as a first step to identify any local structure of B atoms in a-Si.

MD simulations showed that Si bond switching can influence the diffusion processes when increasing the temperature. Any further study of diffusion in a-Si will have to separate the influence of the changes in the network from the diffusion processes of B atoms. As a first attempt, we have obtained from our NEBM simulations the order of magnitude for energy barriers (~0.6 eV).

## ACKNOWLEDGMENTS

The authors thank D. A. Drabold for providing the a-Si structure and for the very helpful discussions. We also thank The Ohio Supercomputer Center for the computing time. I. Santos, L. Pelaz and L.A. Marqués thank the support of the Spanish DGI under project TEC2005-05101 and the JCyL Consejería de Educación y Cultura under project VA070A05.

# REFERENCES

1. International Technology Roadmap for Semiconductors, http://public.itrs.net.
2. B. L. Crowder, J. F. Ziegler and G. W. Cole, *"Ion Implantation in Semiconductors and Other Materials"*, edited by B.L. Crowder (Plenum, New York, 1973), p. 257.
3. R. T. Hodgson, V. Deline, S. M. Mader and J. C. Gelpey, *Appl. Phys. Lett.* **44**, 589 (1984).
4. T. O. Sedgwick, S. A. Cohen, G. S. Oehrlwin, V. R. Deline, R. Kalish and S. Shatas, *"VLSI Science and Technology 1984"*, edited by K.E. Beam and G. Rozgonyi (The Electrochemical Society, Pennington, NJ, 1984), vol. 7.
5. W. K. Hofker, H. W. Werner, D. Oosthoek and N. J. Koeman, *Appl. Phys.* **2**, 265 (1973).
6. P. A. Stolk, H. J. Gossmann, D. J. Eagleshman, D. C. Jacobson and J. M. Poate, *Appl. Phys. Lett.* **66**, 568 (1995).
7. V. C. Venezia, R. Duffy, L. Pelaz, M. Aboy, A. Heringa, P. B. Griffin, C. C. Wang, M. J. P. Hopstaken, Y. Tamminga, T. Dao, B. J. Pawlak and F. Roozeboom, *International Electron Devices Meeting Technical Digest* 2003, 489-492 (2003).
8. S. Solmi, E. Landi and F. Baruffaldi, *J. Appl. Phys.* **68**, 3250 (1990).
9. G. L. Olson and J. A. Roth, *Mater. Sci. Rep.* **3**, 1 (1989).
10. P. Fahey, P. B. Griffin and J. P. Plummer, *Rev. Mod. Phys.* **61**, 289 (1989).
11. E. Tarnow, *J. Phys. Condens. Matter.* **4**, 5405 (1992).
12. J. Zhu, T. Diaz de la Rubia, L. Yang, C. Mailhiot, G. Gilmer, *Phys. Rev. B* **54**, 4741 (1996).
13. B. Sadigh, T. J. Lenosky, S. K. Theiss and M. J. Caturla, *Phys. Rev. Lett.* **83**, 4341 (1999).
14. W. Windl, M. M. Bunea, R. Stumpf, S. T. Dunham and M. P. Masquelier, *Phys. Rev. Lett.* **83**, 4345 (1999).
15. M. J. Caturla, M. D. Johnson and T. Diaz de la Rubia, *Appl. Phys. Lett.* **72**, 2736 (1998).
16. L. Pelaz, G. H. Gilmer, H. J. Gossmann and C.S. Rafferty, *Appl. Phys. Lett.* **74**, 3657 (1999).
17. W. Luo, P. B. Rasband, P. Clancy and B. W. Roberts, *J. Apply. Phys.* **84**, 2476 (1998).
18. X. Y. Liu, W. Windl and M. P. Masquelier, *Appl. Phys. Lett.* **77**, 2018 (2000).
19. P. A. Fedders and D. A. Drabold, *Phys. Rev. B* **56**, 1864 (1997).
20. F. Wooten, K. Winer, and D. Weaire, *Phys. Rev. Lett.* **54**, 1392 (1985).
21. N. Mousseau and G. T. Barkema, *Phys. Rev. B* **61**, 1898 (2000).
22. G. Kresse and J. Hafner, *Phys. Rev. B* **47**, 558 (1993); **49**, 14251 (1994); G. Kresse and J. Furthmüller, *Mater. Sci.* **6**, 15 (1996), *Phy. Rev. B* **54**, 11169 (1996).
23. J. P. Perdew and Y. Wang, *Phys. Rev. B* **45**, 13244 (1992).
24. W. Windl, *Phys. Stat. Sol. B* **241**, 2313 (2004).
25. J.S. Lannin, L.J. Pilione and S.T. Kshirsagar, *Phys. Rev. B* **26**, 3506 (1982).
26. C. Rotaru, S. Nastase and N. Tomozeiu, *Phys. Stat. Sol A* **171**, 365 (1999).
27. H. Jónsson, G. Mills and K. Jacobsen, in *"Classical and Quantum Dynamics in Condensed Phase Simulations"*, ed. by B.J. Berne et al. (World Scientific, Singapore, 1998), p. 385.
28. V. C. Venezia, R. Duffy, L. Pelaz, M. J. P. Hopstaken, G. C. J. Maas, T. Dao, T. Tamminga and T. Graat, *Mat. Sci. and Eng. B* **124-125**, 245 (2005).

Mater. Res. Soc. Symp. Proc. Vol. 1070 © 2008 Materials Research Society     1070-E05-08

# Atomistic Simulation and Subsequent Optimization of Boron USJ Using Pre-Amorphization and High Ramp Rates Annealing

Julien Singer[1,2], François Wacquant[3], Davy Villanueva[1], Frédéric Salvetti[1], Cyrille Laviron[2], Olga Cueto[2], Pierrette Rivallin[2], Martín Jaraíz[4], and Alain Poncet[5]

[1]NXP Semiconductors, 850 rue Jean Monnet, Crolles, 38926, France
[2]CEA-Leti Minatec, 17 rue des Martyrs, Grenoble, 38054, France
[3]STMicroelectronics, 850 rue Jean Monnet, Crolles, 38926, France
[4]University of Valladolid, Valladolid, 47011, Spain
[5]Lyon Nanotechnologies Institute (INL), Villeurbanne, 69621, France

## ABSTRACT

This study presents the use of atomistic process simulations to optimize $p^+/n$ ultra-shallow junctions fabrication process. We first bring to the fore that a high concentration of interstitials close to the boron profile decreases the sensibility of boron diffusion to thermal budget. Preamorphization of the substrate is thus necessary to decrease boron diffusion by thermal budget reduction; the latter is obtained by the use of a high ramp rate annealing system instead of the classical lamp-type rapid thermal annealing. At the same time we show that the use of high ramp rates does not enhance boron activation, the substrate being preamorphized or not. So high ramp rates anneal can improve sheet resistance/junction depth trade-off only with preamorphization implant. Experiments are performed that confirm the predictions of our simulations. Further discussions explain activation path of boron throughout the temperature cycle, as a function of amorphous depth, for both lamp-type and high ramp rate annealing tool.

## INTRODUCTION

Due to transistor size reduction, it is more and more important to introduce high concentrations of active dopants in Source and Drain extensions, while controlling their diffusion. One possible way to reach the best sheet resistance/junction depth ($R_s/X_j$) trade-off is to use high temperature (T), short time annealing, such as rapid thermal annealing (RTA), or fast-ramp spike anneals. Millisecond anneals (flash, LASER) provide good activation levels with low diffusion, but have to be coupled with more classical anneals. Otherwise extended defects remain and junctions can be too abrupt, leading to unacceptable leakage levels [1].

This work presents the use of atomistic simulations in order to understand the effect of aggressive ramps and to determine how to tune them for $R_s/X_j$ trade-off optimization of boron (B) ultra-shallow junctions (USJ). Atomistic simulations are carried out using the kinetic Monte Carlo (kMC) method. We use DADOS, the code developed by the University of Valladolid [2]. One of the main advantages of kMC is to give the right trends even without specific calibration, and to guide the user in understanding the observed phenomena. Indeed all of the implemented mechanisms are always taken into account simultaneously, and all atomistic data are available at any time during the simulation.

## RESULTS AND DISCUSSION

To facilitate interpretation of simulations and favor physical comprehension instead of experiment fitting, we first simulate very simple structures, with square-shaped profiles of B and interstitials (I's). Only then do we proceed to carry out simulations closer to experimental conditions.

## Simulations of simplified structures with square-shaped profiles

The structure used in these first simulations is described in Figure 1. Although simplicity is expected from these simulations, it is necessary to keep in mind the following details in order to interpret the results, and afterwards link simulation to experiments. The square-shaped B profile has a concentration similar to the maximum one of the B as-implanted profile, that is to say $4x10^{21}$ cm$^{-3}$. It is located near the surface so as to imitate USJ conditions. Since the position of I's away from the B profile is justified only if preamorphization implant (PAI) is performed, we also put an amorphous layer in our structure, from the surface to the I's square-shaped profile. Indeed if no amorphous layer is introduced all B will be active, but during solid phase epitaxial regrowth (SPER) B atoms are shared between active (potentially mobile) and clustered (immobile) B. A square-shaped profile of I's was used, with a concentration varying from $1x10^{17}$ to $1x10^{22}$ cm$^{-3}$. Its position corresponds to the highly damaged region, called end of range region. It is made variable since changing the energy of the PAI leads to different amorphous depths. The width of 10 nm is close to the I's peak found just behind the a/c interface after implant. Annealing at constant T around 1000°C is performed, with negligible ramp-up.

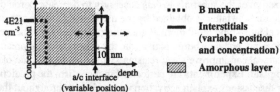

**Figure 1.** Simple structure used in atomistic simulations showing amorphous layer and square-shaped concentration profiles of B and I's. I's concentration and position are variable.

In the following we will refer to $X_j$ as the depth where B concentration reaches $1x10^{18}$ cm$^{-3}$. We look at the $X_j$ increase with an increase of thermal budget (ThB). For this purpose, an anneal time increase at constant T is applied and the resulting $X_j$ raise is measured in %. This indicator informs on the sensitivity of $X_j$ to ThB, depending on the conditions. We look at the effect of I's concentration and position. An example of the diffused profiles is shown in Figure 2.

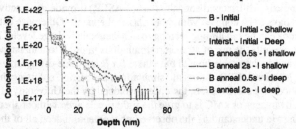

**Figure 2.** Example of B profiles after annealing of 0.5 and 2s. Note the significant effect of ThB when I's are placed at 70 nm in depth, while almost no change is observed when they are 10 nm from surface, close to B profile. Initial profiles of B and I's are also shown for reminder.

First, we observe that increasing I's concentration largely reduces this sensitivity. The trend is clearly shown by Figure 3 (a). B diffusion is due to the formation of boron-interstitial (BI) pairs. In equilibrium conditions, I's are *thermally* generated. If T is raised, the concentration of I's increases; if time is increased, more I's will have been generated during annealing (same concentration). So when ThB increases, the total number of I's generated during annealing goes up, so does the number of BI pairs; thus B diffusion increases. When I's are introduced, the resulting I's concentration is higher than the equilibrium one. The concentration of BI pairs depends on the I's concentration [I]. It is well known that the supersaturation, defined as $[I]/[I]^{equil}$, leads to B transient enhanced diffusion [3]. As the I's injection level increases, the concentration of BI pairs depends more on this injection, and less on the I's thermal generation. B diffusion is in turn less impacted by ThB. In addition, the effect of I's position was investigated. As a result we observe that the sensitivity to ThB is lowered when I's are put closer to the B profile, as plotted in Figure 3 (b). This shows that after their concentration, the position of the I's with regard to the B profile is the second parameter to take into account. This is because supersaturation decreases as we go away from this highly damaged region. Indeed diffusing I's jump from one site to its neighbor, following a random walk. The profile of mobile I's is then approximately a gaussian that broadens during annealing. In summary we show that for these conditions the sensitivity of $X_j$ to ThB is largely reduced when I's are present at high concentration around the B profile.

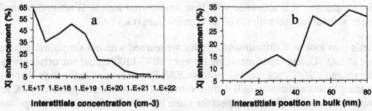

**Figure 3.** $X_j$ increase due to ThB as a function of I's initial concentration (a) and position (b). $X_j$ becomes unaffected by a ThB increase when I's are more numerous or closer to B profile.

## Simulations with experimental parameters and subsequent experiments

B implantation at an energy of 1 keV and a dose of $5 \times 10^{14}$ cm$^{-2}$ was simulated. RTA at T between 1040 and 1100°C involved classical lamp-based RTA, and advanced thermal conduction tool. For the latter we use Levitor [4], which allows T ramps above 600 K/s, while remaining in the same range of anneal times and T. Total ThB is then much lower with Levitor. T profiles of lamp-based and Levitor anneals are presented in Figure 4.

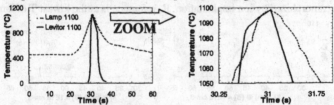

**Figure 4.** Comparisons of T ramps of lamp-based and thermal conduction tools (here 1100°C).

Let's consider first B activation. Figure 5 taken from our simulations represents the B active dose as a function of the T during annealing. Without PAI (a), for both annealing types the time spent at high T allows equilibrium activation at the maximum temperature ($T_{max}$). Although equilibrium is not reached during ramp-up, for any annealing type the time spent at high T leads to the same activation. When PAI is performed, although the activation path is different, once again the same activation level is reached with both annealing types, for a given $T_{max}$. Actually, the activation mechanism consists of over-equilibrium activation after SPER, and subsequent de-activation so as to tend to equilibrium, as shown again by atomistic simulations (b). We will explain this in detail below. So B activation depends on $T_{max}$ and on the time spent at this T, with and without PAI.

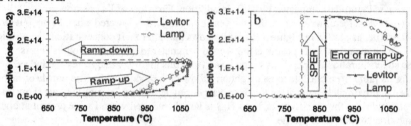

**Figure 5.** Comparison of B activation with lamp and Levitor anneal: B active dose evolution with T cycle without (a) and with (b) 80 keV germanium (Ge) PAI.

Let's now look at B diffusion using what we learned with our simplified simulations. The latter were at 1000°C, and experiments are between 1040-1100°C; but our trends are the same within this range (1080°C, not shown). Without PAI, I's are located in the high B concentration region, leading to low sensitivity of B diffusion to the ThB, as we saw previously. $X_j$ are then similar with both anneals. Thus we expect the same $R_s/X_j$ trade-off with both lamp-based and thermal conduction tools. When PAI is performed, I's remain only beyond initial a/c interface, away from the B profile. So the B diffusion is expected to be more sensitive to the variation of the ThB. In other words, the lower total ThB of Levitor anneal should allow $X_j$ reduction. Therefore we expect the combination of PAI with Levitor anneal to improve the $R_s/X_j$ trade-off.

This is confirmed by realistic atomistic process simulations. We simulated the implantation of B at 1 keV and at a dose of $5 \times 10^{14}$ cm$^{-2}$, preceded or not by a 80 keV Ge amorphizing implant. Annealing was simulated either with lamp or Levitor T cycles, both at different T. Levitor anneal improves $R_s/X_j$ trade-off only if associated with PAI (Figure 6). Experiments based on this process design show exactly the predicted trend (Figure 7): thermal conduction tool with extremely high ramp rates is able to improve $R_s/X_j$ trade-off only if the substrate is preamorphized by ion implantation. $R_s$ reference is at the lowest T of lamp anneals.

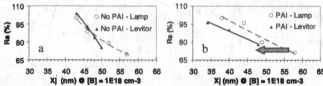

**Figure 6.** Results of atomistic simulations giving $R_s/X_j$ trade-off. No improvement is obtained by Levitor anneal (a), unless PAI is performed (b).

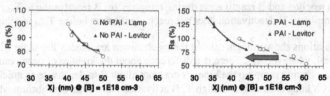

**Figure 7.** Experimental results showing $R_s/X_j$ trade-off improvement with Levitor anneal only if associated with PAI.

## Further discussion on boron activation as a function of the amorphous depth

With PAI, it is interesting to look at the activation path of B as a function of the amorphous depth ($X_{a/c}$). First, SPER leads to over-activation; then deactivation occurs with a given kinetics; finally, reactivation happens. Let's now go into detail.

Figure 8 represents the simulated evolution active B dose during lamp-type annealing. When silicon recrystallizes, the solid solubility limit of B is above the equilibrium [5]. As the Ge PAI energy decreases, $X_{a/c}$ decreases. Recrystallization, which happens during the ramp-up, is then shorter in time, thus in T. This does not change the quantity of active B (after SPER) while all the B as-implanted profile is contained within the amorphous layer, that is to say, down to 15 keV (a). If $X_{a/c}$ is even more lowered, only part of the B profile is below the a/c interface. Thus the quantity of active B decreases with the Ge implant energy (b), for an equivalent solubility limit (concentration). This above-equilibrium activation level tends to reach its equilibrium, which results in deactivation as time passes and T increases (c). Since T increases though, the solid solubility increases. Soon the equilibrium is reached, before the end of the ramp-up, and the active B dose starts increasing again with T. On the activation curve, we see a minimum at the T where the equilibrium is reached (d). When $T_{max}$ is reached, active B diffuses by kick-out and boron-interstitial clusters dissolve to maintain the active B concentration at its equilibrium value. That's why the quantity of active B increases slightly (e). Finally, no significant evolution is observed during ramp-down. A particular attention should be paid to the case of high energy Ge implant (80 keV or above), where the kinetics of deactivation does not allow to reach equilibrium before the end of the ramp-up (f). This is because recrystallization is complete at higher T: the time needed to recrystallize is higher when the amorphous layer is deeper. Finally for any $X_{a/c}$ equilibrium is reached at $T_{max}$, if not before. In summary B activation depends on $T_{max}$ and on the time spent at this T, the substrate being preamorphized or not, and whatever $X_{a/c}$.

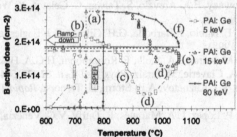

**Figure 8.** T evolution of the dose of active B during the T cycle of lamp-type RTA at 1060°C. (a) Complete recrystallization at different T depending on $X_{a/c}$. (b) B profile not entirely included

within the amorphous layer: the active dose after SPER is not maximum. (c) B deactivation. (d) Equilibrium reached and B reactivation during ramp-up. (e) B reactivation at $T_{max}$. (f) High energy Ge implants: B deactivation does not reach equilibrium before $T_{max}$ is achieved.

Simulations show in Figure 9 that the mechanisms are exactly the same with lamp and Levitor anneal, whatever $X_{a/c}$. There is a difference about the kinetics of deactivation before reaching $T_{max}$. During ramp-up the kinetics does not allow B to deactivate as quickly with Levitor as with lamp anneal. But at high T, B activation reaches its equilibrium whatever the annealing tool. Then the conclusion is the same as previously: activation of B only depends on the value of $T_{max}$ and on the time spent at $T_{max}$.

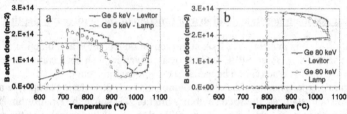

**Figure 9.** Comparison of deactivation and reactivation of B during the T cycle of lamp-type and Levitor RTA at 1060 °C. With Levitor equilibrium is not reached during ramp-up but at $T_{max}$ instead, either for low (a) and high (b) PAI energy.

## CONCLUSION

In this study we present a typical approach that can be followed using atomistic simulations in order to optimize USJ. We first demonstrate that we do not take advantage of Levitor on $R_s$: activation of B is the same as with lamp-type annealing, the substrate being preamorphized or not, and whatever $X_{a/c}$. In order to affirm this, we used atomistic simulations to investigate in detail the activation path of B during temperature cycle. On the contrary, the total ThB lowering allowed by Levitor with respect to lamp-type anneal makes it possible to reduce $X_j$, only if PAI is performed. Thus it is necessary to preamorphize the substrate to take advantage of the high ramp rates of Levitor for $R_s/X_j$ trade-off. Our conclusions were finally confirmed by experiments.

## REFERENCES

[1] S.B. Felch, D.F. Downey, E.A. Arevalo, S. Talwar, C. Gelatos, Y. Wang, *Ion Impl. Tech. Conf. Proc.*, 167 (2000).
[2] M. Jaraíz, L. Pelaz, E. Rubio, J. Barbolla, G.H. Gilmer, D.J. Eaglesham, H.J. Gossmann, J.M. Poate, *Mat. Res. Soc.* (1998).
[3] N.E.B. Cowern, G. Mannino, P.A. Stolk, F. Roozeboom, H.G.A. Huizing, J.G.M. van Berkum, F. Cristiano, A. Claverie, M. Jaraíz, *Phys. Rev. Lett.* **82** (22), 4460 (1999).
[4] E.H.A. Granneman, V.I. Kuznetov, A.B. Storm, H. Terhorst, *Rapid Thermal Processing Conf. Proc.*, 1 (2001).
[5] M. Aboy, L. Pelaz, P. López, L.A. Marqués, R. Duffy, V.C. Venezia, *Appl. Phys. Lett.* **88**, 191917 (2006).

# Modeling and Simulation

Mater. Res. Soc. Symp. Proc. Vol. 1070 © 2008 Materials Research Society          1070-E06-01

# Atomistic Simulation Techniques in Front-End Processing

Luis A. Marqués, Lourdes Pelaz, Iván Santos, Pedro López, and María Aboy
Department of Electronics, University of Valladolid, ETSI Telecomunicación, Campus Miguel
Delibes s/n, Valladolid, 47011, Spain

## ABSTRACT

Atomistic process models are beginning to play an important role as direct simulation approaches for front-end processes and materials, and also as a pathway to improve continuum modeling. Detailed insight into the underlying physics using *ab-initio* methods and classical molecular dynamics simulations will be needed for understanding the kinetics of reduced thermal budget processes and the role of impurities. However, the limited sizes and time scales accessible for detailed atomistic techniques usually lead to the difficult task of relating the information obtained from simulations to experimental data. The solution consists of the use of a hierarchical simulation scheme: more fundamental techniques are employed to extract parameters and models that are then feed into less detailed simulators which allow direct comparison with experiments. This scheme will be illustrated with the atomistic modeling of the ion-beam induced amorphization and recrystallization of silicon. The model is based on the bond defect or IV pair, which is used as the building block of the amorphous phase. It is shown that the recombination of this defect depends on the surrounding bond defects, which accounts for the cooperative nature of the amorphization and recrystallization processes. The implementation of this model in a kinetic Monte Carlo code allows extracting data directly comparable with experiments.

## INTRODUCTION

Silicon processing is facing an increasingly high level of complexity as CMOS technology is pushed closer to its limits. In particular, front-end processing is trying to extend the use of conventional and well established techniques, such as ion implantation and annealing, into the nanometer regime [1]. With further reduction of the devices size, new effects or effects that were neglected so far become relevant. Their experimental characterization is a complex task, firstly because the realization of test lots is extremely expensive, and secondly because usually these effects occur simultaneously which makes difficult the interpretation of measurements. In this situation the use of simulation tools can be very helpful [2].

Most process simulators used in industrial applications are based on continuum methods, where the physics of the system is formulated as a series of differential equations for each particle type considered to be relevant in the process. Typically they are continuity equations, where each particle gain or loss is formulated in terms of its generation and recombination rates and the diffusion flux [3,4]. In the equations there are a number of parameters such as binding energies, diffusivities, capture radii, etc., that have to be provided. The numerical solution of the set of partial differential equations requires spatial and temporal discretization. Continuum simulators are fast and allow the consideration of big sample sizes by adjusting the grid used for the spatial discretization. However, this advantage is reduced as the device size shrinks to nanometric scale. The atomistic nature of the material arises and complex physical interactions show up. The use of a very refined grid and the addition of new equations slow down the resolution of the problem using continuum methods. Then atomistic simulation techniques become a good alternative even for industrial applications [5].

In this paper we show how atomistic simulation techniques can be used to get physical insight on some of the aspects related to the front-end processing of Si. In particular, we will focus on the modeling of the ion-beam induced amorphization and subsequent recrystallization of the Si substrate, which has been defined as a key challenge in the successive editions of the International Technology Roadmap for Semiconductors [2].

## ATOMISTIC SIMULATION TECHNIQUES

In atomistic simulation methods the system under study is described as a set of interacting particles. Depending on the accuracy used to describe the interactions it is possible to distinguish several techniques. In the so-called *ab-initio* methods, the Schrödinger equation is solved for the set of nuclei and electrons which constitute the system under study. They provide an accurate description of the interactions based on the electron distribution of the atoms, with no parameters [6-9]. However, these methods are computationally very expensive, and thus they can only handle systems of a few hundred atoms and are limited to extremely short times. On the other hand, classical molecular dynamics (MD) simulations describe atomic interactions by empirical force laws which include several parameters chosen by fitting to experimental data or *ab-initio* calculations [10,11]. It is possible to simulate systems containing millions of atoms for times of the order of nanoseconds, but at the expense of losing the electronic description of the system [12,13]. The *tight-binding* (TB) technique, based on the method of linear combination of atomic orbitals [14], is an alternative half-way between *ab-initio* and classical MD. It is a non-parameter-free approximation [15] that allows the use of intermediate sizes and times while keeping the electronic description of the system. In summary, these simulation methods allow a full description of the system dynamics at the atomic level. However, they are computationally expensive techniques, and size and time scales accessible to them are still orders of magnitude far from actual processing scales.

In order to bridge the gap between these detailed atomistic techniques and experiments several simulation methods can be used. These methods maintain the atomic-level description, but in order to reach experimental scales they must renounce to keep the full dynamics of the system. In the so-called *binary collision approximation* (BCA) the implantation process is simulated by calculating collisions between energetic atoms and target atoms by assuming that they interact in pairs [16]. Thousands of cascades can be easily calculated to give enough statistical resolution in the generation of dopant profiles. However, damage profiles in amorphizing conditions are not so well predicted, precisely because multiple interactions are not taken into account in the calculation. On the other hand, *kinetic Monte Carlo* (kMC) codes allow the simulation of the annealing step [1-5]. Unlike MD, the vibrational movement of the Si lattice atoms is not simulated, only the dynamics of defects and dopants is followed, which allows the simulation of system sizes typical of today's devices. Parameters that define the dynamics of defects, such as diffusivities, binding energies, capture radii, etc., must be specified. The simulation time-step is variable. It may go from $10^{-9}$ s for the diffusion of point defects to $10^{-3}$ s, or even longer, for the emission of defects from stable clusters. Generally fast events tend to disappear quickly leaving slower events that raise the time-step. This allows to easily access macroscopic times, and so to the simulation of actual processing.

No individual technique can be used alone for the simulation of full Si processing. Each one gives information at a different scale level, so all of them have to be used in a hierarchical or multi-scale scheme. Fundamental techniques such as *ab-initio* and TB can be employed to study defect configurations and energetics, material and electronic properties, and to optimize empirical force potentials along with experimental data. Classical MD in turn can be used to determine interaction and diffusion mechanisms involving defects, or to study the damage morphology obtained from individual implantation cascades as well as its annealing behavior. All these data along with experimental measurements will define the relevant events to be considered in the kMC simulator. The BCA code is used to calculate the implantation cascades. The coordinates of dopants and generated defects are fed to the kMC simulator at time intervals determined by the dose-rate. This procedure is repeated until the specified dose is reached. Afterwards, subsequent anneals can be also simulated.

In the next sections we will show how to apply this hierarchical simulation scheme to develop an atomistic model for the ion-beam induced amorphization and subsequent recrystallization of Si, processes of technological relevance.

## MODELING DAMAGE GENERATION

As it has been already mentioned, implantation is usually simulated using BCA codes, by calculating collisions between the ion and its closest target atom. A target atom is displaced from its position when it receives in a collision an energy higher than the displacement threshold, *Ed*, conventionally 15 eV for Si [17,18]. In that situation, the target atom can become a recoil leaving behind a vacancy and creating a subcascade. For energy transfers below *Ed* no recoiling atoms are formed and energy is assumed to be lost to phonons [19]. Then, BCA describes damage in terms of these created interstitial-vacancy pairs, also called *Frenkel Pairs* (FP). In turn, MD simulations naturally take into account multiple interactions among atoms, and all energy transfers can contribute to possible displacement events. Damage is described in terms of atoms displaced from perfect lattice positions and the corresponding empty lattice sites.

Figure 1 represents the percentage of cascades producing a certain number of FP for BCA and displaced atoms for MD in 1 keV implants of B, Si and Ge ions into Si. BCA produces the same average number of Frenkel pairs and distributions are very similar. However, in MD the amount of damage increases with ion mass, in agreement with experiments [20].

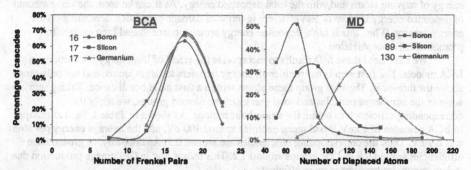

**Figure 1.** Percentage of cascades generating a certain number of FP in BCA and displaced atoms in MD for 1 keV B, Si and Ge cascades into Si. Average numbers of FP and displaced atoms created by each ion are also shown.

In a more detailed analysis of energy transfers in BCA during the cascade, we find that only 23% of deposited energy is used to generate Frenkel pairs (see Table I). The remaining percentage, 77%, is not employed in displacement processes but in energy transfers to atoms below the threshold. If we group them within a first neighbor distance, we find that mean group size and mean group energy increases with ion mass. Consequently, it seems that these low energy transfers usually ignored in BCA simulations could establish the difference in damage morphology between the three ions.

**Table I.** Average results obtained from BCA simulations of 1 keV B, Si and Ge implantations into Si: percentages of deposited energy employed in the generation of FP and in energy transfers below $Ed$, mean size and mean energy of groups of target atoms which receive energies below $Ed$.

|    | FP     | BELOW $Ed$ | MEAN SIZE | MEAN ENERGY (eV) |
|----|--------|------------|-----------|------------------|
| B  | 22.0 % | 78.0 %     | 15.2      | 92.4             |
| Si | 23.3 % | 77.7 %     | 29.6      | 180.8            |
| Ge | 24.0 % | 76.0 %     | 50.5      | 311.4            |

Our aim is to develop an improved BCA model able to provide a description of damage equivalent to the prediction given by MD calculations. For that purpose, we have carried out MD simulations to study the damage generation processes at energy transfers below the displacement threshold [21,22]. We give a certain amount of kinetic energy to a number of atoms located in a sphere in the center of the simulation cell with velocities in random directions. The initial energy of moving atoms varies between 0 and 20 eV per atom, around and below the displacement threshold for Si, and the total deposited energy between 50 and 500 eV, of the order of energies found in BCA simulations. We define the efficiency of damage production as the number of final displaced atoms per initial moving atom.

Figure 2 shows the obtained efficiency of damage production as a function of the initial energy of moving atoms for different total deposited energies. In BCA simulations, the efficiency is zero below the displacement threshold and one over it. However, in MD the efficiency can be quite high even below the threshold. This efficiency increases with initial energy of moving atoms and with the total deposited energy. As it can be seen, the same amount of deposited energy is more or less efficient in terms of damage generation depending on the energy density. When this is high, deposited energy remains concentrated long enough to promote local amorphization.

We have used these MD results to incorporate the effect of low energy interactions in BCA models. The first step is to consider all energy transfers to target atoms, and not only those above the threshold. Then we group these atoms within a first neighbor distance. Taking into account the number of particles and total energy of the formed groups, we apply the corresponding efficiency to obtain the amorphous regions. As shown in Table I, formed groups in BCA cascades of 1 keV B had mean energies around 100 eV, and the average energy per atom was 6.1 eV. Then the corresponding efficiencies are around 0.5. Analogously, Si groups have efficiencies around 0.8 and Ge groups around 1.2. This means that in Ge damage production due to low-energy interactions is more efficient.

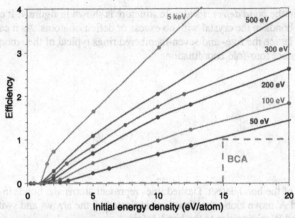

**Figure 2.** Efficiency of damage production as a function of the initial energy of moving atoms for different total deposited energies. Dashed line represents the efficiency of BCA simulations.

We have applied this model to the simulations described in figure 1 and table I. Figure 3 represents the distributions of cascades as a function of the amount of generated damage within this improved BCA model. The average number of displaced atoms for each ion is also shown. In the improved BCA, we can see that now ions produce different number of displaced atoms, with a distribution very similar to that obtained with MD and shown in figure 1 [20].

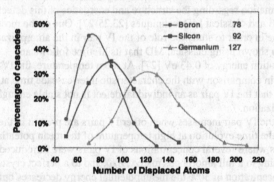

**Figure 3.** Percentage of cascades generating a certain number of displaced atoms obtained from simulations of 1 keV B, Si and Ge implants into Si using the improved BCA model. Average numbers of displaced atoms are also shown.

## MODELING THE AMORPHOUS PHASE

In order to simulate amorphization and recrystallization processes at the atomic level in Si it is essential to identify the defect that can be used to describe the amorphous phase. We have based our atomistic model in the *bond defect*. Its atomic structure is shown in figure 4. It consists of a local rearrangement of bonds in the crystal with no excess or deficit of atoms. As it can be seen, it introduces in the Si lattice the five- and seven-membered rings typical of the amorphous phase while maintaining perfect four-fold coordination.

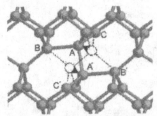

**Figure 4.** Atomic structure of the bond defect. Dashed lines represent atoms and bonds in the perfect lattice. Atoms A and A' move along the directions indicated by the arrows and switch their bonds with atoms B and B', giving rise to the bond defect.

Tang and coworkers encountered the bond defect when studying self-diffusion and interstitial-vacancy recombination in Si using TB techniques [23]. They found that when a vacancy approaches a dumbbell interstitial, a metastable defect structure is generated instead of having immediate interstitial-vacancy recombination. For this reason the bond defect is also known as *IV pair*. Using classical MD techniques, Stock and coworkers observed that the bond defect can be also generated as a result of a pure ballistic process [24]. Thus the IV pair can be a primary defect generated by irradiation, with no need of pre-existing interstitials and vacancies in the lattice for its formation. Information regarding the structure and energetics of this defect has been extracted using *ab-initio*, TB and classical MD techniques [23,25-27]. One of the most important parameters to determine in order to study the role of the IV pair in the amorphization process is its stability. It has been shown using classical MD that its lifetime follows an Arrhenius behavior with an activation energy of 0.43 eV [27]. At room temperature, the IV pair lifetime is about 3 µs, very short in comparison with the characteristic inter-cascade time at typical dose-rates. This indicates that the IV pair as an individual defect is not stable enough to accumulate and promote amorphization.

However, the stability of the IV pair increases when other IV pairs are present in the lattice [27]. In figure 5 we show the time evolution at high temperature of the mean potential energy per atom in crystal lattices where several concentrations of IV pairs were introduced. Atomic energy levels corresponding to a full amorphous matrix, $E_{AM}$, and to a perfect crystal, $E_C$, are also shown. For an initial concentration of 30% the mean potential energy decreases only to the amorphous level. In fact, the obtained structure is identical to a pure amorphous matrix. On the other hand, in lattices with starting concentrations of 10, 20 and 25% full recrystallization occurred. In the case of 10%, the evolution is exponential, with a decay time similar to the lifetime of an isolated IV pair at the same temperature. Thus 10% is a concentration so low that IV pairs do not interact strongly with each other. They recombine one by one, and the overall crystallization behavior is the same as when you have just one IV pair. On the other hand, for higher concentrations the evolution of the potential energy per atom shows plateaus followed by steps, indicated by arrows. This behavior has been also observed in experiments on the recrystallization kinetics of amorphous pockets created by ion irradiation [28]. In these cases, IV pairs interact strongly with each other and form more stable structures, responsible for the plateaus in the curves. Recrystallization then requires the collective movement of several atoms which produces a sudden decrease in the potential energy per atom. All these results indicate that amorphization can be achieved without the intervention of any additional defect, and also that amorphous pocket characterization can be studied by IV pair accumulation.

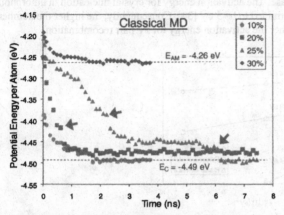

**Figure 5.** Time evolution of the potential energy per atom in lattices with 10, 20, 25 and 30% of IV pairs during annealing at high temperature. Dashed lines indicate the mean potential energy per atom in amorphous and crystal Si, $E_{AM}$ and $E_C$, respectively. Arrows indicate plateaus followed by steep decreases.

To get some insight about the influence of the damage morphology on the recrystallization behavior, we performed MD studies in crystal lattices where we introduced an IV pair concentration of 8%. In one set of simulations IV pairs were distributed randomly in the lattice, separated each other by a distance of at least 4 Å. In other set of simulations IV pairs were arranged in a sphere with a radius of 12 Å [29]. The lattice with scattered IV pairs would represent damage generated by light ion implantation, while the system with concentrated IV pairs would represent damage generated by heavy ion implantation [30]. Both types of systems were annealed at several temperatures, and in all cases the scattered damage disappeared long before the concentrated damage. This indicates that even though the amount of IV pairs was the same, the dynamics of the recrystallization process is totally different. In figure 6 we show the recrystallization velocities for both types of systems as a function of temperature. They have Arrhenius behavior. The activation energy for the scattered damage is very close to that corresponding to isolated IV pair recombination, 0.43 eV. This indicates that a 4 Å separation among IV pairs was long enough to prevent their interaction. In the case of the concentrated damage the activation energy is higher and recrystallization dynamics slower. Recrystallization starts from the amorphous/crystal (a/c) interface, as it is also observed in experiments [31,32]. This is because IV pairs in contact with crystalline atoms are less stable than IV pairs near the center of the sphere. However, their strong interaction with IV pairs in the amorphous side of the interface make them more stable than if they were isolated. This effect is increased in the case of a planar a/c interface, whose recrystallization velocity obtained by MD is also plotted in figure 6.

It shows even higher activation energy for recrystallization, 2.44 eV [29], in very good agreement with the experimental value, 2.7 eV [33]. IV pairs that lie on the planar interface are surrounded by more IV pairs than those on the amorphous sphere. In the limit, as already stated, a pure amorphous matrix would be described by IV pairs completely surrounded by other IV pairs. The recombination of such an IV pair would be equivalent to the generation of a crystal embryo in the amorphous phase. The activation energy for crystal nucleation in amorphous Si has been experimentally determined to be 5 eV [34]. Consequently, the higher the number of surrounding IV pairs, the higher the activation energy for IV pair recombination.

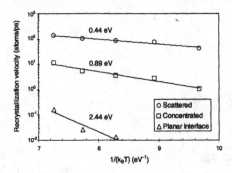

**Figure 6.** Arrhenius plot of the recrystallization velocity in lattices with scattered and concentrated damage. The recrystallization velocity of a planar a/c interface is also shown. Lines are best fits to each data set. Activation energies are also shown.

## MODELING AMORPHIZATION AND RECRYSTALLIZATION

As it has already mentioned, results obtained from these fundamental simulation methods cannot be directly compared with experimental observations. However, based on these results it is possible to develop an atomistic model simple enough to be implemented in a kMC simulator able to reach experimental scales [35]. The IV pair is used as the elementary unit to describe the amorphous material. IV pairs form when Si self-interstitials and vacancies are within the capture radius of each other, and also directly during the collision cascade. Each IV pair is locally characterized by the number of surrounding IV pairs. Its recombination rate decreases as the number of IV neighbors increases. This local description of the disordered zones allows the model to capture any damage morphology that may arise from irradiation cascades, as well as the characteristic regrowth behavior observed in the experiments. The regrowth of amorphous regions starts by IV pairs placed in interfaces with crystalline Si. Each amorphous region regrows at a rate that depends on the local coordination of defects. A continuous amorphous layer is just a particular case of an amorphous pocket. In the planar interface all IV pairs have on average the same number of neighbors. When one of these IV pairs recombines it leaves a "hole", a small crystalline zone, at the interface. The surrounding IV pairs are left with one less neighbor and therefore they recombine faster. The first IV pair recombination acts thus as a *triggering event* for the layer-by-layer recrystallization.

This simple model based on the IV pair quantitatively captures the kinetic features related to the ion-induced amorphization and recrystallization in Si, and it is able to give information of technological relevance. One of the most complex phenomena to reproduce is the critical equilibrium between amorphization and recrystallization depending on dose-rate and temperature [36]. As it can be seen in figure 7, given a dose-rate value there is a critical temperature above which it is not possible to amorphize. Dose-rate specifies the time interval between implant cascades. For low dose-rates, generated damage may anneal out long before the next cascade arrives into the same target region. For high dose-rates, generated damage may overlap with damage created by previous cascades favoring the formation of more stable structures and thus its accumulation. It is important to note that the critical regime between amorphization and recrystallization is determined by a narrow temperature window. If no special care is taken to maintain fixed the wafer temperature during implantation, small variations may lead to very different damage profiles.

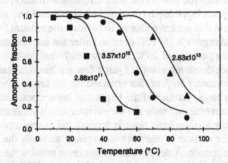

**Figure 7.** Amorphous fraction versus substrate temperature for 1 MeV Si implants to a dose of $10^{15}$ cm$^{-2}$ with several dose-rates (in cm$^{-2}$s$^{-1}$). Solid symbols correspond to the experimental data of Ref. 36, and solid lines represent our simulation results.

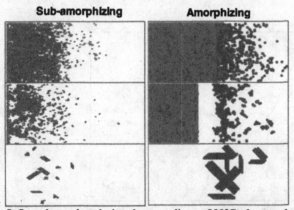

**Figure 8.** Snapshots taken during the annealing at 800°C of a sample implanted with 5 keV $10^{14}$ cm$^{-2}$ Si ions (sub-amorphizing) and 5 keV $10^{15}$ cm$^{-2}$ Si ions (amorphizing). IV pairs are plotted as green circles, whereas Si interstitials and vacancies are indicated by blue and red circles, respectively. Top images show the as-implanted damage profiles.

Whether the implantation is amorphizing or not has important consequences regarding the amount of residual damage left in the lattice. Snapshots shown in figure 8 correspond to the annealing of sub-amorphizing and amorphizing implants. In the former case, the as-implanted profile shows crystalline islands surrounded by large amorphous regions. Upon annealing, amorphous regions recrystallize, releasing to the lattice the unbalanced atoms they contain. These defects accumulate around the mean projected range of the implant, and the amount of retained Si self-interstitials is approximately the implanted dose (+1 model [37]). In the annealing of an amorphizing implant, defects contained in the continuous amorphous layer are swept towards the surface as the a/c interface advances. A band of extended defects is formed only at the end-of-range region, and the amount of stored self-interstitials is significantly lower than the implanted dose.

It is also important to accurately predict the position of the a/c interface in amorphizing conditions, since differences in the amorphous layer depth translate into significant variations in the amount of residual damage. To illustrate this we have plotted in figure 9 the amount of residual damage after solid-phase epitaxial regrowth (SPER) obtained after the implantion of 12 keV Si ions to a dose of $8 \times 10^{14}$ cm$^{-2}$ at a temperature of 20°C, using two different dose-rates. Although the differences in the a/c interface depth are small, of just a few nm, residual damage obtained for the higher dose-rate is approximately 50% of that obtained for the lower dose-rate. Since the implant dose is the same in both cases, profiles of the implanted ion are practically the same. Upon anneal, generated damage recombine and only the excess interstitials generated by the implanted ion survive. When a continuous amorphous layer is formed, the excess atoms contained in the layer are swept towards the surface as regrowth takes place, and only the excess interstitials beyond the initial a/c interface remain. A deeper amorphous layer implies the removal of a larger amount of interstitials, and hence a reduced residual damage. Wafer temperature plays a similar role on the a/c interface depth and the remaining residual damage. If temperature rises during implantation, especially due to the use of high dose-rates, dynamic annealing may lead to an amorphous layer a few nanometers thinner and consequently to a larger amount of residual damage.

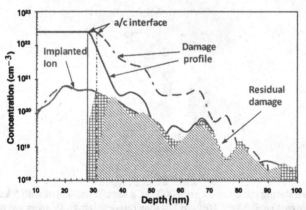

**Figure 9.** Simulated damage profile after implant and residual damage after SPER for a 12 keV $8 \times 10^{14}$ cm$^{-2}$ Si implant for two dose-rates: $2.5 \times 10^{11}$ cm$^{-2}$s$^{-1}$ (solid lines) and $8.3 \times 10^{14}$ cm$^{-2}$s$^{-1}$ (broken lines). Profiles of the implanted ion and positions of the a/c interface are also shown.

# CONCLUSIONS

In this paper we have shown how atomistic simulation techniques can be used to develop Si-processing models with predictive capabilities. These models have to be physically based and accurate enough to give not only qualitative but also quantitative information. The atomistic simulation techniques have to be organized in a hierarchical scheme in order to be able to reach the size and time scales typical of actual Si processing. To illustrate this simulation scheme, we have presented a fully atomistic model for the ion-beam induced amorphization and recrystallization in Si to be used in front-end processing simulators. It is based on the IV pair as the elementary unit to describe the amorphous phase. According to classical MD simulations, the model considers that the IV pair recrystallization rate depends on the local density of surrounding IV pairs. This simple model is able to reproduce most of the experimental observations. It consistently encompasses ion-induced damage ranging from individual defects to full amorphous layers, and thus amorphization needs not to be specified as an input condition but it is the result of the simulation itself. Consequently, it allows to extend the kMC approach to high implant doses, and therefore the atomistic simulation of the fabrication of nanometer-sized Si devices.

# ACKNOWLEDGMENTS

This work has been supported by the Spanish DGI under project TEC2005-05101.

# REFERENCES

1. M. Jaraiz, P. Castrillo, R. Pinacho, I. Martín-Bragado and J. Barbolla, *Simulation of Semiconductor Processes and Devices SISPAD 01*, ed. D. Tsoukalas and C. Tsamis (Springer-Verlag, 2001), p.10.
2. International Technology Roadmad for Semiconductors, http://public.itrs.net/.
3. M.E. Law and S.M. Cea, *Comput. Mat. Sci.* **12**, 289 (1998).
4. C. Rafferty and R.K. Smith, *CMES-Comp. Model. Eng.* **1**, 151 (2000).
5. L. Pelaz, M. Jaraiz, G.H.Gilmer, H.-J. Gossmann, C.S. Rafferty, D.J. Eaglesham and J.M. Poate, *Appl. Phys. Lett.* **70**, 2285 (1997).
6. P.E. Blochl, E. Smargiassi, R. Car, D.G. Laks, W. Andreoni, and S.T. Pantelides, *Phys. Rev. Lett.* **70**, 2435 (1993).
7. E. Smargiassi and R. Car, *Phys. Rev.* **B53**, 9760 (1996).
8. B. Shadigh, T.J. Lenosky, S.K. Theiss, M.-J. Caturla, T. Diaz de la Rubia and M.A. Foad, *Phys. Rev. Lett.* **83**, 4341 (1999).
9. W. Windl, M.M. Bunea, R. Stumpf, S.T. Dunham and M.P. Masquelier, *Phys. Rev. Lett.* **83**, 4345 (1999).
10. F.H. Stillinger and T.A. Weber, *Phys. Rev.* **B31**, 5262 (1985).
11. J. Tersoff, *Phys. Rev.* **B38**, 9902 (1988).
12. L.A. Marqués, L. Pelaz, P. Castrillo and J. Barbolla, *Phys. Rev.* **B71**, 085204 (2005).
13. G.H. Gilmer, T. Diaz de la Rubia, D.M. Stock, and M. Jaraiz, *Nucl. Instrum. Methods Phys. Res.* **B102**, 247 (1995).

14. J.C. Slater and G.F. Coster, *Phys. Rev.* **94**, 1498 (1954).
15. P. Alippi and L. Colombo, *Phys. Rev.* **B62**, 1815 (2000).
16. M.T. Robinson and I.M. Torrens, *Phys. Rev.* **B9**, 5008 (1974).
17. J.M. Hernández-Mangas, J. Arias, L. Bailón, M. Jaraíz and J. Barbolla, *J. Appl. Phys.* **91**, 658 (2002).
18. SRIM documentation, http://www.srim.org.
19. L.A. Marqués, M. J. Caturla, T.D. de la Rubia and G.H. Gilmer, *J. Appl. Phys.* **80**, 6160 (1996).
20. I. Santos, L.A. Marqués, L. Pelaz, P. López, M. Aboy and J. Barbolla, *Mater. Sci. Eng.* **B124-125**, 372 (2005).
21. I. Santos, L.A. Marqués, L. Pelaz and P. López, *Nucl. Instrum. Methods Phys. Res.* **B255**, 110 (2007).
22. I. Santos, L.A. Marqués and L. Pelaz, *Phys. Rev.* **B74**, 174115 (2006).
23. M. Tang, L. Colombo, J. Zhu and T. Diaz de la Rubia, *Phys. Rev.* **B55**, 14279 (1997).
24. D.M. Stock, B. Weber and K. Gärtner, *Phys. Rev.* **B61**, 8150 (2000).
25. F. Cargnoni, C. Gatti and L. Colombo, *Phys. Rev.* **B57**, 170 (1998).
26. S. Goedecker, T. Deutsch and L. Billard, *Phys. Rev. Lett.* **88**, 235501 (2002).
27. L.A. Marqués, L. Pelaz, J. Hernández, J. Barbolla and G.H. Gilmer, *Phys. Rev.* **B64**, 45214 (2001).
28. S.E. Donnelly, R.C. Birtcher, V.M. Vishnyakov and G. Carter, *Appl. Phys. Lett.* **82**, 1860 (2003).
29. L.A. Marqués, L. Pelaz, M. Aboy, L. Enríquez and J. Barbolla, *Phys. Rev. Lett.* **91**, 135504 (2003).
30. M.-J. Caturla, T. Diaz de la Rubia, L.A. Marqués and G.H. Gilmer, *Phys. Rev.* **B54**, 16683 (1996).
31. L. Csepregi, E.F. Kennedy, J.W. Mayer and T.W. Sigmon, *J. Appl. Phys.* **49**, 3906 (1978).
32. A. Battaglia, F. Priolo and E. Rimini, *Appl. Phys. Lett.* **56**, 2622 (1990).
33. G.L. Olson and J.A. Roth, *Mater. Sci. Reports* **3**, 1 (1988).
34. Y. Masaki, P.G. LeComber and A.G. Fitzgerald, *J. Appl. Phys.* **74**, 129 (1993).
35. L. Pelaz, L.A. Marqués, M. Aboy and J. Barbolla, *Appl. Phys. Lett.* **82**, 2038 (2003).
36. P.J. Schultz, C. Jagadish, M.C. Ridgway, R.G. Elliman and J.S. Williams, *Phys. Rev.* **B44**, 9118 (1991).
37. M. D. Giles, *J. Electrochem. Soc.* **138**, 1160 (1991).

Mater. Res. Soc. Symp. Proc. Vol. 1070 © 2008 Materials Research Society          1070-E06-03

# Ab-Initio Modeling of Arsenic Pile-Up and Deactivation at the Si/SiO2 Interface

Naveen Gupta, and Wolfgang Windl
Materials Science and Engineering, The Ohio State University, Columbus, OH, 43210-1178

## ABSTRACT

Two recent papers by Pei *et al*. and Steen *et al*. have shown that the observed pile-up of arsenic at Si/SiO2 interfaces surprisingly does not seem to involve point defects as a major factor, causes local distortions that strain the Si in the pile-up region locally, and that the segregated arsenic atoms are deep donors. In this paper, we use ab-initio modeling to study possible configurations for high As concentrations that may fulfill these criteria. We find for a simple model structure that As nearest neighbors become stable in Si in the vicinity of the interface. We also have studied dopant deactivation using bulk-Si models. Even without invoking point defects explicitly and starting from a purely substitutional arrangement, we find that the energetically most favorable configurations are most stable in the neutral charge state, indicating that high enough concentrations of arsenic atoms make them electrically inactive and hence result in dopant dose loss.

## INTRODUCTION

Arsenic is a group-V element commonly used as a donor dopant in silicon MOS (metal oxide semiconductor) devices, the currently most widely used technology for fabrication of integrated microelectronic circuits. Ion implantation is the preferred route for placing the As dopants into Si, which requires subsequent annealing to heal the implantation damage. During annealing of implanted Si, the As atoms diffuse and concentrate at the Si/SiO2 interface. Since it has been found experimentally that As diffuses in Si with the help of both interstitials and vacancies [1], it has been speculated that this trapping of As at the interface is due to a local supersaturation of point defects generated during the implantation process [2]. However, recent measurements involving anneals in oxidizing and inert ambients found negligible influence of the ambient on the segregated dose under near-equilibrium conditions, indicating that point defects may not play a dominating role in the segregation process [3]. Reference [3] also reported that the segregated arsenic atoms were deep donors with an electrical activity that increased eventually to full electrical activation for high sheet concentrations of the segregated atoms.

Initially, it was believed that As piles up in a narrow region on the oxide side of the interface [5]. However, later characterization by x-ray photoelectron spectroscopy (XPS) [6] and medium energy ion scattering (MEIS) [7] experiments showed that at least half of the pileup is in the first monolayer on the Si side of the interface, while recent atomic-resolution electron-energy loss spectroscopy (EELS) and grazing-incidence X-ray fluorescence spectroscopy measurements [3,8] indicate that most or even all of the As is located on the Si side. The strain observed in the Z-contrast images of [8] also suggests a significant concentration of local distortions within 3 nm from the interface.

The question of As deactivation was first examined with atomistic calculations for the case of bulk Si, where Ramamoorthy et al. [9] argued that in heavily doped Si, As would bind to vacancies to form $As_nV_m$ complexes ($n = 1, 2, 3; m = 1, 2, 3$). Subsequent atomistic simulation studies to explain segregation of As and P at the Si/SiO$_2$ interface by Dabrowski et al. [2] suggest dopant pairing and trapping at interface dangling bonds, enabled by a binding energy of 2 eV for vacancies at the interface. Unfortunately, later studies by Ravichandran and Windl [10] were unable to confirm the high binding energy of vacancies to the interface. Also, impurities such as

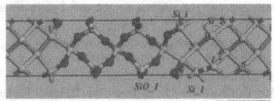

**Figure 1.** Si-SiO$_2$ interface structure from Ref. [15] with labels for the different positions of As atoms considered in our calculations. Positions 1,2,3,4,6 show positions of As atoms when they do not form a monolayer type structure. SiO_X (X = i, 1) refers to monolayers of As parallel to the interface on the oxide side, where As replaces the Si atoms in that layer. Si_X ( X = 1,2,3,i) refers to monolayers of As parallel to the interface on the Si side, where the As atoms replace the Si atoms in that layer.

H suggested in [10] to possibly enhance As pile-up could not be confirmed experimentally [3]. Despite these efforts, As pile-up in the absence of intrinsic point defects or impurities, which experiment seems to suggest [3], has not been studied in the past and is the topic of the present paper. As a starting point, we studied the energetics of As monolayer formation at various distances from the Si/SiO$_2$ interface. We also performed calculations for the case of bulk Si with high As concentrations but no intrinsic point defects with the goal of examining the energetics of submicroscopic precipitates and ordered structures in the vicinity of the interface.

## COMPUTATIONAL METHOD

The total energies of the various structures were calculated using the density functional method [11] as implemented in the Vienna Ab-initio Simulation Package (VASP) [12]. The electron-ion interactions were described by projector-augmented wave (PAW) potentials [13] for the interface studies and ultrasoft pseudopotentials [14] for the bulk studies. The generalized-gradient approximation (GGA) was used for the exchange and correlation energies. We used the Si/SiO$_2$ interface structure by Buczko et al. [15], which has been found previously to excellently combine the features of a realistic interface into a relatively small periodic structure [16]. The structure has seven layers of silicon and 8.37 Å of SiO$_2$. The unit cell dimensions are 5.43 Å × 5.43 Å × 19.55 Å after zero-pressure structural relaxation of basis and lattice. Although the SiO$_2$ strains the silicon by about 10%, we do not expect the strain to change the calculated energetics in a significant way, since we found the relaxation volume [17] of substitutional As to be zero within our numerical accuracy. We chose periodic boundary conditions to eliminate surface effects.

For our calculations we also need to consider that dopant atoms in semiconductors can assume different charge states. Since ab-initio calculations do not take into account the work done to remove electrons from or put electrons into the system, we follow the approach of Ref. [18] and express the total energy $E$ of a system with charge $Q$ as a function of the Fermi energy, $E(E_F,Q) = E_0 + E_F Q$, where $E_0$ is the energy of the charged system and $E_F$ is the sum of the valance band maximum and the (relative) Fermi energy, which can vary between 0 and the Si band gap. The latter can be either taken to have the theoretical value of 0.64 eV, or can be corrected to the experimental value as outlined in Ref. [19].

To check the stability of As in systems with different stoichiometry, we compare its chemical potential in the different configurations, which can be calculated by

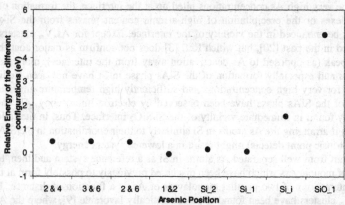

**Figure 2.** Relative energy per As atom for various configurations ("random" and monolayers) of two arsenic atoms in the simulation cell shown in Fig. 1. The labels for the different arsenic positions are shown in Fig. 1.

$$\mu_{As} = \left( E_{charged} + QE_F - N_{Si}\mu_{Si} \right) / N_{As}, \quad (1)$$

where $N_{As}$ ($N_{Si}$) is the number of As (Si) atoms, $E_{charged}$ is the energy of the cell after allowing for band gap and finite size corrections [19], $Q$ is the charge of the system, $E_F$ is the chemical potential of the electrons, and $\mu_{As}$ and $\mu_{Si}$ are the chemical potentials of As and Si, respectively. The former can be calculated if we assume that Si in our simulation cell is in equilibrium with bulk silicon and thus its chemical potential is equal to that of bulk Si.

## RESULTS

As a first step, we studied the energetics of As monolayers at various distances from the interface. In our chosen simulation cell, each Si layer contains two atoms, which means that a monolayer can be realized with two As atoms in the cell. For comparison, we also created "disordered" structures where we randomly selected the positions for the two As atoms. Figure 1 shows the various positions of As atoms considered in our calculations. Our results (Fig. 2) indicate a sharp reduction in energy of ~4 eV or more for the structures with As on the Si side of the interface. This is in qualitative agreement with experiments [3,6,7] and calculations for isolated arsenic atoms which show that As is more stable on the Si [10] side of the interface. On the Si side of the interface, we find the minimum energy with As at positions 1 and 2, which are nearest neighbors and thus in different layers. This is insofar remarkable as in bulk, As nearest-neighbors (pairs as well as extended distributions) are energetically unfavorable compared to As atoms surrounded by all Si atoms. Arsenic monolayers on the silicon side have energies slightly higher than that, e.g., 0.13 eV for the case of As 1. The other non-monolayer As configurations have higher energies (see Fig. 2).

For the very high As concentrations piled up at the interface, the formation of certain As/defect complexes or the precipitation of high-arsenic content phases from the Si-As phase diagram might be enhanced in the vicinity of the interface. Except for $As_nV_m$ clusters which have been studied in the past [20], but which Ref. [3] does not confirm as major contributors to the interface peak (as opposed to As deactivation away from the interface), more generalized ordering effects and especially formation of the SiAs phase in Si have not been studied previously. Indeed, for very high concentrations and sufficiently high temperature-annealing, larger precipitates of the SiAs phase have been observed by electron microscopy [4]. The SiAs phase was mainly found in immediate vicinity of the Si/SiO$_2$ interface. Thus, in the following we will examine if arranging the As atoms in Si similarly to their coordination in the SiAs phase (but without intrinsic point defects) might result in a lowered system energy.

We start from well separated As atoms in Si as a reference system and then, to expand upon the idea of monolayers which have been discussed previously to possibly form at the interface, study the interactions of two parallel monolayers of As as a function of distance. Furthermore, since $As_nV_m$ clusters have been found to be energetically favorable [9], where the As atoms are threefold coordinated, we also consider an arrangement of As similar to two adjacent top layers of a 2×1 reconstructed silicon surface which has also been found to be energetically favorable in the case of an As terminated surface [20]. Instead of surface termination by As, we rather consider a continuous structure with two parallel reconstructed arsenic layers. Finally, to study the energetics of the SiAs phase in Si, we construct cells with a distribution of As atoms similar to their arrangement in SiAs.

These structures were realized in the configurations shown in Fig. 3, which are, specifically, (i) As atoms away from each other ("random") , (ii) two As monolayers next to each other ("As 1"),

(iii) two As monolayers separated from each other ("As 2"), (iv) two neighboring reconstructed As monolayers with three-fold coordinated arsenic ("reconstructed"), (v) a one-dimensional (continuous only in one dimension) SiAs-type arrangement of As ("SiAs 1D"), and (vi) a layer of SiAs-type arrangement of As ("SiAs layer").

The cell size for the first four structures (i - iv) is 5.42 Å × 17.20 Å × 5.42 Å with 20 Si and

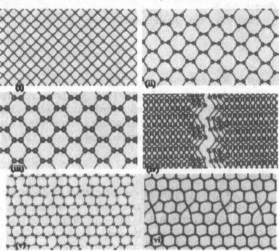

**Figure 3.** Various configurations for arsenic in perfect silicon used in our study. (i) arsenic atoms away from each other ("random"); (ii) two arsenic monolayers next to each other ( As_1); (iii) two arsenic monolayers separated from each other (As_2); (iv) arsenic reconstruction leading to the formation of interlayer of three fold coordinated arsenic ("reconstructed"), (v) one-dimensional (because it is continuity only in one dimension) SiAs type arrangement of arsenic (SiAs_1D); and (vi) layer of SiAs type arrangement of arsenic (SiAs_layer).

4 As atoms. Structures 'v' and 'vi' have dimensions of 18.81 Å × 19.94 Å × 3.85 Å. Structure 'v' has 64 Si and 6 As atoms, while structure 'vi' has 58 Si and 12 As atoms.

To compare the stability of As in the different structures, we calculate the As chemical potential in the different configurations using Eq. (1). Since As is a donor dopant, we examine the positive charge states besides the neutral one, specifically, charges of +1, +2, +3 and +$n$, where $n$ is the number of As atoms in the simulation cell. In Fig. 4, we compare the chemical potential for As in the different charge states. For all Fermi energies, either the +$n$ or the neutral charge states result in the lowest As chemical potential, whence we will exclude the

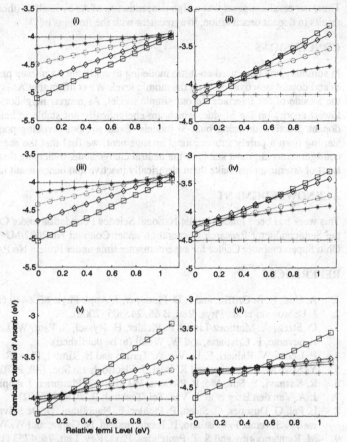

**Figure 4.** Chemical potential of arsenic in structures (i – vi) with different charge in the simulation cell (neutral (magenta crosses), +1 (red stars), +2 (green circles), +3 (black diamonds), and +$n$ (number of As atoms in the system; blue squares)) .

other charge states from the following discussion. For structures (i) and (iii), where all As atoms are fourfold coordinated by Si atoms, we find the As atoms to be fully ionized throughout the whole Fermi level range. For all remaining structures – including the two with the lowest energy – we find the neutral charge state to have the lowest energy in the $n$-type and high-temperature regime with ionization levels below midgap. This means that the As atoms are not ionized, thus electrically inactive, and contribute to the dose loss, as postulated in [3]. Therefore, even for these structures which do not explicitly include clustering with point defects, we surprisingly find that agglomeration of As atoms by itself results in dose loss. This includes the structure with two neighboring monolayers where As atoms are nearest neighbors to each other (structure (ii)). As we have shown above, As nearest neighbors are stabilized in the vicinity of the interface.

Thus, our results suggest that a high concentration of As by itself (without intrinsic point defects) results in dopant deactivation, in agreement with the findings of [3].

## CONCLUSIONS

In summary, we have used ab-initio modeling to study the As pile-up process at $Si/SiO_2$ interfaces and dopant deactivation from the atomic level. We confirm that As is most likely to pile up on the Si side of the interface. In our simple model, As nearest neighbors were found to have the lowest energy on the Si side, which are energetically not stable in bulk. We also have studied dopant deactivation using bulk-Si models. Even without invoking point defects explicitly and starting from a purely substitutional arrangement, we find that the energetically most favorable configurations are most stable in the neutral charge state, indicating that high enough concentrations of arsenic atoms make them electrically inactive and hence result in dopant dose loss.

## ACKNOWLEDGMENT

This work has been funded by the National Science Foundation under Contract No. 0244724 and the Semiconductor Research Corporation under Contract No.2002-MJ-1018. We also thank the Ohio Supercomputer Center for supercomputer time under Project No.PAS0072.

## REFERENCES

1. A. Ural, P. B. Griffin, and J. D. Plummer, J. Appl. Phys. **85**, 6440 (1999).
2. J. Dabrowski et al., Phys. Rev. B **65**, 245305 (2002).
3. C. Steen, A. Martinez-Limia, P. Pichler, H. Ryssel, S. Paul, W. Lerch, L. Pei, G. Duscher, F. Severac, F. Cristiano, and W. Windl (to be published).
4. F. Iacona, V. Raineri, F. La Via, A. Terrasi and E. Rimini, Phys. Rev. B **58**, 10990 ( 1998).
5. Y. Sato, M. Watanabe, and K. Imai, J. Electrochem. Soc. **140**, 2670 (1993).
6. R. Kasnavi, Y. Sun, Mo, P. Pianetta, P. Griffin, J. Plummer, J. Appl. Phys. **87**, 2255 (2000).
7. J. A. Van den Berg et al. J. Vac. Sci. Technol. B **20**, 974 (2002).
8. L. Pei, G. Duscher, C. Steen, P. Pichler, E. Napolitani, D. De Salvador, A. M. Piro, A. Terrasi, F. Severac, F. Cristiano, L. Ravichandran, N. Gupta, and W. Windl (to be published).
9. M. Ramamoorthy and S. T. Pantelides, Phys. Rev. Lett. **76**, 4753 (1996).
10. K. Ravichandran And W. Windl, Appl. Phys. Lett. **86**, 152106 (2005).
11. W. Kohn and L. J. Kelly, Phys. Rev. **140**, A1133 (1965).
12. G. Kresse and J. Furthmüller, VASP the Guide (Vienna University of Technology, Vienna, 1999), [http://cms.mpi.univie.ac.at/vasp/vasp/vasp.html].
13. P. Blöchl, Phys. Rev. B **50**, 17953 (1994); G. Kresse and J. Joubert, *ibid.* **59**, 1758 (1999).
14. D. Vanderbilt, Phys. Rev. B. **41**, 8412 (1990).
15. R. Buczko, S. J. Pennycook, S. T. Pantelides, Phys. Rev. Lett. **84**, 943 (2000).
16. W. Windl , T. Liang, S. Lopatin, and G. Duscher, Mat. Sci. & Eng. B **114**, 156 (2004).
17. M. S. Daw, W. Windl, N. N. Carlson, M. Laudon, and M. P. Masquelier,Phys. Rev. B **64**, 045205 (2001).
18. W. Windl, M. M. Bunea, R. Stumpf, S. T. Dunham, and M. P. Masquelier, Phys. Rev. Lett. **83**, 4345 (1999).
19. W. Windl, Physica Status Solidi B **241**, 2313 (2004).
20. M. Ramamoorthy, E. L. Briggs, and J. Bernhole, Phys. Rev. B **60**, 8178 (1999).

Mater. Res. Soc. Symp. Proc. Vol. 1070 © 2008 Materials Research Society　　　　1070-E06-04

## Atomistic Modeling of {311} Defects and Dislocation Ribbons

Bart Trzynadlowski, Scott Dunham, and Chihak Ahn
Electrical Engineering, University of Washington, Seattle, WA, 98195

## ABSTRACT

*Ab-initio* calculations of dislocation and {311} defect structures in silicon were performed in order to investigate the formation energies as functions of geometry, including the effects of applied strain, with a simple model. Predictions were made concerning the size at which a {311} defect becomes less favorable than a dislocation ribbon, and it was shown that this is affected by applied strain.

## INTRODUCTION

Extended defects in silicon have adverse effects on device performance and it is therefore desirable to better understand their formation and evolution, particularly as a function of stress and strain, which arises both intentionally and unintentionally in modern nano-scale CMOS processes. *Ab-initio* calculations can be used to investigate the properties of extended defect, and this study aims to offer a simple model for the conditions under which dislocation loops become energetically more favorable than {311} defects.

## CALCULATIONS

*Ab-initio* calculations of edge dislocation structures were performed using VASP, a density functional theory (DFT) code, with generalized gradient approximation (GGA) pseudopotentials [1,2]. Both full and partial stacking faults, present in the (111) plane, were simulated in a local coordinate system where the X axis is along the [110] direction, Y is [112], and Z is [111]. In this rotated coordinate system, the minimum repeatable (primitive) cell consists of 12 atoms and has dimensions $a_0/\sqrt{2} \times a_0\sqrt{6}/2 \times a_0\sqrt{3}$, where $a_0$ is the silicon lattice constant. A 180-lattice site volume ("Si180") was used to create the full stacking fault (20 interstitials, Si200) and two dislocation "ribbon" systems (Si188, Si192, with 8 and 12 interstitials, respectively.) In this case, the dislocation line extends infinitely along [110] due to periodic boundary conditions. A 252-lattice site volume (Si252) was used for the Si260 and Si264 (also 8 and 12 interstitials) systems, with the dislocation line extending along [112]. Figure 1 shows Si264 as well as Si134, a {311} defect system, described below.

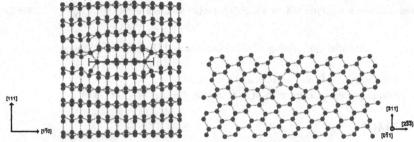

**Figure 1.** Si264 dislocation ribbon structure (left) with the edge dislocations marked, and a single chain of {311} defects, with interstitials marked green.

{311} defect structures were generated in a local coordinate system with X along [011], Y along [233], and Z along [311] [3]. The primitive cell in this orientation contains 44 atoms and has dimensions $a_0/\sqrt{2} \times a_0\sqrt{11}/\sqrt{2} \times a_0\sqrt{11}$. The chains are elongated in the X direction and multiple chains are positioned side-by-side along Y. Single and two-chain systems were simulated in a volume three cells wide (132 lattice sites). Due to periodic boundary conditions, only two interstitials are required to form an infinite chain. Energy as a function of chain length was investigated by creating systems where the periodicity results in an infinite number of chains, but of finite length. Finally, a planar configuration was simulated, where the periodicity results in infinite length as well as number of chains.

## RESULTS

The unstressed formation energy of a defect is defined by Eq. 1, where $E_{N+NI}^D$ is the energy of a system with N lattice sites and $N_I$ defect atoms, and $E_N^{Si}$ is the minimum energy of the equivalent non-defective system. Formation energy is modeled in this work as a function of geometry as in Eq. 1, where $\Delta E_f^A$ is energy per area, $\Delta E_f^i$ is energy per length of a given side, i, $L^i$ is the corresponding side length, and A is area.

$$E_f^D = E_{N+NI}^D - \frac{N_I + N}{N} E_N^{Si} = A \cdot \Delta E_f^A + \sum_i \left( L^i \cdot \Delta E_f^i \right) \tag{1}$$

Eq. 2 expresses the energy of a system as a function of an applied strain state, $\varepsilon$, where $\varepsilon_\alpha$ is the applied strain in direction $\alpha$, and $C^{Si}$ is the elasticity tensor for silicon [4]. The effect of the defect, which will cause an induced strain and a modification of the elasticity tensor, are expressed by $\Delta\varepsilon$ and $\Delta C$. $N_I$ is the number of interstitial atoms in the system that form the defect, and N is the number of lattice sites in the non-defective system. The normalization factor, x, is simply $N_I/N$; $x\Delta\varepsilon$ is the total induced strain, and depends on the volume considered. The volume of the system is $\Omega$ and $E_0$ is the fully-relaxed system energy.

$$E(\varepsilon) = E_0 + \frac{\Omega}{2} \sum_{\alpha,\beta} \left( \varepsilon_\alpha - x\Delta\varepsilon_\alpha \right) \cdot \left( C_{\alpha\beta}^{Si} + x\Delta C_{\alpha\beta} \right) \cdot \left( \varepsilon_\beta - x\Delta\varepsilon_\beta \right) \tag{2}$$

Calculation of the induced strain is accomplished by performing a full relaxation of the simulation volume, allowing expansion in all directions, to find the minimum energy size/shape. An alternative method is to apply strain to the volume (hydrostatic, uniaxial, and/or biaxial), and compute the induced strains [4].

## Dislocation Results

A simple model for rectangular dislocation ribbon formation energies would be to add energy per fault area to energy per core length (Eq. 3.) The result for un-terminated ribbons, as in the simulations performed, neglects one of the edge terms, depending on the orientation.

$$E_f^D = A^{DL} \cdot \Delta E_f^{SF} + 2 \cdot L^{110} \cdot \Delta E_f^{110} + 2 \cdot L^{211} \cdot \Delta E_f^{211} \qquad (3)$$

$A^{DL}$, $L^{110}$, and $L^{211}$ are the area of the dislocation (per $a_0^2$), length of the <110> edge (per $a_0$), and length of the <211> edge, respectively. $\Delta E_f^{SF}$ is a normalized energy per area due to the stacking fault (planar) contribution; we find this value to be very small as expected. $\Delta E_f^{110}$ and $\Delta E_f^{211}$ are energies per edge (core) length. Results are listed in Table I.

**Table I.** Calculated dislocation stacking fault (planar) and core (edge) energies.

| | |
|---|---|
| $\Delta E_f^{SF}$ | 0.01525 eV/interstitial, 0.07044 eV/$a_0^2$ |
| $\Delta E_f^{110}$ | 8.627 eV/$a_0$ |
| $\Delta E_f^{211}$ | 7.283 eV/$a_0$ |

Table II shows the induced strains of the systems studied, rotated back into the standard cubic lattice coordinate system (e.g., subscript 11 represents normal strain along the [100] direction).

**Table II.** Summary of induced strain data for the dislocation ribbon and stacking fault (Si200) systems.

| | Si260 | Si264 | Si188 | Si192 | Si200 |
|---|---|---|---|---|---|
| $\Delta\varepsilon_{11}$ | 0.31717 | 0.29152 | 0.22619 | 0.30554 | 0.30892 |
| $\Delta\varepsilon_{22}$ | 0.31717 | 0.29152 | 0.22619 | 0.30554 | 0.30892 |
| $\Delta\varepsilon_{33}$ | 0.30996 | 0.32155 | 0.32265 | 0.33408 | 0.31915 |
| $\Delta\varepsilon_{23}$ | 0.37539 | 0.36395 | 0.34092 | 0.38595 | 0.34034 |
| $\Delta\varepsilon_{31}$ | 0.37539 | 0.36395 | 0.34092 | 0.38595 | 0.34034 |
| $\Delta\varepsilon_{12}$ | 0.35217 | 0.36716 | 0.34828 | 0.36806 | 0.34452 |

The induced strains can be modeled as a function of the dislocation size as

$$x\Delta\varepsilon(N_1) = N_1 \cdot \left(\frac{\Delta\varepsilon^{SF}}{N}\right) + 2 \cdot L^{DC} \cdot \left(\frac{\Delta\varepsilon^{DC}}{N}\right), \qquad (4)$$

where the total induced strain is given as a function of the defect size. $\Delta\varepsilon^{SF}$ is the normalized induced strain of the stacking fault component of the dislocation. $\Delta\varepsilon^{DC}$ represents the normalized induced strain due to the dislocation core; $L^{DC}$ being the corresponding length of the dislocation core. In our case, for all systems, the core length corresponds to the simulation cell size along that direction. Table III summarizes the extracted parameters for Si252 and Si180-based systems.

**Table III:** Extracted stacking fault and dislocation core contributions of the induced strain of dislocation ribbons. Here, the Z direction is [111].

| Systems | $\Delta\varepsilon_Z^{SF}$ | $\Delta\varepsilon_Z^{DC}$ |
|---------|------------|------------|
| Si180 | 0.995 | $0.193/a_0$ |
| Si252 | 0.996 | $0.175/a_0$ |

$\Delta\varepsilon^{SF}$ obtained from the Si200 system is almost identical to the value extracted from the Si252 systems, with both values very close to one. The induced strains due to dislocation cores are an order of magnitude smaller. Thus, it is possible to estimate the induced strain due to dislocation loops quite accurately by a dilation in the direction normal to the (111) fault plane with magnitude equal to the volume of the excess silicon atoms involved. It was found that stacking faults have little effect on the elasticity tensor, and thus the silicon elasticity tensor can be used to calculate resulting stresses.

## {311} Defect Results

The formation energy of {311} defects can be modeled by Eq. 5 [5], where x and y are the (011) and (233) edge lengths. Simulations of one and two-chain systems (infinitely long), finite-length chains (2, 4, and 6 atoms long), and a planar configuration were performed to extract the energy components in Table IV.

$$E_f^{311}(x, y) = x \cdot y \cdot \Delta E_f^{PL} + 2x \cdot \left( \Delta E_f^{011} + \frac{\Delta E_f^B}{y} \right) + 2y \cdot \Delta E_f^{233} \tag{5}$$

**Table IV.** Planar and edge energies for {311} defects.

| | |
|---|---|
| $\Delta E_f^{PL}$ | 1.00 eV/interstitial |
| $\Delta E_f^{011}$ | $0.37$ eV/$a_0$ |
| $\Delta E_f^{233}$ | $1.00$ eV/$a_0$ |
| $\Delta E_f^B$ | 0.33 eV |

The induced strains for {311} defect systems with varying number of chains are shown rotated back into the conventional coordinate system in Table V.

**Table V.** Induced strains for {311} defects with varying numbers of chains.

| | Si134, one chain | Si136, two chains | Si92, 2 chains (plane) |
|---|---|---|---|
| $\Delta\varepsilon_{11}$ | 1.005273 | 0.804091 | 0.570182 |
| $\Delta\varepsilon_{22}$ | 0.173864 | 0.228955 | 0.303909 |
| $\Delta\varepsilon_{33}$ | 0.173864 | 0.228955 | 0.303909 |
| $\Delta\varepsilon_{23}$ | -0.060136 | 0.074955 | 0.084909 |
| $\Delta\varepsilon_{31}$ | -0.382091 | -0.214364 | -0.077727 |
| $\Delta\varepsilon_{12}$ | -0.382091 | -0.214364 | -0.077727 |

# DISCUSSION

Boninelli *et al.* [6] report observing {311} clusters transform into intermediate rod-like defects lying in {111} planes before developing into dislocation loops. The experimental observations are consistent with these rod-like defects being dislocation ribbons. Comparing the formation energies of {311} defects and ribbons offers insight into how these structures evolve. It is clear that a stacking fault is energetically favorable over a large {311} cluster because its energy is much lower per area. However, due to the high perimeter energies of a dislocation, {311} defects are favored for small sizes. As the dislocation ribbon increases, the ratio between the stacking fault area and the core (perimeter) becomes such that the defect as a whole has a smaller formation energy than an equivalent size {311} cluster. We term the size at which a {311} defect is no longer energetically favorable relative to a dislocation loop or ribbon the "crossover point."

A {311} defect can be considered to be a rectangular structure and optimized for a given area, allowing the aspect ratio to be determined. Eq. 6 is the result, and Eq. 7 relates the area to the number of atoms.

$$E_f^{311}(A^{311}) = \Delta E_f^{PL} \cdot A^{311} + 2 \cdot \Delta E_f^{011} \cdot \sqrt{\frac{\Delta E_f^{233}}{\Delta E_f^{011}} A^{311}} + 2 \cdot \Delta E_f^{233} \cdot \sqrt{\frac{\Delta E_f^{011}}{\Delta E_f^{233}} A^{311}} \quad (6)$$

$$A^{311} = \frac{\sqrt{11}}{8} a_0^2 \cdot N \quad (7)$$

The dislocation ribbon is assumed to have the same length along (110) as a {311} defect with the same number of atoms, but with the (233) width modified appropriately to reflect the difference in density of atoms. Eq. 8 is the formation energy, and Eq. 9 is the relation of area to atoms.

$$E_f^{DR} = \Delta E_f^{SF} \cdot A^{DR} + 2 \cdot \Delta E_f^{110} \cdot \sqrt{\frac{\Delta E_f^{233}}{\Delta E_f^{011}} A^{311}} + 2 \cdot \Delta E_f^{211} \cdot \frac{A^{DR}}{\sqrt{\frac{\Delta E_f^{233}}{\Delta E_f^{011}} A^{311}}} \quad (8)$$

$$A^{DR} = \frac{\sqrt{3}}{8} a_0^2 \cdot N \quad (9)$$

A similar analysis can be performed for dislocation loops. A dodecagon of alternating (110) and (211) edges can be formed. Our results show that the edge energies are fairly similar, and optimizing for aspect ratio results in a hexagon of (211) sides, which reasonably approximates a circular loop structure given our data.

Setting the energies of the dislocation ribbon/loop and {311} defect equal to each other and solving for N yields the crossover points. To include the effects of applied strain on the per-interstitial formation energy, Hooke's Law can be used,

$$\Delta E_f(\varepsilon) = -\frac{1}{2} \sum_{i=1}^{3} \sum_{j=1}^{3} \sum_{k=1}^{3} \sum_{l=1}^{3} \Omega_0 \Delta \varepsilon_{ij} C_{ijkl}^{Si} \varepsilon_{kl} , \quad (10)$$

where $\Omega_0$ is the atomic volume of silicon. This allows the crossover point to be computed for arbitrary strain conditions and, with the induced strain tensors appropriately rotated, any defect orientations. Only the planar components of the strain for both types of defects were considered, for simplicity, because they dominate over the edge terms for large defect sizes.

Fig. 2 is a plot of the defect energies as a function of their size in atoms both without applied strain and with 0.01 uniaxial strain along [100], for the most favorable defect orientations.

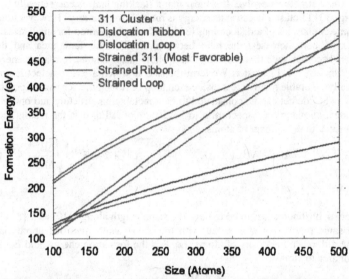

**Figure 2.** Formation energy of extended defects as a function of their size with and without applied strain (uniaxial, 0.01 tensile, along [100].)

## CONCLUSION

This study has demonstrated DFT calculations leading to a simple model for dislocation and {311} defect formation energies as a function of geometry and applied strain. As expected, {311} structures are energetically favorable up to a specific crossover point, and this point is influenced by applied strain.

## REFERENCES

1. G. Kresse and J. Furthmüller, *Phys. Rev. B* **54**, 11169 (1996).
2. D. Vanderbilt, *Phys. Rev. B* **41**, 7892 (1990).
3. J. Kim, J.W. Wilkins, F.S. Khan, and A. Canning, *Phys. Rev. B* **55**, 16186 (1997).
4. M. Diebel, PhD. Thesis, University of Washington, 2004.
5. P. Fastenko, PhD. Thesis, University of Washington, 2002.
6. S. Boninelli, N. Cherkashin, A. Claverie, and F. Cristiano, *Appl. Phys. Let.* **89**, 161904 (2006).

Mater. Res. Soc. Symp. Proc. Vol. 1070 © 2008 Materials Research Society    1070-E06-05

# A Comparison of Intrinsic Point Defect Properties in Si and Ge

Jan Vanhellemont[1], Piotr Spiewak[2,3], Koji Sueoka[4], Eddy Simoen[5], and Igor Romandic[6]

[1]Department of Solid State Sciences, Ghent University, Krijgslaan 281 S1, Ghent, Belgium
[2]Materials Design Division, Faculty of Materials Science and Engineering, Warsaw University of Technology, Woloska 141, Warsaw, Poland
[3]Umicore, Ludwiki 4, Warsaw, Poland
[4]Department of System Engineering, Okayama Prefectural University, 111 Kuboki, Soja, Okayama, Japan
[5]IMEC, Kapeldreef 75, Leuven, Belgium
[6]Umicore EOM, Watertorenstraat 33, Olen, Belgium

## ABSTRACT

Intrinsic point defects determine to a large extent the semiconductor crystal quality both mechanically and electrically not only during crystal growth or when tuning polished wafer properties by thermal treatments, but also and not the least during device processing. Point defects play e.g. a crucial role in dopant diffusion and activation, in gettering processes and in extended lattice defect formation.

Available experimental data and results of numerical calculation of the formation energy and diffusivity of the intrinsic point defects in Si and Ge are compared and discussed. Intrinsic point defect clustering is illustrated by defect formation during Czochralski crystal growth.

## INTRODUCTION

Intrinsic point defects determine to a large extent the semiconductor crystal quality both mechanically and electrically not only during crystal growth or when tuning polished wafer properties by thermal treatments, but also and not the least during device processing. They play a crucial role in dopant diffusion and activation, in gettering processes and in extended lattice defect formation. An in-depth knowledge of the point defects properties is therefore essential for successful defect engineering and process and yield control.

Thanks to 50 years of intensive research, a huge database and know how is available for Si. For Ge the situation is less favourable as few (and mostly very old) experimental data exist for the vacancy [1 and references therein] and data are even lacking for the self-interstitial. Ge receives a lot of attention at the moment for its application as active layer in advanced devices in view of the much higher carrier mobility compared to Si and of its compatibility with Si processing. As it is often used as a thin epitaxial layer on the Si substrate, it is useful to compare (point) defect properties in both materials [2]. In the present paper, the available experimental data are compared with results of numerical calculation of the formation energy and diffusivity of the intrinsic point defects in Si and Ge [3,4]. Intrinsic point defect recombination and clustering are also discussed.

# EXPERIMENTAL AND THEORETICAL ASSESSMENT OF INTRINSIC POINT DEFECT SOLUBILITY AND DIFFUSIVITY

## Experimental results for silicon

In Si, the most reliable quantitative data on the solubility and diffusivity of intrinsic point defects were obtained on the basis of metal diffusion experiments on one hand and of intrinsic point defect cluster formation during Czochralski (Cz) crystal growth on the other hand. Originally the data obtained from both approaches differed considerably in part due to the very different temperature window which was used. Recently Voronkov et al. [5] made an attempt to unify the results and finally obtained for the vacancy $V$ and interstitial $I$ diffusivity $D_{V,I}$ and solubility $C^{eq}_{V,I}$, respectively

$$D_V = \left(0.002 cm^2 s^{-1}\right) e^{-\frac{0.38 eV}{kT}}, C^{eq}_V = \left(2.4 \times 10^{26} cm^{-3}\right) e^{-\frac{3.95 eV}{kT}}$$

$$D_I = \left(0.0095 cm^2 s^{-1}\right) e^{-\frac{0.3 eV}{kT}}, C^{eq}_I = \left(7 \times 10^{28} cm^{-3}\right) e^{-\frac{4.85 eV}{kT}}$$

(1)

## Experimental results for germanium

For Ge, quenching from high temperatures followed by annealing at lower temperatures was used to study the vacancy solubility and diffusivity. In addition, the study of the diffusion and/or precipitation of fast diffusing dopants such as Cu and Zn can provide indirect information on intrinsic point defects. Finally, also ab initio calculations can shed more light on the formation and migration energy and the electrical activity of the intrinsic point defects.

Quenching of n-type Ge from temperatures above 800°C can result in p-type material, pointing to the creation of acceptors. Annealing at 500°C recovers the original n-type, indicating out-diffusion/annihilation of the quenched-in thermal acceptors assumed to be related to vacancies. It is clear that quenching experiments can only yield indirect (and incomplete) information on single vacancies as diffusion is so fast that vacancy clustering during quenching is difficult to avoid.

Self-diffusion in Si and Ge is the slowest diffusion process and provides direct evidence of the existence of an equilibrium concentration of intrinsic point defects. Giese et al. [6] showed that the self-diffusion coefficient in Ge is dominated by vacancies which is different from the behavior in Si where there is an important interstitial contribution to the self-diffusion.

Figure 1(left) illustrates that the self-diffusivity in Si and Ge is very similar when plotted as a function of the temperature normalized to the melt temperature $T_m$. Figure 1(right) reveals however the wide spread of reported vacancy solubilities which is partly due to the difficult separation of the diffusivity and solubility as mostly the product $C_{I,V}D_{I,V}$ is assessed by the experiments.

**Thermal equilibrium concentration of vacancies in Ge** Most of the available information on vacancy solubility and diffusivity is based on the quenching experiments performed during the 1950's and 1960's [1]. From electron irradiated and low temperature annealed oxygen-rich Ge samples, a migration barrier for the single vacancy of about 0.2eV has been estimated by Whan already in 1965 [7]. More than a decade later, the diffusivities of vacancies and self-interstitials have been determined in the temperature range between 420 and 620K from photo-stimulated electron emission experiments by Ershov *et al.* [8], yielding a vacancy migration energy of 0.52 ± 0.05eV and 0.42 ± 0.04eV for the uncharged and single negatively charged state, respectively. For the neutral self-interstitial, a migration energy of 0.16 ± 0.03eV was derived. Hiraki [9] obtained a vacancy migration energy of 1.2eV based on precipitation kinetics of Cu in vacancy-rich intrinsic Ge samples obtained by quenching. Shaw [10] derived migration energies of 0.7 and 1.0eV for the uncharged and single negatively charged vacancy, respectively.

In contrast to Si, self-interstitials do not play a major role in diffusion processes in Ge, except possibly after ion-implantation. There are also no indications for extended interstitial-type defects in Ge after Cz crystal growth as is the case for slow pulled Si crystals. Similar as for Si, so called "crystal originated pits" or COP's can be observed on polished Ge wafers, related to void formation by vacancy clustering during Cz growth.

In the quenching experiments reported in literature, the measured vacancy concentration is mainly due to negatively charged vacancies for which an estimate of the equilibrium concentration $C_V^{eq}$, is obtained as shown in Figure 1(right). It should be noted that the measured equilibrium concentrations for charged vacancies are valid for close to intrinsic material (Fermi level at half bandgap). High n- or p-type dopant concentrations will most likely change the respective concentrations of the various charge states of the vacancy.

## Simulation of intrinsic point defect formation energy

Density functional theory (DFT) was used to calculate the formation energy of intrinsic defects in Si and Ge as a function of the Fermi level [3,4]. Figure 2 shows the calculated formation energy of the vacancy $V$ and the self-interstitial $I$ in Si and Ge, respectively, for different charge states as a function of the Fermi level position.

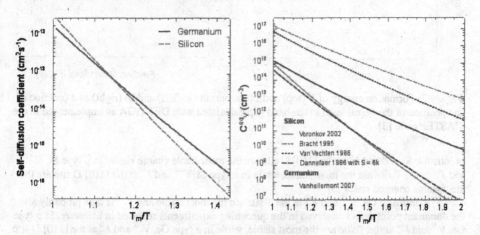

**Figure 1:** Self-diffusion coefficient (left) and vacancy solubility (right) in Si and Ge as a function of the temperature normalized with respect to the melt temperature [1, 2 and references therein]. No data are available for the self-interstitial in Ge.

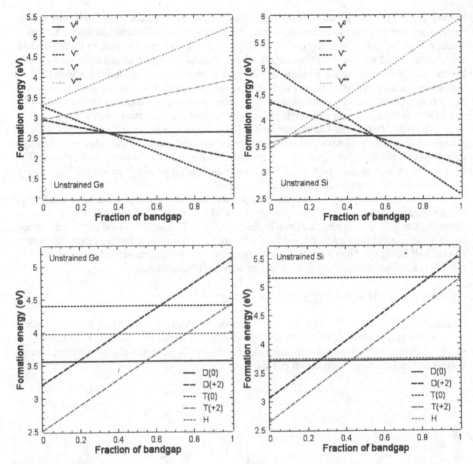

**Figure 2:** Formation energy of $V$ (top) and I (bottom) in Ge (left) and Si (right) as a function of the position of the Fermi level in the band gap calculated with DFT-GGA as implemented in the CASTEP code [3].

In intrinsic Si, $V^0$ and $I^0$ at the [110] D site are the most stable charge states. In p type Si, $V^{+2}$ and $I^{+2}$ at the T site are the most stable, while in n type Si, $V^{-2}$ and $I^0$ at the [110] D site are the most stable charged states [3].

In intrinsic Ge, $V^{-2}$ and $I^{+2}$ at the T site are the most stable and $V^{-2}$ is thus probably also the dominant point defect observed in the quenching experiments reported in literature. In p type Ge, $V^0$ and $I^{+2}$ at the T site are the most stable, while in n type Ge, $V^{-2}$ and $I^0$ at the [110] D site are the most stable charged states [3].

## INTRINSIC POINT DEFECT RECOMBINATION AND CLUSTERING

In processes such as crystal pulling, plastic deformation and ion implantation both types of intrinsic point defects are generated and/or coexist. An important parameter controlling the concentration of both point defects is the recombination factor $k_{IV}$ which is given by the well-known expression

$$k_{IV} = 4\pi a_c (D_V + D_I) e^{-\frac{(\Delta G)_{rec}}{kT}},$$ (2)

with $a_c$ the capture radius of the order of 1 nm and $(\Delta G)_{rec}$ the free energy barrier against recombination.

Assuming an Arrhenius-type behavior, (2) can also be written as [11]

$$k_{IV} = k_m e^{\frac{E_{rec}}{kT_m}\left(\frac{T_m}{T}-1\right)}.$$ (3)

$k_m$ is a constant and $E_{rec}$ the activation energy for recombination. Applying (2) and (3) for data obtained from grown-in defect distributions in floating zone and Czochralski grown Si crystals [12,13] yields the $k_{IV}$ dependence on the temperature normalized to the melt temperature as shown in Fig. 3. It is clear that a large uncertainty still exists with respect to the value of $k_{IV}$. As for Ge, data for the self-interstitial are lacking, it is even not possible to estimate the recombination factor.

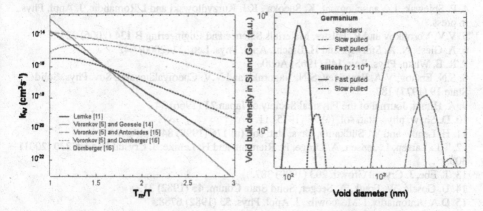

**Figure 3:** $k_{IV}$ as a function of the temperature normalized with respect to the melt temperature. The parameters used are Lemke: (3) with $k_m = 20/C^{eq}_V$ and $E_{rec} = 1.25$ eV; Voronkov: (1) and (2) with $(\Delta G)_{rec}$ from [14], [15] and [16]; Dornberger: [16].

**Figure 4:** Simulated void size/density distributions in Si [17] and Ge [18] illustrating the impact of the crystal pulling speed. The simulations also show the larger void size in Ge which is associated with a four orders of magnitude lower volume density.

Cz Si and Ge crystals can contain large vacancy clusters that lead to the formation of COPs ("crystal-originated particles") observed on polished wafer surfaces. During the solidification process, thermal equilibrium concentrations of both intrinsic point defects are introduced at the melt/solid interface. They are transported axially by thermal diffusion due to the thermal gradient $G$ and by the crystal itself that is moving away from the melt with the pulling speed $v$. These two transport mechanisms and the recombination of the intrinsic defects determine the dominant intrinsic point defect in the cooling crystal and thus, also the type of extended defects formed by point defect clustering. Si crystals pulled with a $v$ over $G$ ratio larger than a critical value are vacancy-rich, while below the critical value the crystal is interstitial-rich. Fig. 4 shows simulation results of voids formed in Si and Ge crystals [17,18]. A transition from vacancy- to interstitial-rich material has not yet been reported for Ge.

## ACKNOWLEDGMENTS

Part of this research was supported by the Polish Ministry of Science and Higher Education under contract no. N507 011 31/0315 and by the European Space Agency through ESTEC contract no. 19633/06/NL/GLC.

## REFERENCES

1. J. Vanhellemont, P. Śpiewak and K. Sueoka, J. Appl. Phys. **101** (2007) 036103.
2. J. Vanhellemont and E. Simoen, J. Electrochem. Soc. **154** (2007) H572.
3. K. Sueoka, P. Śpiewak and J. Vanhellemont, submitted for publication in J. Electrochem. Soc.
4. P. Śpiewak, J. Vanhellemont, K. Sueoka , K.J. Kurzydłowski and I. Romandic, J. Appl. Phys., in press.
5. V.V. Voronkov and R. Falster, Materials Science and Engineering B **134** (2006) 227.
6. A. Giese, N. A. Stolwijk and H. Bracht, Appl. Phys. Lett. **77** (2000) 642.
7. R. E. Whan, Phys. Rev. **140** (1965) A690.
8. S.N. Ershov, V.A. Panteleev, S.N. Nagornkh and V.V. Chernyakhovskii, Sov. Phys. Solid State **19** (1977) 187.
9. A. Hiraki, Journal of the Physical Society of Japan **21** (1966) 34.
10. D. Shaw, phys. stat. sol. (a) **72** (1975) 11.
11. H. Lemke and W. Südkamp, phys. stat. sol. (a) **176** (1999) 843.
12. T.L. Larsen, L. Jensen, A. Lüdge, H. Riemann and H. Lemke: J. Crystal Growth **230** (2001) 300.
13. T. Ebe, J. Crystal Growth **203** (1999) 387.
14. U. Gosele, W. Frank, A. Seeger, Solid State Comm. **45** (1982) 31.
15. D.A. Antoniadis, I. Moscowitz, J. Appl. Phys. **53** (1982) 6788.
16. E. Dornberger, T. Sinno, J. Esfandyari, J. Vanhellemont, R.A. Brown, W. von Ammon, High Purity Silicon V, **98-13**, Electrochem. Soc. Inc., 1998, p. 170.
17. J. Vanhellemont, E. Dornberger, D. Gräf, J. Esfandyari, U. Lambert, R. Schmolke, W. von Ammon and P. Wagner, in Proceedings of The Kazusa Akademia Park Forum on The Science and Technology of Silicon Materials, Kazusa Akademia Park, Chiba, Japan, (1997) p.173.
18. S. Hens, J. Vanhellemont, D. Poelman, P. Clauws, I. Romandic, A. Theuwis, F. Holsteyns and J. Van Steenbergen, Appl. Phys. Lett. **87** (2005) 061915.

Mater. Res. Soc. Symp. Proc. Vol. 1070 © 2008 Materials Research Society                    1070-E06-06

# Modeling Evolution of Temperature, Stress, Defects, and Dopant Diffusion in Silicon During Spike and Millisecond Annealing

Victor Moroz[1], Ignacio Martin-Bragado[2], Nikolas Zographos[3], Dmitri Matveev[3], Christoph Zechner[3], and Munkang Choi[2]

[1]TCAD, Synopsys, 700 East Middlefield Road, Mountain View, CA, 94043
[2]Synopsys, Mountain View, CA, 94043
[3]Synopsys, Zurich, Switzerland

## ABSTRACT

The bulk CMOS devices continue to be the dominant player for the next few technology nodes. This drives the increasingly contradicting requirements for the channel, source/drain extension, and heavily doped source/drain doping profiles. To analyze and optimize the transistors, it has become necessary to simultaneously analyze effects that have been previously decoupled. The temperature gradients, combined with stress engineering techniques such as embedded SiGe and Si:C source/drain and stress memorization techniques, create non-uniform stress distributions which are determined by the layout patterns. The interaction of implant-induced damage with dopants, stress, and defect traps defines the dopant activation, retention of useful stress, and junction leakage. This work reviews recent trends in modeling these effects using continuum and kinetic Monte Carlo methods.

## INTRODUCTION

Millisecond annealing is being introduced as a performance boost option in addition to the conventional spike anneal, and is expected to replace the spike anneal within the next two technology nodes. This work explores several facets of the spike and millisecond annealing techniques that impose considerable side effects on the process flow, employing physics-based simulation tools which have been calibrated to silicon data. The optical simulations in this work are performed using EMW simulator [1], while the mechanical stress, heat transfer, defect formation, and dopant diffusion are simulated using Sentaurus Process [2]. In addition to the conventional continuum simulation approach, some of the simulations are performed with kinetic Monte Carlo (KMC) approach [3,4].

## HEAT ABSORPTION

Spike and millisecond annealing techniques exploit electromagnetic waves to heat the silicon wafers. Consider a scanning laser that can be employed for millisecond annealing as well as for the expected future shrinkage of the thermal budget into microsecond and nanosecond timeframe. A bare silicon wafer without any pattern or a wafer with features that are larger than the illumination wavelength will be heated uniformly according to the specific heat absorption coefficients of the materials involved. However, feature sizes that are comparable to or smaller than the illumination wavelength create diffraction patterns as shown on Fig. 1, indicative of non-uniform heating of the wafer.

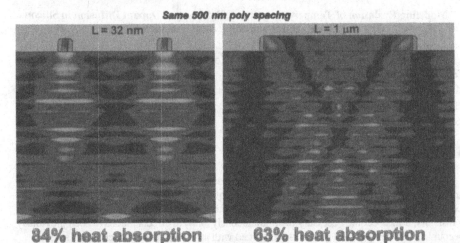

**Figure 1. Light diffraction patterns in silicon wafer and the polysilicon gates with nitride spacers. Coherent illumination with λ=810nm is assumed. The peak light intensity is shown as red, and the minimum light intensity is shown as blue, corresponding to 0.1 of the peak value.**

It can be seen in Fig. 1 that the narrow 32 nm wide polysilicon lines act as lenses, focusing the light into the silicon beneath. The polysilicon gates are heated the most with 84% of the illumination being converted into heat. In contrast the 1 μm wide polysilicon lines are much cooler and the overall heat absorption is lower at 63%. Temperature-wise, the structure with the 32nm polysilicon lines, on the left is about 200°C hotter than the structure on the right.

In order to investigate the effect of the polysilicon pitch on the heat absorption, the polysilicon line width was fixed and the pitch varied. Figure 2 shows how heat absorption responds to the polysilicon pitch for two fixed polysilicon widths. The apparent oscillations in the absorption behavior correlate with the wave length and can be attributed to wave diffraction and interference. It is clear from the optical analysis that the heat absorption is a strong function of the polysilicon pattern.

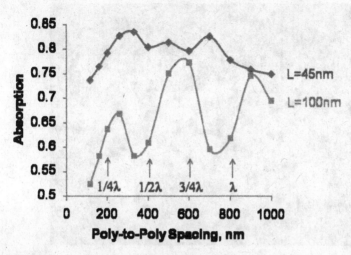

Figure 2. Heat absorption coefficient for two different polysilicon widths as a function of poly-to-polysilicon spacing for the same illumination conditions as on Fig. 1.

## HEAT TRANSFER

Millisecond annealing is only possible due to the fast heat transfer from the wafer surface to the entire wafer thickness. The heat front travels over 100 um-ms$^{-1}$ in silicon and polysilicon, and about a third of that value in the oxide. Figure 3 depicts the temperature distribution around a transistor surrounded by STI and several dummy gates. The temperature distribution is calculated at the temperature peak, i.e. at the end of laser scan. At that point, the temperature non-uniformity is the largest and therefore it represents the worst case scenario. Still, the temperature gradient is less than 1°Cum$^{-1}$, therefore it is reasonable to assume a uniform temperature distribution.

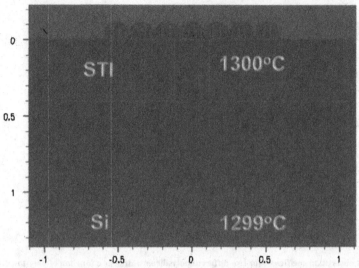

**Figure 3. Temperature distribution near the wafer surface at the peak temperature, just after the laser scan.**

It is evident from Figures 1 through 3, that the fast heat transfer will result in an approximately uniform temperature within a 100 um radius on the wafer surface. Therefore, any pattern differences on a smaller scale will not introduce temperature non-uniformities. Large areas of distinct pattern density (over about 100 microns radius) will lead to non-uniform temperature.

There are several ways to improve temperature uniformity over the wafer surface. One method is to cover the wafer with a heat absorption layer. Another it to generate a dummy polysilicon fill using dense placement of narrow polysilicon lines [5]. The high sensitivity of heat absorption to the polysilicon pattern shown on Fig. 2 provides high flexibility for automatic creation of such dummy fill.

In terms of practical applications, the layout blocks with memory are so regular that they are definitely uniformly heated. The logic is usually regular enough to have only minor non-uniformities. Significant temperature differences can only be expected in SoC applications, and can be fixed with the dummy polysilicon fill.

## STRESS INDUCED BY MILLISECOND ANNEALING

The inherent nature of millisecond annealing is such that the first few tens of microns of silicon are heated to a very high temperature, whereas the bulk of the wafer below is much cooler. This vertical temperature gradient creates thermal mismatch stress, which is expected to be compressive in the top hot part of the wafer. Figure 4 illustrates stress distributions in a silicon-dominant versus an STI-dominant layout at the peak annealing temperature, when the thermal mismatch stress is the highest. It can be seen that the transistor channels indeed experience moderate compressive stress in silicon-dominant environment. However, the stress sign changes to tensile in an STI-dominant environment. Neither of the two stress levels is dangerous by itself, but it can push a wafer with built-in stress due to stress engineering over the limit and create dislocations or cracks.

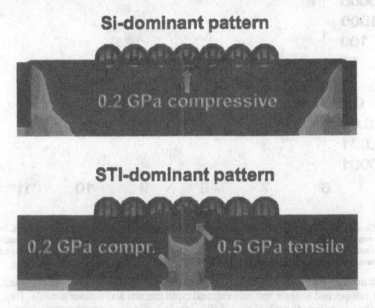

Figure 4. Stress distribution around wafer surface at the peak temperature.

Simulation of combined millisecond annealing-induced stress and the built-in stress from stress engineering can be instrumental in locating potential problems. Similar to the temperature non-uniformity described above, here introduction of a dummy STI fill would mitigate the problem. Specifically, more regular silicon/dummy STI pattern would reduce stress variations in the silicon islands.

## DISSOLVING IMPLANT DAMAGE WITH MILLISECOND ANNEALING

The resultant thermal budget from millisecond annealing produces almost diffusionless junctions and allows higher active dopant concentrations to be quenched in compared to the conventional spike anneal. However, the technique is limited in its ability to dissolve the implant damage as illustrated in Fig. 5, for several As and Si implants. The symbols show both calculated and measured data points, while the lines illustrate the apparent trends. It can be seen that millisecond annealing is barely capable of annealing damage created by a shallow 5 keV $1 \times 10^{15}$ cm$^{-2}$ arsenic implant, but is unable to remove damage generated by a deeper 40 keV $3 \times 10^{15}$ cm$^{-2}$ arsenic implant. Therefore, the transition from spike anneal to the millisecond anneal imposes limitations on the implant energy that can be used without introducing residual defects.

271

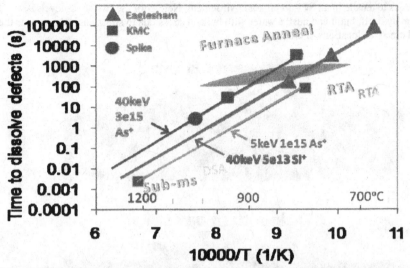

Figure 5. Different time scales needed to dissolve implant-induced defects at different temperatures. Notice the gap between RTA and Sub-ms anneal, where damage annealing can only be done by multiple millisecond anneals, for example, by performing several laser scans. Also notice the problems to dissolve defects using ms-anneal when the implant energy is bigger than 10keV.

Additionally, there is a significant time and temperature gap between the operating regimes for millisecond anneals and conventional rapid thermal and furnace anneals. This difference can be attributed to the distinct differences in heat transfer mechanisms mentioned above. The inherent nature of the millisecond anneal does not allow the time scale of the anneal to be increased more than several milliseconds. Otherwise, the fast solid-state heat transfer would result in a uniformly distributed temperature profile and therefore limit the speed of wafer cooling to slow heat radiation, as in conventional techniques.

One approach to bridging the thermal budget gap is to combine spike and millisecond anneals. Another is to apply the millisecond anneal multiple times. Each additional scan of the laser beam doubles the original millisecond thermal budget and can, theoretically, bridge the gap An alternative approach that does not increase the thermal budget is to introduce impurities that are known to trap the interstitials (like C, F or N) and therefore suppress the defect formation. The following examples illustrate usage of both approaches.

Figure 6 shows the sheet resistance of a boron junction in a sample annealed for 1 minute at different temperatures, after either 1 or 10 laser scans at 1150°C. The sample was preamorphized with a 5keV, $1x10^{15}$ cm$^{-2}$ Ge implant and subsequently implanted with $1x10^{15}$ cm$^{-2}$, 500 eV B. A single scan of the laser was not enough to fully dissolve the damage: the residual defects deactivated the boron by creating Boron Interstitial Clusters (BICs) in the 700°C to 900°C temperature range. In contrast, ten laser scans completely annealed the implant damage and stabilize the boron sheet resistance. The symbols are the experimental data from [6] and the lines represent the results of KMC simulation using the advanced calibration parameter set without any parameter fine-tuning. The agreement between measurement and simulation is excellent, which indicates that all important mechanisms are accurately captured in the KMC model.

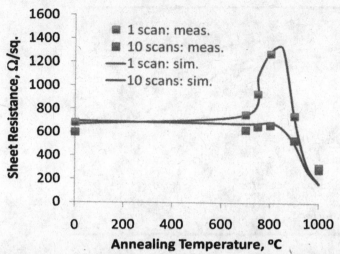

**Figure 6. Differences in boron activation in a sample annealed for 1 minute after 1 or 10 laser scans at 1150°C. The symbols are the experimental data from reference [6] and the lines represent the results of KMC simulation using the advanced calibration parameter set without any parameter fine-tuning.**

In the case of halo implants, deactivation via clustering is not an issue because of the lower dopant concentration. However, the residual defect levels due to higher implant energies are important since they are likely to survive the annealing process. These residual defects introduce mid-gap energy levels, leading to thermal carrier generation and therefore junction leakage. Figure 7 illustrates the synergy between a carbon co-implant and a laser scan. The laser scans helps to cut damage annealing time by a factor of 2.5, allowing for much smaller dopant diffusion in the next regular annealing step.

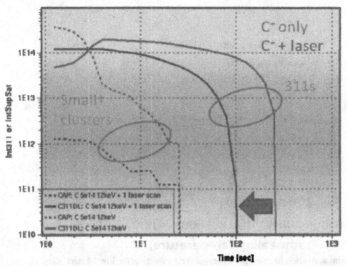

**Figure 7.** Effect of a laser scan on defect dissolution. The figure shows the simulation results of a mid temperature annealing of a preamorphized sample including carbon ($5 \times 10^{15}$ cm$^{-2}$ PAI, 12 keV) followed by an Arsenic implant ($10^{13}$ cm$^{-2}$ halo, 20 keV) with or without a laser scan right after the implants.

The carbon co-implant is vital for the successful reduction of annealing time after the laser scan. Without it the extra laser scan does not seem to improve the damage dissolution, and multiple laser scans can actually make it worse.

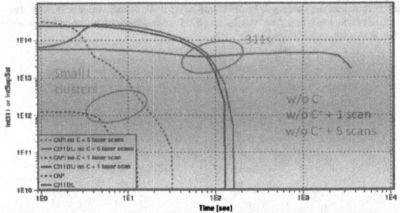

Figure 8 illustrates this point by representing the defect time evolution under the same conditions as Fig. 7, but without the carbon co-implant. One additional scan (red curves) does significantly change the dissolution time (blue curves) but 5 scans increases the thermal budget required to dissolve the defects 60-fold. This is due to the formation of stable dislocation loops, as the {311}-type defects evolve with additional thermal budget, which are more difficult to anneal out.

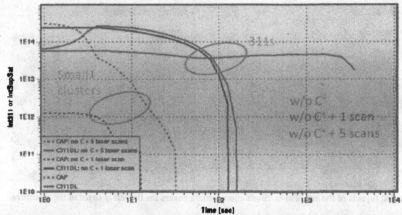

**Figure 8. Similar implant and annealing conditions to those used in Fig. 7, except that there is no carbon co-implant and several laser scans are evaluated in addition to the zero scan and one scan options.**

Carbon plays ~~such~~ an important role in this scenario because the halo implants are non-amorphizing; rather they are deep and introduce damage well below the channel surface. Consequently, the additional thermal budget given by the laser will lead to further Ostwald ripening and defect evolution rather than defect dissolution, given that the silicon surface that facilitates interstitial recombination is quite far. In this case, the presence of carbon will tell us whether the interstitials released during the laser scan will be trapped and accumulated in form of harmless silicon-carbon clusters, or will be trapped by other extended defects growing into the stable dislocation loops.

## IMPACT OF LAYOUT ON THRESHOLD VOLTAGE

The mechanical stress in the transistor channel arising from STI and other stress sources is the primary reason behind stress-enhanced carrier mobilities. Another major layout-induced effect is the threshold voltage shift according to the size and shape of the active area [7]. This effect is caused by the transient-enhanced diffusion (TED) rather than stress [8,9].

Figure 9 shows schematically two different layouts of the source/drain. On the left a traditional simple rectangular, or 'I' shaped layout is shown, and on the right, the increasingly common 'H' shaped layout is depicted. Also shown is the quarter of the structure that was simulated, taking advantage of the structure symmetry to fasten the simulation. KMC simulations were employed to investigate the impact of these different layouts on the interstitial population and ultimately on the threshold voltage.

Simulation domain                    Simulation domain

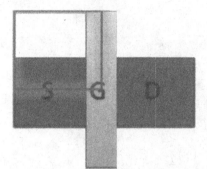

Figure 9. Different layouts of the active area. Simulation domain covers the top left quarter of the structure.

Figure 10 shows the distribution of interstitial jumps integrated over the annealing time after an amorphizing arsenic source/drain implant and an activation anneal. This distribution is proportional to the amount of TED seen by the source/drain and by the halo dopants. It can be seen that the H-shaped transistor exhibits higher TED levels in the channel. The clean (green colored) area is due to amorphous layer recrystallization.

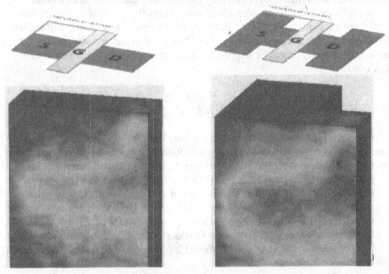

Figure 10. Distribution of interstitial jumps integrated over the annealing time. This distribution is proportional to TED.

Figure 11 shows the amount of TED along the line connecting the middle of the silicon/STI edge behind the source to the middle of the channel (i.e. the lower edge of the simulation domain rectangle as shown on Fig. 9). The "H" structure exhibits about 30% larger

TED in the channel, due to the interstitials from additional implant damage in the "H" structure coming around the corner towards the channel and enhancing dopant diffusion there.

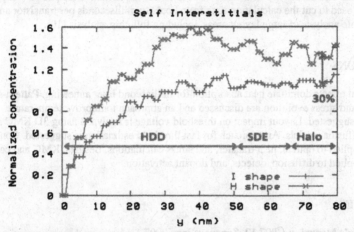

**Figure 11. Normalized distribution of the interstitial jumps/TED along a line going from source/STI edge to the middle of the channel.**

This enhanced TED in the "H" structure leads to larger lateral source/drain diffusion, and therefore shorter effective channel length. Moreover, it increases halo TED and affects the channel doping level, which translates into a shift of the threshold voltage, illustrated in Fig. 12. Both the poly-to-STI distance and the "H" feature size lead to comparable threshold shifts of around 40 mV, which is large enough to make a major impact on transistor performance.

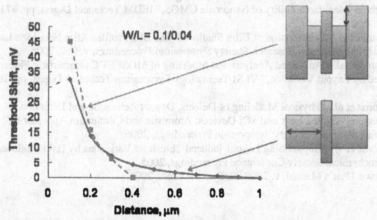

**Figure 12. Impact of active area shape and size on threshold voltage obtained using 3D process and device simulators, calibrated to a 32 nm process.**

The impact of layout on threshold voltage described above is obtained using 3D process and device simulators that are calibrated to a 32 nm process. Typical physics-based simulation time for such 3D structures is several hours of CPU time. An alternative simplified compact model can be used to cut the calculation time from hours to milliseconds per transistor and therefore enable analysis of large layout areas, including full-chip analysis [10].

## CONCLUSIONS

Optical effects dominate heat absorption for millisecond laser annealing. Pattern-specific heat transfer and stress evolution are discussed and an approach to improve temperature uniformity is suggested. Layout impact on threshold voltage is analyzed using 3D KMC and continuum diffusion models. An approach for fast threshold estimation is suggested. Continuum models are applied to optical, heat transfer, and stress calculations. Besides, KMC and compact models are applied to diffusion, defects, and dopant activation.

## REFERENCES

[1] EMW User's Manual, v. 2007.12, Synopsys Inc., 2007.
[2] Sentaurus Process User's Manual, v. 2007.12, Synopsys Inc., 2007.
[3] M. Jaraiz et al., "Atomistic Front-End Process Modelling: A Powerful Tool for Deep-Submicron Device Fabrication," SISPAD 2001 Proceedings, Springer-Verlag, pp. 10 – 17, 2001.
[4] Ignacio Martin-Bragado et al. "Modeling charged defects, dopant diffusion and activation mechanisms for TCAD simulations using kinetic Monte Carlo," Nuclear Instruments and Methods in Physics Research B, v. 253, pp. 63-67, 2006.
[5] K. Kuhn, "Reducing Variation in Advanced Logic Technologies: Approaches to Process and Design for Manufacturability of Nanoscale CMOS," IEDM Technical Digest, pp. 471-474, 2007.
[6] J. Sharp et al., "Deactivation of Ultra Shallow B and BF2 Profiles After Non-Melt Laser Annealing," Materials Research Society Symposium Proceedings, v. 912, 2006.
[7] H. Tsuno et al., "Advanced Analysis and Modeling of MOSFET Characteristic Fluctuation Caused by Layout Variation," VLSI Technology Symposium Technical Digest, pp. 204-205, 2007.
[8] V. Moroz et al., "Physical Modeling of Defects, Dopant Activation and Diffusion in Aggressively Scaled Bulk and SOI Devices: Atomistic and Continuum Approaches," Materials Research Society Symposium Proceedings, 2006.
[9] V. Moroz et al., "Suppressing Layout-Induced Threshold Variations by Halo Engineering," Electrochemical Society Conference Proceedings, 2005.
[10] Seismos User's Manual, v. 2008.03, Synopsys Inc., 2008.

Mater. Res. Soc. Symp. Proc. Vol. 1070 © 2008 Materials Research Society

# F+ Implants in Crystalline Si: The Si Interstitial Contribution

Pedro Lopez[1], Lourdes Pelaz[1], Ray Duffy[2], P. Meunier-Beillard[2], F. Roozeboom[3], K. van der Tak[4], P. Breimer[4], J. G. M. van Berkum[4], M. A. Verheijen[4], and M. Kaiser[4]

[1]Electricidad y Electronica, Universidad de Valladolid, E.T.S.I. Telecomunicacion. Campus Miguel Delibes s/n, Valladolid, 47011, Spain
[2]NXP Semiconductors, Leuven, 3001, Belgium
[3]NXP Semiconductors, Eindhoven, 5656, Netherlands
[4]Philips Research Laboratories Eindhoven, Eindhoven, 5656, Netherlands

## ABSTRACT

In this work the Si interstitial contribution of F+ implants in crystalline Si is quantified by the analysis of extended defects and B diffusion in samples implanted with 25 keV F+ and/or 40 keV Si+. We estimate that approximately 0.4 to 0.5 Si interstitials are generated per implanted F+ ion, which is in good agreement with the value resulting from the net separation of Frenkel pairs obtained from MARLOWE simulations. The damage created by F+ implants in crystalline Si may explain the presence of extended defects in F-enriched samples and the evolution of B profiles during annealing. For short anneals, B diffusion is reduced when F+ is co-implanted with Si+ compared to the sample only implanted with Si+, due to the formation of more stable defects that set a lower Si interstitial supersaturation. For longer anneals, when defects have dissolved and TED is complete, B diffusion is higher because the additional damage created by the F+ implant has contributed to enhance B diffusion.

## INTRODUCTION

The co-implantation of F+ and B+ in pre-amorphized Si, followed by solid phase epitaxial regrowth, has been proved to be beneficial for ultrashallow junction formation since a remarkable reduction of B diffusion can be achieved [1]. The beneficial effect of F has been attributed to the formation of fluorine-vacancy complexes ($F_nV_m$) during recrystallization. This results in a regrown layer rich in vacancies (V´s) which act as annihilation centers for Si interstitials (I´s) injected from the end of range (EOR) damage. This hypothesis is supported by the high affinity of F with vacancies, as indicated by theoretical calculations [2], and by the direct observation by transmission electron microscopy (TEM) of bubbles in the high concentration region of a F profile [3].

The case of F+ implantation in crystalline Si (c-Si) has been less studied and the available information is sometimes contradictory. There is not even agreement about the possible beneficial effect of F. Both B diffusion enhancement [4] and reduction [5, 6] by the presence of F have been reported. The clarification of the role of F in c-Si is gaining relevance since in some cases amorphization is not advisable due to the high leakage currents caused by residual EOR defects [7] or to the poor crystal quality after regrowth as observed in FinFET structures [8].

An important difference between $F^+$ implantation in amorphous and crystalline Si is related to the defect balance. In amorphizing conditions, the introduction of F in the amorphous layer could favor the formation of $F_nV_m$ complexes during recrystallization but the corresponding Si I´s contained in the amorphous layer (excess Si atoms) may be swept to the surface during regrowth. Thus, the presence of $F_nV_m$ complexes in the regrown layer represents an excess of V´s that is not accompanied by an excess of Si I´s. In crystalline Si, Si I´s and V´s are generated in pairs by the $F^+$ implant, and therefore, the populations of both point defects can be considered to be balanced (although they could be spatially separated). If some V´s were trapped in $F_nV_m$ complexes the corresponding Si I´s would exist.

We have designed experiments to quantify the amount of damage generated by a $F^+$ implant in c-Si and to clarify whether the presence of F may have a beneficial effect on B diffusion even in non-amorphizing conditions.

## EXPERIMENT

In our experiments, two B doped layers were grown by chemical vapor deposition at depths of 120 and 440 nm to act as diffusion markers. B peak concentration was low, approximately $2.5 \times 10^{18}$ cm$^{-3}$, to minimize the formation of boron-interstitial clusters. $10^{14}$ and $5 \times 10^{14}$ cm$^{-2}$ $F^+$ was implanted at 25 keV whose mean projected range ($R_p$) (~56 nm) is located close to the position of the shallow B spike. The two non-amorphizing F doses were chosen to introduce a different amount of F and damage in the samples. A 40 keV $5 \times 10^{13}$ cm$^{-2}$ Si$^+$ implant, whose $R_p$ is similar to that of the $F^+$ implants, was also performed alone or combined with one of the $F^+$ implants to study the effect of F with additional damage. Note that the Si dose is half the lower F dose ($10^{14}$ cm$^{-2}$) to reproduce the ratio between the B and F dose in BF$_2^+$ implants. A sample with no implants was used as a reference for equilibrium B diffusion. All samples were annealed at 850 °C for 18, 180 or 1800 s. Secondary ion mass spectrometry (SIMS) characterization was used to analyze B and F profiles and TEM images to detect the presence of extended defects.

## DISCUSSION

SIMS B profiles for the samples implanted with 25 keV $10^{14}$ cm$^{-2}$ $F^+$, Si$^+$, and $F^+$ plus Si$^+$ after annealing at 850 °C for 18 or 180 s are plotted in Fig. 1. F profiles as implanted and after annealing are also included. After 18 s anneal, B diffusion is reduced when $F^+$ and Si$^+$ are co-implanted, compared to the sample only implanted with Si$^+$. The situation changes as the anneal proceeds. After 180 s the advantage of $F^+$ co-implantation is no longer observed and a higher B diffusion appears when $F^+$ and Si$^+$ are co-implanted compared to only a Si$^+$ implant. B SIMS profiles for 1800 s (not shown) are similar to those at 180 s.

Previous works have reported a reduced B diffusion when $F^+$ is implanted in c-Si, by using a B spike as a diffusion marker located in the shallow part of a high energy $F^+$ implant [5, 9]. In those conditions the reduction of B diffusion can be attributed to an excess of V´s close to the surface generated by the $F^+$ implant, as also observed for other ions [10]. In our experiments, both B spikes are deeper than $R_p$ and, therefore, they are located in the Si interstitial-rich region of the damage profile. The reported reduction on B diffusion at short anneal times cannot either be caused by a chemical F-B bond or the formation of boron-interstitial clusters, because this effect is also observed in the deep B spike that is hardly covered by F or implant damage.

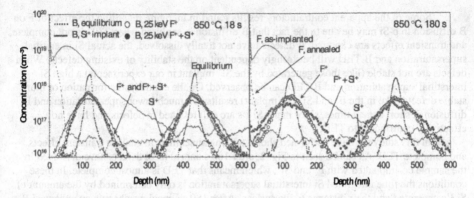

**Figure 1.** B SIMS profiles after anneal at 850 °C for 18 or 180 s for the samples implanted with 25 keV $10^{14}$ cm$^{-2}$ F$^+$, 40 keV $5\times10^{13}$ cm$^{-2}$ Si$^+$ and co-implanted with F$^+$ and Si$^+$. B spikes in equilibrium conditions (no implants) and F profiles as-implanted and after anneal are also shown.

TEM characterization was used to analyze the presence of extended defects in F-enriched samples. It is known that a 40 keV $5\times10^{13}$ cm$^{-2}$ Si$^+$ implant produces small Si I´s clusters and {113} defects that set a high Si I´s supersaturation but dissolve quickly [11]. In the sample implanted only with Si$^+$ no defects are observed in TEM images after annealing at 850 °C for 18 s (not shown). This indicates that TED is complete and only equilibrium B diffusion is observed in subsequent annealing times. In Fig. 2 we show the plan-view TEM images corresponding to the sample co-implanted with F$^+$ and Si$^+$, after annealing for 18 or 180 s. After 18 s anneal a high density of defects is observed. The additional damage generated by the F$^+$ implant favors the growth of more stable defects, setting a lower Si I´s supersaturation. This can explain the reduced B TED observed for short anneal times when F$^+$ and Si$^+$ are co-implanted, compared to only a Si$^+$ implant.

**Figure 2.** Plan-view TEM images for 25 keV $10^{14}$ cm$^{-2}$ F$^+$ plus 40 keV $5\times10^{13}$ cm$^{-2}$ Si$^+$ after annealing at 850 °C for 18 (a) and 180 s (b).

Some of the apparent contradictory results found in literature regarding the effect of F on B diffusion in c-Si may be due to the fact that B diffusion is analyzed when TED is not complete, and transient effects are captured. If defects have not totally dissolved, the actual Si interstitial supersaturation and B TED will be strongly dependent on the stability of existing defects. When defects are not stable (like those generated by the $Si^+$ implant in our experiments) a high Si interstitial supersaturation and B TED can be observed. On the contrary, the formation of more stable defects (like in the $F^+$ and $Si^+$ co-implant) results in a much lower supersaturation and B diffusion at short anneal times, since many Si I's are still retained in defects and have not effectively contributed to TED.

In this study we have performed short and long annealings to analyze transient effects and complete TED. As can be seen in Fig. 2, after 180 s anneal most defects have dissolved in the sample co-implanted with $F^+$ and $Si^+$, which means that TED is almost complete. In these conditions the time integrated Si interstitial supersaturation is only determined by the amount of Si I's generated and their distance to the surface. After 180 s anneal we observe an enhanced B diffusion by $F^+$ co-implantation with $Si^+$, compared to only a $Si^+$ implant (note that $F^+$ and $Si^+$ implants have similar $R_p$ and defects are located at approximately the same depth), because a much larger amount of Si I's (those produced by both the $F^+$ and $Si^+$ implant) have contributed to B TED.

An estimate of the effective number of Si I's generated by the $F^+$ implant can be derived from the analysis of B diffusion when TED is complete. After 180 s, the averaged value of B diffusivity multiplied by time (which is proportional to the time integrated free Si interstitial concentration) in the deep B spike for the $10^{14}$ $cm^{-2}$ $F^+$ implant alone is about 0.7 times that of the $Si^+$ implant, while for the $F^+$ and $Si^+$ co-implantation this value equals 1.7 times compared to Si. Therefore, the contribution of this $F^+$ implant is about 0.7 times that of the $Si^+$ implant, although the implanted F dose is twice the Si dose. Considering that the effective "+n" factor for the $Si^+$ implant is approximately +1.3 [12], this indicates that the number of effective Si I's per implanted $F^+$ ion in c-Si is about +0.5.

A quantification of the Si interstitial damage produced by the $F^+$ implant and its net contribution to B diffusion can also be directly obtained from TEM images. For this purpose we implant 25 keV $5\times10^{14}$ $cm^{-2}$ $F^+$, because the larger F dose favors the formation of bigger and more stable defects, as shown in the TEM image included in Fig. 3, taken after annealing at 850 °C for 18 s. We have used several plan-view TEM images from a sample whose thickness was

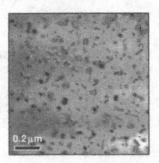

**Figure 3.** Plan-view TEM image for 25 keV $5\times10^{14}$ $cm^{-2}$ $F^+$ after annealing at 850 °C for 18 s.

larger than $R_p$ to ensure that the whole band of defects is present within the TEM sample. For the calculation of the Si interstitial concentration stored in {113} defects and dislocation loops we have followed the method described elsewhere [13, 14]. We have determined that approximately $2\times10^{14}$ Si I´s/cm$^2$ are stored in {113} defects and dislocation loops for the $5\times10^{14}$ cm$^{-2}$ F$^+$ implant after 18 s anneal (typical error is 20%). Since some defects may have already dissolved or may not be visible by TEM, this calculation is a low limit. Thus, we experimentally estimate that approximately 0.4 Si I´s per implanted F$^+$ ion are stored in defects.

Theoretical calculations reveal that F atoms tend to remain interstitial rather than to occupy a substitutional position [15], which may explain the low +n value experimentally obtained. According to this hypothesis, we have run a MARLOWE simulation of the 25 keV $5\times10^{14}$ cm$^{-2}$ F$^+$ implant considering that F does not become substitutional. The simulation results included in Fig. 4 reveal that once Frenkel pairs have locally recombined, there are approximately 0.4 Si I´s per implanted F$^+$ ion in the interstitial-rich region, which is in good agreement with the values estimated from experiments (0.4-0.5). Therefore, the net separation between Si I´s and the corresponding V´s in a F$^+$ implant in c-Si accounts for the presence of Si interstitial-type defects and the observed enhanced B diffusion.

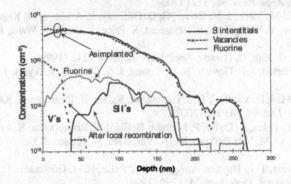

**Figure 4.** MARLOWE simulations of the 25 keV $5\times10^{14}$ cm$^{-2}$ F$^+$ implant considering that F atoms do not become substitutional. The as-implanted Si interstitial and vacancy distribution, as well as the F$^+$ profile, are shown. Si I´s and V´s profiles once Frenkel pairs have locally recombined are also plotted.

## CONCLUSIONS

From the analysis of damage evolution and B diffusion in 25 keV F$^+$ implants in c-Si we have estimated that approximately 0.4 to 0.5 Si I´s are generated per implanted F$^+$ ion, in good agreement with the value resulting from the net separation of Frenkel pairs. The excess Si I´s generated by F$^+$ co-implants contribute to the formation of a larger amount of more stable extended defects, which set a lower but longer-lasting Si interstitial supersaturation. As a result, at short anneal times a reduction of B diffusion is observed in F$^+$ co-implanted samples, but if the anneal is complete (which is desirable to reduce junction leakage), B diffusion is enhanced by the $F^+$ co-implant. Our results clearly indicate that $F^+$ co-implantation in c-Si causes additional Si interstitial defects and it has an overall negative effect on junction formation.

## ACKNOWLEDGMENTS

This work has been funded by the Spanish DGI under project TEC2005-05101 and the JCyL Consejería de Educación y Cultura under project VA070A05.

## REFERENCES

1. G. Impellizzeri, S. Mirabella, F. Priolo, E. Napolitani, A. Carnera, *J. Appl. Phys.* **99**, 103510 (2006).
2. G. M. Lopez, V. Fiorentini, G. Impellizzeri, S. Mirabella, E. Napolitani, *Phys. Rev. B* **72**, 045219 (2005).
3. S. Boninelli, A. Claverie, G. Impellizzeri, S. Mirabella, F. Priolo, E. Napolitani, F. Cristiano, *Appl. Phys. Lett.* **89**, 171916 (2006).
4. T. Noda, *J. Appl. Phys.* **96**, 3721 (2004).
5. H. A. W. El Mubarek, J. M. Bonar, G. D. Dilliway, P. Ashburn, M. Karunaratne, A. F. Willoughby, Y. Wang, P. L. F. Hemment, R. Price, J. Zhang, P. Ward, *J. Appl. Phys.* **96**, 4114 (2004).
6. J. Park, Y. -J. Huh, H. Hwang, *Appl. Phys. Lett.* **74**, 1248 (1999).
7. S. D. Brotherton, J. P. Gowers, N. D. Young, J. B. Clegg, J. R. Ayres, *J. Appl. Phys.* **60**, 3567 (1986).
8. R. Duffy, M. J. H. Van Dal, B. J. Pawlak, M. Kaiser, B. Degroote, E. Kunnen, E. Altamirano, *Appl. Phys. Lett.* **90**, 241912 (2007).
9. P. Lopez, L. Pelaz, R. Duffy, P. Meunier-Beillard, F. Roozeboom, K. van der Tak, P. Breimer, J. G. M. Van Berkum, M. A. Verheijen, M. Kaiser, *J. Vac. Sci. Technol. B* **26**, 377 (2008).
10. V. C. Venezia, T. E. Haynes, A. Agarwal, L. Pelaz, H. -J. Gossmann, D. C. Jacobson, D. J. Eaglesham, *Appl. Phys. Lett.* **74**, 1299 (1999).
11. D. J. Eaglesham, P. A. Stolk, H. -J. Gossman, J. M. Poate, *Appl. Phys. Lett.* **65**, 2305 (1994).
12. L. Pelaz, G. H. Gilmer, M. Jaraiz, S. B. Herner, H. -J. Gossmann, D. J. Eaglesham, G. Hobler, C. S. Rafferty, J. Barbolla, *Appl. Phys. Lett.* **73**, 1421 (1998).
13. N. Cherkashin, P. Calvo, F. Cristiano, B. de Mauduit, A. Claverie, *Mater. Res. Soc. Symp. Proc.* **810**, 103 (2004).
14. J. K. Listebarger, K. S. Jones, J. A. Slinkman, *J. Appl. Phys.* **73**, 4815 (1993).
15. M. Diebel, S. T. Dunham, *Mater. Res. Soc. Symp. Proc* **717**, C4.5.1 (2002).

Mater. Res. Soc. Symp. Proc. Vol. 1070 © 2008 Materials Research Society

# Concentration-Dependence of Self-Interstitial and Boron Diffusion in Silicon

Wolfgang Windl
Materials Science and Engineering, The Ohio State University, Columbus, OH, 43210-1178

## ABSTRACT

In this paper, we discuss the accuracy of ab-initio calculations for self-interstitial and boron diffusion in silicon in light of recent experimental data by de Salvador et al. and Bracht et al. Mapping the experimental data onto the activation energy vs. Fermi level representation commonly used to display ab-initio results, we show that the experimental results are consistent with each other. While the theoretical LDA value for the boron activation energy as a function of the Fermi level agrees well with experiment, we find for the self-interstitial in line with other calculations an underestimation of the experimental values, despite using total-energy corrections.

## INTRODUCTION

The diffusion mechanism of boron has not been understood for a very long time despite considerable research interest [1]. It is widely accepted by now that boron diffuses nearly exclusively with the help of Si self-interstitials ($I$s) [2], *i.e.*, the mobile entity is thought to be a B atom paired with an $I$. As concerning the diffusion mechanism, first *ab-initio* modeling for neutral systems using the drag method had suggested that a kick-out mechanism with long-range low-barrier interstitial migration would be the dominant mechanism [3,4], in contrast to previous perception. Later, we used the nudged elastic band method (NEBM) [5] implemented into VASP [6] to re-examine the minimum-energy barrier diffusion path for $I$-assisted, charge-state dependent B diffusion within both LDA and GGA, using 64-atom supercells [7] with new results (see below). Although this work has been performed several years ago, very recent experiments by De Salvador et al. [8] and Bracht et al. [9] have re-examined the charge states of both the boron atom and interstitials involved, which helps to re-evaluate the simulation results.

## DIFFUSION EQUATIONS

Especially interesting is the simultaneous determination of the interaction rate $g$ between B and $I$ and the mean free path $\lambda$ in [8], from which the diffusivity $D$ can be calculated, $D = g\lambda^2$ [10,11]. Under thermodynamic equilibrium conditions in a homogenous material, the impurity transport can be described by [10]

$$\partial C_{BI}/\partial t = D_{BI}\nabla^2 C_{BI} - rC_{BI} + gC_{Bs}, \qquad \partial(C_{Bs} + C_{BI})/\partial t = D_{BI}\nabla^2 C_{BI}, \qquad (1)$$

where $C_{Bs}$ and $C_{BI}$ are the concentrations of the ground state, substitutional boron $B_s$, and the mobile boron interstitial pair, $BI$. Since for B diffusion in Si the Frank-Turnbull mechanism ($B_s \rightarrow BI + V$) is energetically much less favorable than the kick-out mechanism ($B_s + I \rightarrow BI$), the generation rate $g$ is given by the forward reaction rate [12] of the kick-out mechanism times self-interstitial concentration,

$$g = 4\pi a_c^{BI} D_I C_I, \qquad (3)$$

where $a_c^{BI}$ is the B-$I$ capture radius. $g$ is thus directly proportional to the self-interstitial transport coefficient, $D_I C_I$.

The mean free path can be calculated from $\lambda = \sqrt{D_{BI}/r}$ [10]. With $\beta = 1/(k_B T)$, $D_{BI} = D_0^{BI} \exp(-\beta E_m^{BI})$, $C_I = C_0^I \exp(-\beta E_f^I)$, and $D_I = D_0^I \exp(-\beta E_m^I)$, we define the recombination rate as [13] $r = (D_I/a_{Si}^2)\exp(-\beta E_b^{BI})$, where prefactors $D_0$, migration energies $E_m$, formation energy $E_f^I$, average Si spacing $a_{Si}$ and B-$I$ binding energy $E_b^{BI}$ have been used. With that, we get $\lambda = a_{Si}\sqrt{D_0^{BI}/D_0^I}\exp(-\beta(E_m^{BI} - E_b^{BI} - E_m^I)/2)$. Introducing the Si atom density $C_{Si} = 1/a_{Si}^3 = 5\times10^{22}\,\text{cm}^{-3}$ which defines the average Si spacing $a_{Si}$, we finally obtain $D_B = (4\pi a_c^{BI} C_0^I)/(a_{Si}C_{Si})D_0^{BI}\exp(-\beta(E_f^I - E_b^{BI} + E_m^{BI}))$. Defining the activation energies $E_a^B = E_f^I - E_b^{BI} + E_m^{BI}$ and $E_a^I = E_f^I + E_m^I$ and the prefactors $D_{B0} = 4\pi a_c^{BI} C_0^I D_0^{BI}/(a_{Si}C_{Si})$ and $(D_I C_I)_0 = D_0^I C_0^I$, we can also write

$$D_B = D_{B0}\exp(-\beta E_a^B), \tag{10}$$
$$g = 4\pi a_c^{BI}(D_I C_I)_0 \exp(-\beta E_a^I), \tag{11}$$
$$\lambda = \sqrt{D_{B0}/4\pi a_c^{BI}(D_I C_I)_0}\,\exp(-\beta(E_a^B - E_a^I)/2). \tag{12}$$

Since all energies and the capture radius $a_c^{BI}$ in Eqs. (10-12) have been calculated in the past, we will first report those results, and then compare them to the experimental results of [8] and [9].

## THEORETICAL RESULTS

Equations (10-12) are controlled by the activation energies for B and $I$ diffusion. 216-atom, $2\times2\times2$ k-point GGA calculations for the Si self-interstitial with the corrections discussed in [14] suggest that only tetrahedral $T^{++}$ and split-interstitial $X^0$ are stable (mid-gap formation energies are 3.74 and 3.70 eV, respectively), whereas $T^+$ and negative charge states have for no Fermi-level position the lowest energy (Fig. 1). For the migration energies, we find values of 1.03, 0.02, and 1.14 eV for $X^0$, $T^+$, and $T^{++}$, respectively. Although the barrier for $T^+$ diffusion may seem small, experiments have suggested that self-interstitials in Si are mobile at temperatures of 4.2 K [15] and even 0.5 K [16], which would be easiest possible for the small migration barrier found for the positive charge. The activation energy for the transport coefficient $C_I D_I$ at midgap is given by 4.73, 4.32, and 4.88 eV, for $X^0$, $T^+$, and $T^{++}$, respectively, making $T^+$ the dominant mobile species. These val-

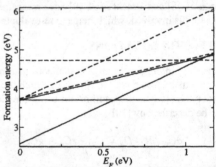

**Figure 1.** GGA formation energies (solid) and activation energies (dashed lines) for the Si self interstitial with (in order of increasing slope) 0, +, and ++ charge, using 217-atom supercells and the corrections from [14].

ues are somewhat lower than recent values of 4.96, 4.82, and 5.02 eV [9], extracted from dopant and self-diffusion experiments in Si isotope multilayer structures.

Concerning B diffusion, It is by now well established that a B atom needs to capture a self-interstitital to form some kind of B$I$ pair (which in our definition also includes an interstitial B atom, B$_i$). The results for the energetics of boron diffusion have been published in [7]. These activation energies from LDA and GGA calculations, along with experimental data, are shown in Fig. 2.

Finally, Beardmore *et al.* [17] calculated the B$I$ capture radius $a_c^{BI}$ from kinetic Monte Carlo simulations based on ab-initio calculated interaction energies up to the 11$^{th}$ neighbor shell to be 4.6 Å at 900 °C. Assuming a similar temperature dependence as found in [17] for As$V$ and P$V$, the capture radius for B$I$ at 700 °C might be somewhat higher, around 5.4 Å. However, due to the lack of hard data, we use in the following section the 900 °C value of 4.6 Å.

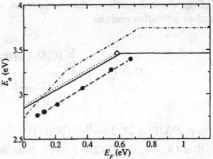

**Figure 2.** Activation energy for B diffusion. Solid and dot-dashed lines, theoretical LDA and GGA curves from [7]. The diamond is an experimental value from [9], the circles values calculated from $D$ vs. $p$ values from [8] (see text), both with rescaled energies to match the high-temperature intrinsic Fermi energy to the zero-temperature value. The dashed line is a fit to the values from [8], the dotted line corresponds to the +1 slope suggested in [9].

## COMPARISON TO EXPERIMENT

The results of the zero-temperature calculations described in the previous section were reported as functions of the Fermi energy. In order to compare the experimental values from [8] to these results, it is helpful to define the carrier-concentration dependence for Eqs. (10-12) in a somewhat unusual form by allowing concentration dependence of prefactor and activation energy. For the example of the diffusion constant, this results in $D(p) = D_0(p)\exp(-\beta E_a(p))$ and thus

$$E_a(p) = -k_B T \ln(D(p)/D_0(p)) \tag{14}$$

(with analogous expressions for $g$ and $\lambda$, respectively). Thus, $E_a(p)$ can be calculated, provided a value/function for the prefactor $D_0(p)$ is available. To plot $E_a$ as a function of the Fermi level $E_F$ instead of $p$, we use the familiar $E_F(p) = E_F^i - k_B T \ln(p/n_i)$. For the necessary parameters to convert the experimental results of [8] at 700 °C, we use $n_i = 1.0 \times 10^{18}\,\text{cm}^{-3}$ [18] and $E_F^i = E_g/2 + 3/4\ln(m_p^*/m_n^*) = 0.42$ eV (using Thurmond's expression for $E_g$ [19] and Jain and Van Overstraeten's effective masses $m^*$ [20]).

Variations in $D_0$ with Fermi energy, as are typically present in the widely used expression $D = D^0 + (p/n_i)D^+ + \cdots$, are not important for boron diffusion which is dominated by a single charge state, but have to be taken into account for self-interstitials. For the interstitial prefactor

$(C_ID_I)_0$, which depends on the charge state, we use a Boltzmann average with Fermi-level dependent activation energies,

$$(C_ID_I)_0(E_F) = \frac{\sum_{Q=0}^{2}(C_ID_I)_0(Q)\exp\{-\beta[E_a^I(Q)+Q(E_F-E_F^i)]\}}{\sum_{Q=0}^{2}\exp\{-\beta[E_a^I(Q)+Q(E_F-E_F^i)]\}}. \tag{16}$$

with $(C_ID_I)_0(0) = 2732$ cm$^2$/s, $(C_ID_I)_0(1) = 69.9$ cm$^2$/s, $(C_ID_I)_0(2) = 469$ cm$^2$/s, $E_a^I(0) = 4.96$ eV, $E_a^I(1) = 4.82$ eV, and $E_a^I(2) = 5.02$ eV from Ref. [9]. For B, we use $D_{B0} = 0.87$ cm$^2$/s [9].

**Boron Diffusion Coefficient:** In Fig. 2, we show the theoretical activation energies from LDA and GGA calculations as a function of the Fermi energy from [7]. Also, we include the experimental data from [8] and [9]. The energies of [8] have been determined from $D$ vs. $p$ data according to the previously described procedure. There is some uncertainty concerning the necessary prefactor $D_0$ [Eq. (14)], which we assumed to be equal to the most recent value of 0.87 cm$^2$/s [9]. Taking a value of ~4 cm$^2$/s instead would result in a best fit to the theoretical LDA values. Knowledge of the temperature dependence of the ionization energies is necessary to compare zero-temperature ab-initio calculations to high-temperature data. Since it is hard to extrapolate from the temperature dependence of high-temperature data (*e.g.*, in [9]) to zero, we simply chose for the B diffusion coefficient to scale the experimental energies to match the theoretical intrinsic Fermi level at zero temperature.

The values from [8] are lower than both LDA and GGA results. However, the most recent experimental value for the B activation energy in intrinsic Si is higher at 3.46 eV [9] and very close to especially the LDA result. Independent of this, the most important result of the carrier-concentration dependent diffusivity values from [8] is the shape of the underlying curve. It is straight with a fitted slope of 1.09 (close to the expected value of 1) and extends into the *n*-type regime, which was experimentally achieved by counterdoping with phosphorous. Reference [9] comes to the same conclusion. This behavior is in agreement with the theoretical LDA curve in the *p*-regime, which predicts a diffusion coefficient proportional to $p/n_i$ in that region. Although it extends this behavior into the *n*-type regime, the theoretical $p/n_i$ regime ends at a lower Fermi energy than experiment finds. This situation might be different when scaling the calculations to high-temperatures with potential ionization-energy shifts, which we did not attempt.

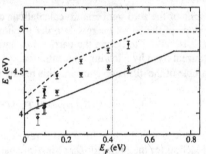

**Figure 3.** Activation energy for the generation rate, calculated as described in the text, and scaled to the high-temperature band gap with fixed intercepts (solid line); from an analysis of dopant and self-diffusion in silicon isotope multilayer structures (dashed line) [9]; and calculated with Eq. (14) from the values for the generation rate in [8] with either an averaged prefactor from [9] (empty circles) or the Fermi-level dependent prefactor from Eq. (16) (full circles).

In either case, the main LDA prediction of [7] that a neutral B$I^0$ pair dominates B diffusion in Si is confirmed. However, the GGA curve displays higher values with larger discrepancies to experiment and shows a $(p/n_i)^2$ behavior for Fermi energies close to the valence band edge (*i.e.*, high *p*-type doping), different from the experimental curve. On the other hand, its linear regime extends to higher Fermi energy values, in agreement with experiment.

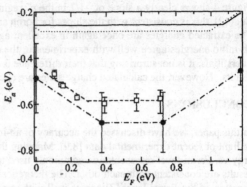

**Figure 4.** Activation energy for the mean free path. Solid (dotted) lines are calculated from one half the difference between B and *I* activation energies from LDA (GGA) calculations. Solid dots and dot-dashed lines are calculated from the experimental values from [9]. Open squares are calculated according to Eq. (14) from the mean free path values in [8] using the Fermi-level dependent prefactor from Eq. (16).

**Generation Rate:** According to Eq. (3), the generation rate *g* is proportional to the self-interstitial transport coefficient. We can thus use an equation analogous to Eq. (14) to extract the activation energy for Si interstitial diffusion from the *g* results of [8], provided we know the prefactor for the *I* transport coefficient and the capture radius for the B*I* reaction. For the latter, we take the value of 4.6 Å from [17], for the former, Eq. (16) with the values from [9]. To demonstrate the influence of the prefactor, we also use a constant, Fermi-level independent prefactor from averaging the values from [9]. Figure 3 shows the results of this procedure in comparison to the theoretical activation energies and the results from [9]. For the comparison with the theoretical calculations, the latter have been scaled to the 700 °C band gap, leaving the relative ionization levels fixed.

The theoretical GGA results and the values from [9] agree in the sense that they both predict $I^0$, $I^+$, and $I^{++}$ as dominant mobile species, depending on the Fermi level. The low theoretical value for the activation energy of $I^+$ ($T^+$) causes the ionization energies to be close to the band edges.

Using the constant prefactor, the experimental points from [8] follow a parabolic shape (open circles in Fig. 3), similar to the plot against $p/n_i$ in [8], which then could be fitted with a constant and a plus-two term. However, when using the Fermi-level dependent prefactor from Eq. (16), one can clearly distinguish two linear regions with different slopes of potentially ~1 and ~2 (with the exception of the point closest to the valence band edge which is somewhat lower). Thus, from this picture, one could argue that the slopes for single and double positive charge states of the self-interstitial can be seen. The activation energies extracted from the generation rate are very similar to the experimental results of [9] when using the Fermi-level dependent prefactor and thus higher than the calculated results, and vice versa when using the constant prefactor.

**Mean Free Path:** As shown in Eq. (12), the mean free path $\lambda$ can be calculated from the boron diffusion and self-interstitial transport coefficients as well as the B-*I* capture radius. Its energy dependence is given by one half the difference between the activation energies for boron and self-interstitial transport, respectively. In order to extract the energy dependence of the mean free path data in [8], we use Eq. (14) along with Eq. (12) to define the necessary prefactor from the data described above. The prefactor for self-interstitial transport is again calculated as a function of the Fermi level using Eq. (16).

Figure 4 shows clearly a slope of $-1/2$ in the $p^+$ regime and zero closer to mid-gap. Considering Eq. (12), this is consistent with the slopes for boron and self-interstitial diffusion in Figs. 2 and 3. The extracted energies are once again in excellent agreement with the findings of [9]. Since the ab-initio energies agree well with experiment in the case of boron, but seem to be low for self-interstitials, it is not surprising that their difference is considerably higher than both experimental results. However, the calculated charge states are once again in agreement with experiment.

## CONCLUSIONS

In this paper, we have discussed the accuracy of ab-initio calculations for $I$ and B diffusion in Si in light of recent experimental data [8,9]. Mapping the experimental data onto the activation energy vs. Fermi level representation commonly used to display ab-initio results, the experimental results are consistent with each other. The theoretical activation energy for boron diffusion as a function of the Fermi level [7] agrees well with experiment. For the self-interstitial, we find in line with other calculations an underestimation of the experimental values, despite using total-energy corrections. Since the predicted activation energies not only for boron, but also for other interstitial diffusers is in general predicted much closer to experiment than the self-interstitial energies, it indeed might be plausible that point-defect interactions and reactions could be the reason for the discrepancy between experiment and theory, as previously speculated in [9].

## REFERENCES

1. P. Pichler, *Intrinsic Point Defects, Impurities, and Their Diffusion in Silicon* (Springer, 2004).
2. A. Ural, P. B. Griffin, and J. D. Plummer, J. Appl. Phys. **85**, 6440 (1999).
3. J. Zhu, T. Diaz de la Rubia, L. H. Yang, C. Mailhiot, and G. H. Gilmer, Phys. Rev. B **54**, 4741 (1996); J. Zhu, Comput. Mater. Sci. **12**, 309 (1996).
4. C. S. Nichols, C. G. Van de Walle, and S. T. Pantelides, Phys. Rev. B **40**, 5484 (1989).
5. H. Jónsson, G. Mills, and K. W. Jacobsen, in *Classical and Quantum Dynamics in Condensed Phase Simulations*, edited by B. J. Berne *et al.* (World Scientific, Singapore, 1998), p. 385.
6. G. Kresse and J. Hafner, Phys. Rev. B **47**, 558 (1993); **49**, 14251 (1994); G. Kresse and J. Furthmüller, Comput. Mater. Sci. **6**, 15 (1996); Phys. Rev. B **54**, 11169 (1996).
7. W. Windl *et al.*, Phys. Rev. Lett. **83**, 4345 (1999).
8. D. De Salvador, E. Napolitani, S. Mirabella, G. Bisognin, G. Impellizzeri, A. Carnera, and F. Priolo, Phys. Rev. Lett. **97**, 255902 (2006).
9. H. Bracht, H. H. Silvestri, I. D. Sharp, and E. E. Haller, Phys. Rev. B **75**, 035211 (2007).
10. N. E. B. Cowern *et al.*, Phys. Rev. Lett. **65**, 2434 (1990).
11. N. E. B. Cowern *et al.*, Phys. Rev. Lett. **67**, 212 (1991).
12. T. R. Waite, Phys. Rev. **107**, 463 (1957).
13. C. S. Rafferty *et al.*, Appl. Phys. Lett. **68**, 2395 (1996).
14. W. Windl, Phys. Stat. Sol. (B) **241**, 2313 (2004).
15. G. D. Watkins, *A Review of EPR Studies in Irradiated Silicon*, in *Radiation Damage in Semiconductors* (Dunod, Paris, 1964), p. 97.
16. P. S. Gwozdz and J. S. Koehler, Phys. Rev. B **6**, 4571 (1972).
17. K. M. Beardmore, W. Windl, B. P. Haley, and N. Grønbech-Jensen, in Computational Nanoscience and Nanotechnology 2002 (Appl. Comput. Res. Soc., Cambridge, 2002), p. 251.
18. F. J. Morin and J. P. Maita, Phys. Rev. **96**, 28 (1954).
19. C. D. Thurmond, J. Electrochem. Soc. **122**, 1133 (1975).
20. R. K. Jain and R. J. Van Overstraeten, IEEE Trans. Electron Devices **ED-21**, 155 (1974).

Mater. Res. Soc. Symp. Proc. Vol. 1070 © 2008 Materials Research Society

# An Alternative Approach to Analyzing the Interstitial Decay from the End of Range Damage During Millisecond Annealing

Renata Camillo-Castillo[1,2], Mark E Law[3], and Kevin S Jones[2]

[1]IBM, 1000 River Street, Essex Junction, VT, 05452

[2]Department of Materials Science and Engineering, University of Florida, Gainesville, FL, 32611

[3]Department of Electrical and Computer Engineering, University of Florida, Gainesville, FL, 32611

## ABSTRACT

Flash-assist Rapid Thermal Processing (RTP) presents an opportunity to investigate annealing time and temperature regimes which were previously not accessible with conventional annealing techniques such as Rapid Thermal Annealing. This provides a unique opportunity to explore the early stages of the End of Range (EOR) damage evolution and also to examine how the damage evolves during the high temperature portion of the temperature profile. However, the nature of the Flash-assist RTP makes it is extremely difficult to reasonably compare it to alternative annealing techniques, largely because the annealing time at a given temperature is dictated by the FWHM of the radiation pulse. The FWHM for current flash tools vary between 0.85 and 1.38 milliseconds, which is three orders of magnitude smaller to that required for a RTA to achieve similar temperatures. Traditionally, the kinetics of the extended defects has been studied by time dependent studies utilizing isothermal anneals; in which specific defect structures could be isolated. The characteristics of Flash-assist RTP do not allow for such investigations in which the EOR defect evolution could be closely tracked with time. Since the annealing time at the target temperature for the Flash-assist RTP is essentially fixed to very small times on the order of milliseconds, isochronal anneals are a logical experimental approach to temperature dependent studies. This fact presents a challenge in the data analysis and comparison. Another feature of Flash-assist RTP which makes the analysis complex is the ramp time relative to the dwell time spent at the peak fRTP temperature. As the flash anneal temperature is increased the total ramp time can exceed the dwell time at the peak temperature, which may play a significantly larger role in dictating the final material properties. The inherent characteristics of Flash-assist RTP have consequently required the development of another approach to analyzing the attainable experimental data, such that a meaningful comparison could be made to past studies. The adopted analysis entails the selection of a reference anneal, from which the decay in the trapped interstitial density can be tracked with the flash anneal temperature, allowing for the kinetics of the interstitial decay to be extracted.

## INTRODUCTION

One of the challenges that the semiconductor industry continues to face in scaling CMOS devices is the formation of ultra shallow, highly doped junctions. Current MOS process flows employ a pre-amorphization implant which serves to distort the crystal structure such that there is reduced channeling of the subsequently implanted dopant atoms down the interstitial rows, thereby allowing for shallow junction formation. Although the ion implantation process offers a number of advantages, the inherent damage to the crystal structure must be repaired, which is usually achieved by a thermal anneal. Such anneals serve to reduce the mobility degradation in final device structures by the crystallographic defects[1] and should simultaneously activate the implanted dopant atoms, whilst minimizing their diffusion. The key to achieving these objectives is minimization of the total thermal budget imparted to the wafer. In recent years numerous techniques have been employed to realizing this end. Most recently millisecond annealing has come to the forefront in the semiconductor industry in its attempt to continue to drive Moore's law by reducing the total thermal budget imparted to the wafer.

Commercially known as Flash-assist rapid thermal process (RTP), millisecond annealing is designed to operate within the time gap between spike rapid and laser thermal processing techniques. The process offers three main advantages over conventional RTA systems which stem from the differences in the heating technology. Tungsten filament lamps are utilized in conventional RTAs compared to the water-walled arc lamps in flash systems. The arc lamps, which are very high quality optical sources, deliver greater power; have a faster response time and deliver short wavelength radiation which is more effective in heating the silicon substrate. High pressure argon plasma in an arc lamp when heated to 12000K produces radiation power of $1x10^6$W, enabling very high ramp rates that are four orders of magnitude higher than a conventional RTA, which has a resultant power of $1x10^3$W. The smaller thermal mass of the argon in the arc lamps also enables the arc lamps to respond approximately ten times faster than tungsten lamps[2]. Since the transition from heating to cooling is also a function of the response time of the heat source, a wafer heated by flash techniques will transition much faster from heating to cooling. Finally, over 95% of the arc radiation is below the 1.2μm band gap absorption of silicon compared to 40% for radiation generated by the tungsten lamps[2], hence it is more effective in heating the wafer.

The flash process uses a continuous arc lamp to heat the bulk of the wafer to an intermediate temperature (iRTP), where the dwell time is essentially zero. This heating is slower than the thermal conduction rate through the wafer, thus the entire wafer remains at approximately the same temperature[3]. The iRTP serves as the initial temperature of the flash anneal. Subsequently, a capacitor bank is discharged through an arc lamp which adds additional power to the device side of the wafer, at a rate much faster than the thermal conduction rate. Short time pulses allow for heating of the surface of the wafer to the peak flash temperatures (fRTP), while the substrate never attains these high temperatures. This is possible since the time constant of the flash which is on the order of 1 millisecond is much shorter than the thermal time constant of the wafer (~10-20ms). Therefore a thin slice of the device side of the wafer is heated and cooled rapidly at rates on the order of $1x10^6$ °Cs$^{-1}$. The fast cooling is achieved since the bulk of the wafer acts as a heat sink removing heat from the top layer via conduction much more efficiently and faster than can be accomplished in bulk cooling. The high absorbance of the reactor chamber also complements the cooling rates.

Knowledge of the defect structures which exist in the silicon lattice is crucial to understanding the variations in the silicon interstitial supersaturation and therefore dopant diffusion, during thermal annealing process. It has been previously demonstrated that the silicon interstitial supersaturation in the vicinity of the End of Range (EOR) damage drives the anomalous diffusion or transient enhanced diffusion (TED)[4] of dopant atoms. Additionally the release of interstitials which accompanies dissolution of the EOR damage also contributes to TED. TED is known to occur during the early stages of the annealing process, when the interstitial supersaturation is high. The interstitial supersaturation is also known to be a function of the defect size and also the anneal temperature, being high for smaller defects and lower anneal temperature. Hence, investigations which can provide further insight into the processes that occur in the early stages of the thermal anneal of an implanted wafer, when the defects in the early stages of evolution, are beneficial. Damage annealing in the millisecond time regime presents a unique opportunity to investigate the early stages of the damage evolution process, which were not previously possible with former available annealing processes such as rapid thermal annealing. The high temperatures attainable by the flash annealing technique coupled with the extremely short anneal times, has enabled the defect evolution to be mapped from the early stages in the evolution to mature defect structures, which have been extensively characterized in the past, thus allowing for a more complete picture of the defect evolutionary processes.

## EXPERIMENT

200mm 12ohm-cm (100) n-type Czochralski (CZ) grown silicon wafers were amorphized with a 30keV germanium ion implant at a dose of $1x10^{15}cm^{-2}$. The ion implantations were performed in deceleration mode on an Applied Materials XR80 Leap Implanter at a twist of $27^{o}$ and a tilt of $7^{o}$. The silicon wafers were then annealed by Flash-assist RTP at Vortek technologies in Vancouver, Canada. These thermal anneals comprised heating the bulk silicon wafers by arc irradiation at a heating rate of $150^{o}Cs^{-1}$ to an intermediate temperature (iRTP) of $700^{o}C$, where the dwell time was 0s. The iRTP served as the initial temperature for the flash anneal. The flash anneal (fRTP) which arises by discharging capacitor banks into flash lamps, produced a pulse of radiation with a full width at half maximum ranging 0.85-0.9ms, enabling temperatures of 1100, 1200 and $1300^{o}C$ at a heating rate of $10^{6}\,^{o}Cs^{-1}$. Only the near surface region of the wafers was heated during the fRTP anneal, thus allowing for conductive heat loss through the cooler layers on the backside of the wafer, and rapid cooling rates very similar to the heating rates attained of $10^{6}\,^{o}Cs^{-1}$. Throughout the course of the flash pulse the backside of the wafers experienced an increase in temperature of approximately $100^{o}C$. Once the flashed wafer surface was in thermal equilibrium with the bulk of the wafer, subsequent cooling was dominated by radiative heat loss to the surrounding black environment at a maximum rate of $90^{o}Cs^{-1}$. Plan-view transmission electron microscopy (PTEM) was used to investigate the effects of flash-assist RTP on the end of range (EOR) defect evolution and morphology. PTEM sample preparation entailed cutting a 3mm disc of each sample, chemically mechanically thinning the sample using 15µm alumina slurry and chemically etching the backside of the sample with a 1:3 49%HF: HNO3 solution, until electron transparent regions were evident. PTEM images of the EOR damage in these regions were then captured using a JEOL 200CX microscope, operating at an accelerating voltage of 200keV in weak-beam-dark-field (WBDF), g220 two-beam imaging conditions. The resultant images of the EOR damage were analyzed using the quantification technique of Bharatan[5].Cross-sectional TEM (XTEM) was also performed using a JEOL 200CX TEM operating at an accelerating voltage of 200keV. However the images were taken under g110 bright field conditions. The XTEM samples were prepared via a Focused Ion Beam and the method was used to determine amorphous layer depths and track layer re-growth.

## DISCUSSION

The 30keV germanium implant produced a continuous amorphous layer from the surface to a depth of 50nm. WBDF PTEM images of the EOR damage observed for the Flash-assist RTP are illustrated in Figure 1. Examination of the damage due to the 30keV germanium PAI reveals the presence of small dot-like interstitial clusters of very high density, immediately after the 700°C iRTP anneal. On application of the 1100°C fRTP, some signs of coarsening become apparent. The EOR defects appear somewhat larger and of smaller areal density. This trend is sustained as the fRTP temperature is raised to 1200 and 1300°C, in which cases the defects are observed to coarsen into {311}-type structures and dislocation loops. Defect density as a function of the fRTP anneal temperature is shown in Figure 2a, which exhibits a decreasing defect density with increasing fRTP anneal temperature. Defect densities on the order of $1 \times 10^{11} cm^{-2}$ are attained for anneal temperatures of 700, 1100 and 1200°C. However, the density rapidly plummets two orders of magnitude on application of the 1300°C fRTP anneal. An analysis of the defect size indicates that the defect diameter does not significantly change between 700 and 1100°C averaging between 5 and 7nm, respectively. As the fRTP temperature is increased to 1200°C the predominant {311}-defects present in the microstructure are of an average length of 14nm, whereas, the dislocation loops observed at 1300°C are approximately 18nm in diameter. Quantification of the interstitials trapped in the EOR defects were conducted using the average defect sizes summarized above. As the fRTP anneal temperature was increased the number of interstitials trapped by the EOR defects were observed to decrease. The trapped interstitial dose decreased from values on the order of $1 \times 10^{14} cm^{-2}$ between 700 and 1200°C to $1 \times 10^{13} cm^{-2}$ for the 1300°C fRTP, as shown in Figure 2b. This trend is very similar to that observed for the defect density and demonstrates an exponential decay with anneal temperature.

Figure1. WBDF PTEM images of the EOR defects imaged under $g_{220}$ two-beam conditions of the 30keV $1 \times 10^{15} cm^{-2}$ Ge amorphizing implant. a)700°C iRTP, b) 700°C iRTP,1000°C fRTP, c) 700°C iRTP, 1100°C fRTP, d) 700°C iRTP, 1200°C fRTP, e) 700°C iRTP, 1300°C fRTP

The damage observed in the microstructure can be classified as Type II, which results when an amorphous layer is present and occurs beyond the amorphous-crystalline interface in the EOR region[6]. The evolution of Type II defects from point defects to extrinsic dislocation loops upon annealing is believed to occur via intermediate defect configurations and it is now widely accepted that sub-microscopic interstitial clusters (SMICs)[7,8,9] are the precursors for the formation of {311}-type defects[10,11,12]. The {311}-type defects which are metastable, eventually unfault to form dislocation loops[13]. The fact that different defect structures were identified at different annealing temperatures in this work is not unexpected. Isochronal anneals although a valid experimental approach to temperature dependent studies, do not yield EOR defects in the same phase of their evolution. At a given isochronal annealing time, lower temperature anneals would generate EOR defects in their earlier nucleation, growth and coarsening stages, while high temperature anneals would result in defects further along in their evolution, possibly in the dissolution regime[14]. The evolution of the Type II damage observed in this study concurs with previous findings – the EOR defects are observed to evolve from {311}-type defects into dislocation loops with increasing fRTP anneal temperature. However, there has been no previous evidence of the EOR defects evolving from these dot-like structures to {311}-type defects. The structure of the dot-like interstitial clusters observed after the 700°C iRTP and 1100°C fRTP anneals proved difficult to discern from the PTEM images. The exact configuration of small interstitial clusters has been the center of a number of investigations, yet very little is still known. Recent experimental and theoretical data[15,16,17,18] demonstrates that precise cluster sizes exhibit enhanced stability, indicated by the existence of minima and maxima in the cluster binding energy curve. However, considerable debate remains over the exact sizes of the stable clusters. Cowern[17] found that interstitial clusters which consisted of more than twenty atoms had a similar differential formation energy to the {311}-type defect, suggesting that the interstitial clusters undergo a transition to {311}-type defects at a smaller cluster size. Other investigations [19,20,21] support this idea and suggest that the transition from small interstitial clusters to {311}-type defects occur for interstitial clusters containing eight atoms. Hence the small interstitial clusters observed at 700°C iRTP and the 1100°C fRTP may in fact be {311}-type defects, since the total number of atoms in these structures exceeds 8 atoms and the smallest defect that can be imaged by a conventional TEM is approximately 100 atoms[22]. Other studies[23,24] of lower energy germanium amorphizing implants propose that small interstitial clusters may exhibit a defect morphology very similar to plate-like dislocation loops. These dislocation loops have been shown to be very unstable, dissolving with an activation energy of $1.13 \pm 0.14\text{eV}$[25]. If the dot-like interstitial clusters in these experiments are analogous to the loops observed by Gutierrez and King then any additional thermal budget applied to them should result in a defect dissolution behavior in accordance with their kinetics.

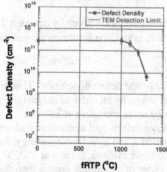

Figure 2a. Defect density as a function of fRTP anneal temperature for a $1\times10^{15}\text{cm}^{-2}$ 30keV Ge amorphizing implant

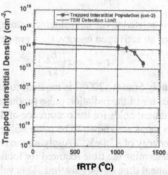

Figure 2b. Trapped Interstitial Density as a function of fRTP anneal temperature for a $1\times10^{15}\text{cm}^{-2}$ 30keV Ge amorphizing implant

In an effort to corroborate that the defect evolution captured during the lower fRTP temperatures, was in its initial stages, post flash thermal anneals were conducted. The experimental selection was based on the notion that the additional thermal budget would coarsen or evolve the EOR damage if it was in its infancy or dissolve the damage if the dissolution adhered to the findings of King[25]. The modification to the original experiment entailed subjecting the wafer which was annealed at a 700°C iRTP, 1100°C fRTP thermal condition a 950°C spike RTA. The thermal profile comprised a ramp up to an intermediate soak temperature of 600°C at a heating rate of 100°Cs$^{-1}$; at which the temperature was held constant for 10 seconds. After which the temperature was increased at the same heating rate to 950°C, where the dwell time was 0 seconds. The 950°C spike RTA evolved the EOR defects from dot-like interstitial clusters, which were present after the 1100°C fRTP anneal, to dislocation loops of much lower density. This critical data provided the evidence needed to substantiate the supposition that at the lower fRTP anneal temperatures, the system was in fact in the early stages of evolution. Robertson[22] demonstrated for amorphizing implants that approximately 75% of {311}-type defects unfaulted to dislocation loops. Hence it can be inferred that a possible evolutionary path from the dot-like interstitials to dislocation loops involved the formation of {311}-type defect. This inference is supported by the existence of the {311}-type defect in the microstructure after the 1200°C fRTP anneal. The evolution of {311}-type defects from dot-like interstitial clusters have not been previously reported and strongly suggest the existence of another defect regime prior to the formation of the {311}-type defect and dislocation loop.

{311}-type defects have been shown to be relatively unstable, dissolving after only 3 minutes at 815°C with an activation energy of 3.7eV[10,11]. These dissolution kinetics indicate that such defect structures can only exist for extremely short anneal times at very high anneal temperatures. Knowledge of the time, t1 required for {311}-type defect dissolution at a given anneal temperature, T1 allows one to calculate the anneal time, t2 required to effect similar defect structures at another temperature, T2 from the ratio of the time constants for decay, $\tau$ in accordance with the relation

$$t_2 = \frac{\tau_1 t_1}{\tau_2}$$
Equation 1

in which $\tau$ is related to the anneal temperature, T by an Arrhenius relation given by equation 2 that includes an activation energy, $E_a$, a pre-exponential factor, $\tilde{\tau}_0$ and the Boltzmann constant, k

$$\tau = \tau_0 \exp\left(\frac{E_a}{kT}\right)$$
Equation 2

Approximately 5ms at 1200°C is required to effect total {311}-type defect dissolution. This time exceeds the size of the radiation pulse produced by the flash lamps by an order of magnitude. A similar argument holds for the occurrence of dislocation loops upon annealing at 1300°C. Dislocation loops are known to be more stable than {311}-type defects and to dissolve with an activation energy of ~5.0eV[26]. Applying the relation in equation 1 and an activation energy, $E_a$ of 5.0eV, to determine the time required for total dissolution of these extrinsic defects at 1300°C, yields an anneal time of roughly 9ms. This value is also an order of magnitude greater than the time for which the wafer was subjected to this temperature during the flash-assist RTP anneals.

The number of interstitials trapped by the EOR defects remained effectively unchanged for the sample subjected to the 700°C iRTP and the 1100°C fRTP anneals. The extremely small defect sizes at these temperatures made it exceedingly difficult to determine the exact area of the dot-line interstitial clusters which were present in the microstructure at this time, introducing a larger error in the analysis. However the values obtained are in agreement with previous studies which examined the number of trapped interstitials for germanium amorphizing implants[23,27]. At the higher fRTP anneal temperatures, the larger defect sizes allowed for the determination of the trapped interstitial density with more confidence. Hence, the reduction in the number of trapped interstitials observed between 1100 and 1300°C fRTP anneal temperatures is believed to be accurate. This reduced number can possibly be accounted for by defect dissolution and interstitial recombination at the surface.

It is important to recognize that a number of simultaneous complex reactions are occurring during the thermal cycle, which impacts the interstitial population in the microstructure. The nature of the Flash-assist RTP is such that the system is in a continuous dynamic state as the time spent at the target temperature is on the order of milliseconds. Hence a significant portion of the thermal anneal is spent in either the ramp up or ramp down stages of the thermal profile. As a consequence, the interstitial supersaturation and diffusivities are directly impacted and are not fixed values. These parameters have a large influence on the interstitial recombination rate, which has been demonstrated to be a di-interstitial mediated mechanism proportional to the square of the interstitial supersaturation[28] [LAW98]. Hence at higher temperatures, the recombination rate should decrease with the reductions in supersaturation levels. Conversely, larger interstitial diffusivities at higher anneal temperatures increases the probability of interstitial recombination. These competing mechanisms dictate the interstitial recombination rate, which varies with the system temperature. At the fRTP anneal temperatures investigated, both the defect and trapped interstitial densities of the dot-like structures were seen to simultaneously fall of as the temperature is increased. These events suggest that the defects may be in a coarsening regime in which a fraction of the interstitials is not re-captured by other defect structures. The likelihood that interstitials are continually lost by recombination at the Si-SiO₂ interface or at the amorphous-crystalline interface during the re-crystallization process is therefore high. Consequently, the system is viewed as a "leaky box" from which interstitials are lost as the EOR defects undergo coarsening.

Interstitial loss from the EOR by defect dissolution is also a likely process to account for the reductions detected. Robertson[22] demonstrated approximately 25% of the {311}-type defects undergo dissolution versus the unfaulting process to form dislocation loops. Furthermore, application of the dissolution kinetics for {311}-type defects[10,11] and dislocation loops[26], established that this process was a viable explanation for the diminished interstitial densities observed at 1200 and 1300°C fRTP temperatures. Both of these processes of interstitial losses via recombination and defect dissolution have therefore been attributed to the fall in trapped interstitial densities.

## A New Approach to Kinetic Analysis of the Interstitial Loss from the EOR damage

The nature of the Flash-assist RTP makes it is extremely difficult to reasonably compare it to alternative annealing techniques, largely because the annealing time at a given temperature is dictated by the FWHM of the radiation pulse. The FWHM for current flash tools vary between 0.85 and 1.38ms, which compares to a FWHM of 2s for a RTA to achieve similar temperatures. Since the annealing time at the target temperature for the Flash-assist RTP is essentially fixed to very small times on the order of milliseconds, isochronal anneals are a logical experimental approach to temperature dependent studies. However, such anneals yield defect morphologies at different stages of the defect evolution. For instance a lower temperature anneal is expected to probe the early stages of the evolution such as nucleation and growth, whereas a high temperature anneal may capture the defects further along in their evolution, possibly in the dissolution regime. Traditionally, the kinetics of the extended defect have been studied by time dependent studies utilizing isothermal anneals [26,10,11,25], in which specific defect structures could be isolated. The characteristics of Flash-assist RTP do not allow for such investigations in which the EOR defect evolution could be closely tracked with time. This fact presents a challenge in the data analysis and comparison, since different defect structures are detected at each fRTP anneal temperature investigated. Another feature of Flash-assist RTP which makes the analysis complex is the ramp time relative to the dwell time spent at the peak fRTP temperature. The ramp up to and ramp down from the desired anneal temperature becomes increasingly important in analyzing an anneal process, as the time spent at the target temperature is decreased. The ramp times are on the same order as the FWHM of the anneal pulse itself. The ratio of the FWHM value to the total ramp time is shown in Table 1, which illustrates that as the fRTP temperature is increased the total ramp time can exceed the dwell time at the peak temperature. This simple calculation emphasizes the importance of accounting for these portions of the thermal profile, which may play a significantly larger role in dictating the final material properties. The inherent characteristics of Flash-assist RTP have consequently required the development of another approach to analyzing the attainable experimental data, such that a meaningful comparison could be made to past studies. The adopted analysis method entailed the selection of a reference anneal, from which the decay in the trapped interstitial density was tracked with the fRTP anneal temperature, allowing for the kinetics of the interstitial decay to be extracted. This analytical procedure as well as the data analysis is discussed in the ensuing sections.

The reductions in trapped interstitial density discussed in the previous section, suggested that the additional thermal budget associated with the fRTP anneal effected the interstitial losses from the EOR damage. In order to extract the kinetics associated with the interstitial decay, it was necessary to estimate the rate at which these interstitials were being lost. This calculation required at least two data points. The experiment yielded the number of trapped interstitials for each fRTP anneal, which served as one data point. However, the initial trapped interstitial concentrations were not available; hence an assumption had to be made. Since the iRTP anneal temperature from which all the fRTP anneals were conducted was not varied for the experiment, it was assumed that the interstitial density obtained for this temperature could serve as an initial trapped interstitial value, as all the wafers were subjected to this temperature prior to the fRTP anneal. This allowed for the decay in the trapped interstitials during the flash portion of the thermal profile i.e. fRTP anneal, to be extracted such that the effect of the flash on the defects could be isolated.

In accordance with kinetic rate theory the interstitial decay was approximated by the relation

$$\frac{\partial C_i}{\partial t} = -\frac{C_i}{\tau} \qquad \text{Equation 3}$$

where $C_i$ is the concentration of trapped interstitials ($cm^{-2}$), $t(s)$ is anneal time and $\tau$ (s) is the captured interstitial lifetime, which is related to the anneal temperature, $T(K)$ by an Arrhenius expression that includes an activation energy, $E_a$.

$$\tau = \tau_0 \exp\left(\frac{E_a}{kT}\right) \qquad \text{Equation 4}$$

The process simulator FLOOPS[29] was utilized to calculate the trapped interstitial density for each fRTP anneal temperature from an initial trapped interstitial density (value for the 700°C iRTP anneal), by fitting the parameters $\tau_0$ and $E_a$. The simulation incorporated the temperature-time variation for each anneal, thus enabling τ to be accurately accounted for in time as the temperature was varied. This facilitated a precise integration of the trapped interstitials with time. The close agreement between the trapped interstitial populations for the simulations and the experiment is also demonstrated in Figure 3a. While the decay rates derived from the fits of the experimental trapped interstitial populations are illustrated in Figure 3b. The interstitial decay rate varied linearly with the inverse fRTP temperature, yielding an activation energy, Ea of 2.1 ± 0.05eV and pre-exponential factor, $K_0$ of $3.3 \times 10^{10}\,s^{-1}$. The interstitial decay rates varied over two orders of magnitude for the fRTP temperatures investigated, from approximately $100s^{-1}$ at 1000°C, compared to $2000s^{-1}$ at 1200°C. The kinetic parameters extracted in this work are pertinent only to the interstitial loss from the EOR defects during the flash portion of the thermal profile (i.e. fRTP anneal), since the root cause for the drop in trapped interstitials could not be pinned to any particular process and more than one defect structure was analyzed to yield the trapped interstitial data presented herein.

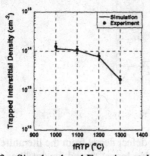

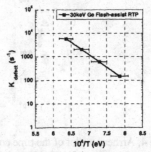

Figure 3a. Simulated and Experimental trapped interstitial densities for a 30keV, $1 \times 10^{15}\,cm^{-2}$ germanium amorphizing ion implant into (100) silicon

Figure 3b. Arrhenius plot of the time constant derived for defect decay extracted from the simulated experimental data, indicating an activation energy, $E_a$ of 2.1 ±0.05eV for dissolution

Comparison of the interstitial decay rates for {311}-type defects[10] and dislocation loops[26,25] demonstrates the much higher interstitial decay rates during the flash anneal compared to former studies as demonstrated in Figure 4. The interstitial decay rates for the flash anneals range two orders of magnitude between $1 \times 10^2$ and $1 \times 10^4 s^{-1}$ for anneal temperatures from 1000 through 1300°C, which are three orders of magnitude larger than the maximum rate previously reported, obtained by Seidel[26] for the interstitial decay from stable dislocation loops. The interstitial decay rates for {311}-type defects and small dislocation loops were much lower. The temperature range of Seidel's experiments coincides with flash temperatures lower than 1200°C. Yet, the interstitial decay rates are vastly dissimilar for this temperature regime, suggesting the interstitial loss is most probably from a defect not similar to the dislocation loop, but suggests the existence of a highly unstable defect structure at these temperatures. This correlates with the dot-defects observed at this temperature during the flash anneal. Examination of the differences in the activation energies for defect dissolution in the literature provides further insight into the characteristics of the proposed highly unstable defect. The extracted activation energy for interstitial loss during the flash anneal is 2.7eV and 1.6eV smaller than the values obtained for dislocation loop and {311}-type defect dissolution, respectively. Seidel extracted an activation energy of approximately 4.8eV[26] for interstitial loss from dislocation loops, which is remarkably similar to the activation energy value for silicon interstitial self-diffusion[30,31] whereas the activation energy for {311}-type dissolution was determined to be 3.7eV[10]. Most recently, King's studies revealed the existence of dot-like dislocation loops which dissolved with an activation energy of 1.13eV[25]. The fact that the activation energy determined for the interstitial decay during the flash anneal is not similar to those previously extracted values, supports the theory that the interstitial loss is not from comparable defect structures. It also clearly indicates that this defect is less stable than the {311} defect and dislocation loops.

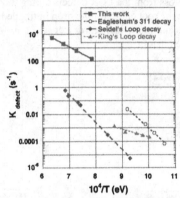

Figure 4. Arrhenius plot of the time constant derived for defect decay from the literature, including decay rates for {311}-type defects[10,11] dislocation loops[26,25] and the dot-like defects observed in this work

The dot-like defects existing in the structure for flash anneal temperatures of 1100°C and lower, signifies that these defects only exist in the early stages of annealing and either evolve into {311}-type defects or dissolve. The decrease in the trapped interstitial concentration between 1100 and 1200°C flash anneal temperatures, suggests that some of these defects dissolved, losing interstitials to either the surface to the bulk of the material. The data implies that those dot-like defects which did not dissolve, must have therefore evolved into the {311}-type defect detected after the 1200°C anneal, since they are no longer observed in the microstructure. This validates the supposition that the dot-like defect is a precursor for the {311}-type defect. Consequently, the extracted kinetics for the decrease in trapped interstitial density as a function of the temperature applies to the dissolution of this dot-like defect structure.

## CONCLUSIONS

The inherent nature of Flash-assist RTP which makes it attractive as a possible annealing technique for ultra shallow-junction formation, presents challenges in evaluating its effectiveness in eliminating the EOR damage. These challenges stem from the very short anneal times, on the order of milliseconds, that Flash-assist RTP offers. These short anneal times makes time dependent studies of the damage evolution impossible. Consequently, the use of isochronal anneals for temperature dependent studies of the EOR defect evolution is necessitated. This experimental approach yields the extended defects at different stages of their evolution, requiring an alternate method to conventional data analysis techniques. The method employed in this work tracks the interstitial decay with anneal temperature by utilizing the initial interstitial density as that for the iRTP anneal, which was not varied for the experiment. Thus the interstitial decay rate over the fRTP portion of the anneal could be isolated, enabling the effectiveness of the flash to be evaluated.

The short anneal times and high temperatures attainable with Flash-assist RTP have enabled investigations of the early stages of the EOR defect evolution, which have revealed the existence of a highly unstable defect structure preceding the {311}-type defect. This defect is believed to be a precursor for the {311}-type defect and is thought to go through one of two evolutionary paths, defect dissolution or evolution into the {311}-type defect. An activation energy of 2.1eV was determined for dissolution of these dot-like defects. This is the first experimental evidence of the existence of such a defect, whose dissolution kinetics differ from previously reported defect structures.

# REFERENCES

1. International Technology Roadmap for Semiconductors. Available: http://public.itrs.net/
2. D. M. Camm, B. Lojek, Proc. 2nd Int. Conf. Advanced Thermal Processing of Semiconductors (RTP 1994), p.259.
3. G.C. Stuart, D.M. Camm, J. Cibere, L. Kaludjercic, S.L. Kervin, B. Lu, K.J. McDonnell, N. Tam. 10th IEEE International Conferenceof Advanced Thermal Processing of Semiconductors, pg. 77 (2002).
4. C. Bonafos, M. Omri, B. de Mauduit, G. BenAssayag, A. Claverie, D. Alquier A. Martinez, D. Mathiot, J. Appl. Phys. 82, 2855 (1997)

5. S. Bharatan, J. Desrouches, and K. S. Jones, Materials and Process Characterization of Ion Implantation, Vol. 4, p. 222. (Ion Beam, 1997).
6. K.S. Jones, S. Prussin, and E. R. Weber, Appl. Phys. A 45, 1 (1988).
7. J.L. Benton, S. Libertino, P. Kringhoj, D. J. Eaglesham and J. M. Poate, J. Appl. Phys. 82 (1), 120 (1997).
8. S. Coffa, S. Libertino, C. Spinella, Appl. Phys. Lett, 76(3), 321 (2000).
9. S. Libertino, J. L. Benton, S. Coffa, and D. J. Eaglesham, Mater. Res. Soc. Symp. Proc. 504, 3, (1998).
10. D.J. Eaglesham, P. A. Stolk, H.-J. Gossmann, and J. M. Poate, Appl. Phys. Lett. 65 (18) (1994).
11. P.A. Stolk, H.-J. Gossmann, D. J. Eaglesham, D. C. Jacobson, C. S. Rafferty, G. H. Gilmer, M. Jaraı´z, J. M. Poate, H. S. Luftman, T. E. Haynes, J. Appl. Phys. 81 (9), (1997).
12. G.Z. Pan, K.N. Tu, A. Prussin, J. Appl. Phys. 71(5), 659, (1997).
13. J. Li and K.S. Jones, Appl. Phys. Lett. 73 (25), (1998).
14. P. Keys. PhD. Dissertation, Department of Materials Science and Engineering. University of Florida, Gainesville, FL (2001).
15. J. Kim, J. W. Wilkins, F. S. Khan, and A. Canning, Phys. Rev. B 40, 10351 (1989).
16. G. H. Gilmer, T. Diaz de la Rubia, D. M. Stock and M. Jaraiz, Nucl. Instr. Meth. Phys. Res. B 102, 247 (1995).
17. N.E.B. Cowern, G. Mannino, P. A. Stolk, F. Roozeboom, H. G. A. Huizing, J. G. M. van Berkum, F. Cristiano, A. Claverie, M. Jaraı´z, Phys. Rev. Letts. 82 (22), 4460 (1999).

18. M. P. Chichkine, M. M. De Souza, and E. M. Sankara Narayanan, Phys. Rev. Lett. 88, 085501 (2002).
19. M.M. De Souza, M.P. Chichkine, E.M. Sankara Narayanan, Mater. Res. Soc. Proc. N0. 610, B11.3.1, (2000).
20. A. Claverie, B. Colombeau, F. Cristiano, A. Altibelli, C. Bonafos, Mater. Res. Soc. Proc., 669, J9.4 (2001).
21. A. Claverie, B. Colombeau, F. Cristiano, A. Altibelli, C. Bonafos, Nucl. Instr. Meth. Phys. Res. B 186 (1-4), 281 (2002).
22. L.S. Robertson, K.S. Jones, L.M. Rubin, J.Jackson, J. Appl. Phys. 87(6), 2910 (2000).
23. A.F. Gutierrez, M.S. Thesis, Department of Materials Science and Engineering, University of Florida: Gainesville (2001)
24. A.C. King. M.S. Thesis, Department of Materials Science and Engineering, University of Florida, Gainesville FL (2003).
25. A.C. King, A. F. Gutierrez, A. F. Saavedra, K. S. Jones and D. F. Downey, J. Appl. Phys. 93(5), 2449 (2003).
26. T.E. Seidel, D.J. Lischerner, C.S. Pai, R.V. Knoell, D.M. Maher, D.C. Jacobson, Nucl. Instr. Meth. Phys. B 7/8, 251, (1985).
27. I. Avci. Ph.D. Dissertation, Department of Electrical and Computer Engineering. University of Florida. (2002)
28. M. E. Law, Y. M. Haddara, and K. S. Jones, J. Appl. Phys. 84, 3555 (1998).
29. FLOOPS, Mark Law, University of Florida, Electrical and Computer Engineering Department (2003).
30. A. Ural, P. B. Griffin, J. D. Plummer. Appl. Phys. Lett., 79(26), 24, (2001).
31. H. Bracht, E.E. Haller, R. Clark-Phelps, Phys. Rev. Lett. 81(2), 393, (1998).

Mater. Res. Soc. Symp. Proc. Vol. 1070 © 2008 Materials Research Society 1070-E06-10

## Dopant Condensation beyond Solubility Limit in the Vicinity of Silicon/Silicide Interface Based on First-Principles Calculations

Takashi Yamauchi, Yoshifumi Nishi, Atsuhiro Kinoshita, Yoshinori Tsuchiya, Junji Koga, and Koichi Kato

Advanced LSI Technology Laboratory, Corporate Research & Development Center,Toshiba Corporation, 1,Komukai-Toshiba-cho, Saiwai-ku, Kawasaki, 212-8582, Japan

### ABSTRACT

We studied B atom condensation behavior around the nickel silicide (NiSi)/Si Schottky interface by first-principles calculations. We found that the Schottky barrier height (SBH) for a hole is dramatically reduced by 0.3eV due to atomic-scale dipole generation across the NiSi/Si interface, in case of substitution of a B atom for a Si atom in a few Si layers around the interface. We also found that this interface dipole generation leads to the remarkable increase in the B solubility limit at the interface. The possibility of interface dipoles comforting Schottky (DCS) barrier height was verified by both the observed B profiles by secondary ion mass sepectroscopy (SIMS) and the measured I-V characteristics for the NiSi/Si Schottky diodes formed through the ion implantation after silicidation (IAS) process. Furthermore, we studied the SBH modulation effect induced by dipoles, which other impurity atoms (As, Mg, Al, In) generate at the NiSi/Si interface. Based on these understandings, we discussed their applicability to the DCS junction, showing a principle of choosing dopants.

### INTRODUCTION

In the trend of scaling down metal-oxide-semiconductor field effect transistors (MOSFETs), reduction of contact resistance at the silicide/Si interface will be essential for higher performance[1]. Nickel silicide (NiSi) is considered as a substitute for a present electrode material in MOSFETs, cobalt silicide (CoSi$_2$), because silicidation temperature can be reduced as compared with the case of the conventional CoSi$_2$[2-4]. Hence, we have focused on the NiSi/Si Schottky interface.

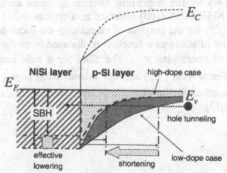

**Figure 1.** Schematic view of a hole transport at the NiSi/Si interface. As the dopant concentration at the interface increases, the stronger band-bending effect of the Si layer leads to the increase in the tunnel probability for a hole passing through the interface.

An ordinary method to increase the dopant concentration at the interface is ion implantation before silicidation (IBS) process. The dopant atoms are consequently condensed around the interface by snowplow effect[5-8], leading to the effective lowering of the Schottky barrier height (SBH) because of the band bending enhancement of the Si layer, as shown in Fig.1. However, this band bending technique does not reduce the SBH in further scaled MOSFETs, because the Fermi level cannot be pinned in the Si layer. In this context, we studied another possibility of SBH modulation technique, based on the first-principles calculations. Throughout our calculations, we found that a large atomic-scale dipole between impurity and silicide atoms across the interface leads to the dramatic reduction of SBH. Also, impurity atoms are expected to be condensed because of a large energy gain at the interfaces. Based on these results, we proposed a novel dipole comforting Schottky (DCS) junction[9]. Furthermore, the thickness of the Si layer interfacing with the NiSi layer can be 1nm or less.

We applied this idea to the actual process through experimental techniques. In this paper, we focused on the P-type DCS junction, since reduction of SBH for P-type Si is normally smaller than for N-type Si. The NiSi/Si Schottky diodes were prepared by ion implantation after silicidation (IAS) process. This is because the calculated results suggest that B implantation after silicidation leads to larger B concentration at the interface than that before silicidation. We evaluated the interface dipoles contribution to the measured SBH reduction. As a result, the B atoms were found to be condensed beyond solubility limits on the interface Si side, and we verified the generated interface dipoles actually reduce the SBH. Furthermore, we explored the other possibility of another type of impurity atoms applicable to the DCS junction. Among some other impurity atoms (As, Mg, Al, In), the calculated SBH modulation due to dipoles generated around these impurity atoms were found to be further enhanced in some cases. With these understandings, we propose a principle for choosing dopants towards ultimate lowering of the contact resistance in ultimately scaled MOSFETs.

## CALCULATION AND EXPERIMENTAL METHODS

Our calculations are based on density functional theory with generalized gradient approximation (GGA) including spin-polarization. All the calculations were performed with (1×1) and (2×2) unit cells consisting of 3 vacuum layers/4NiSi layers/12Si layers in Fig.2. Inversion symmetry is applied to the end of the Si layer to increase the computational efficiency. The calculations were performed using ultrasoft pseudopotentials[10] for the specific atoms with 16-$k$ points for Brillouine zone samplings. We found that the cut-off energies of 25 Ry for the wave functions and 196 Ry for the augmented electron densities are sufficient for converging energies. We chose the NiSi(020)/Si(100) interface as a typical case[11]. We divided this unit cell into 51 equal slices($\Omega$) for the purpose of calculating the local density of states (LDOS), which displays the density of states as a function of distance from the interface, and identified each region by its central coordinate($Z_\Omega$). The SBH for a hole can be evaluated from the difference between the Fermi level ($E_F$) and the valence band edge ($E_v$) of the LDOS at the farthest point from the interface (1.59nm) in the Si layer. We used the crystal structural data for NiSi in the references[12,13].

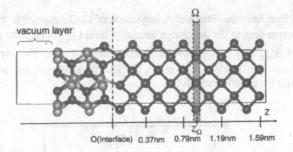

**Figure 2.** Calculated atomic structure of the NiSi/Si interface. Gray and white circles stand for Si and Ni atoms, respectively.

Both IAS and IBS processes were employed for NiSi/Si Schottky diodes, to clarify the behavior of a B atom around the interface. The experimental procedures are illustrated in Fig.3. The I-V characteristics for these diodes were measured at sufficiently low temperature, and then the SBHs were evaluated through reproducing calculated curves with the free-electron model[14].

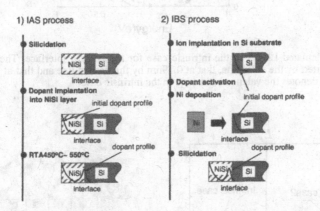

**Figure 3.** Formation procedures of the NiSi/Si Schottky junction. In the IAS process, ion implantation is performed after silicidation, on the other hand, in the IBS process, before silicidation. In the IBS process, Schottky junction is formed with snowplow effect.

## RESULTS AND DISCUSSIONS

Figure4 shows the calculated LDOS for the NiSi/Si interface in the intrinsic case. Since the metallic region penetrates into the Si layer, the Si gap does not fully open in the LDOS at the point less than 0.79nm from the interface. A large peak of LDOS appears in the Si band gap at the interface, corresponding to the the the charge neutrality level ($E_{CNL}$)[15]. The energy difference (~0.15eV) between $E_{CNL}$ and $E_F$ induces the charge transfer through the interface, generating intrinsic dipoles around the interface. As a result, the SBH for a hole is 0.4eV and this value is in good agreement with the experiments[16,17].

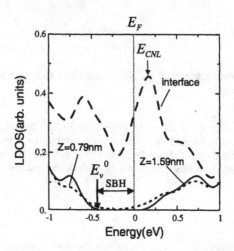

**Figure 4.** Calculated LDOS in the intrinsic case for the NiSi/Si interface. The LDOS at the interface is plotted by the solid line, that at 0.79nm by the dashed line and that at 1.59nm by the dotted line. $E_v^0$ denotes the valence band edge in the intrinsic case.

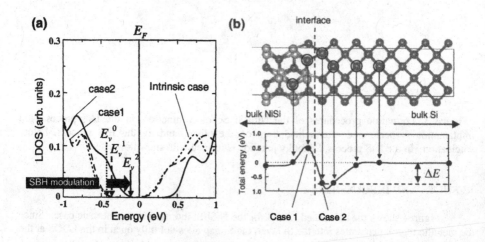

306

**Figure 5.** (a) The calculated LDOS at 1.59nm in both the B-doped case for the NiSi/Si interface. The dotted, dashed and solid lines show the LDOS in the intrinsic case, case1 and case2, respectively. The result in the intrinsic case is the same that in the case that z=1.59nm in Fig.2. (b) The total energy of the interface structure with a B atom at the substitutional site for a Si atom. The circles in the upper figure mean the same as those in Fig.2. Black circles in the lower figure show the calculated total energies of the interface structure with a B atom at the substitutional site for a Si atom.

Figure5(a) shows the calculated LDOS in the B-doped case. When a B atom replaces a Si atom on the NiSi side (case1), the SBH almost remains the same as that in the intrinsic case. On the other hand, when a B atom replaces a Si atom on the Si side (case2), the SBH modulation width ($\Delta\phi_b$) for a hole is 0.3eV. For understanding these results, we calculated the effective charges on the atoms in case2, using the valence charge division with the Wigner-Seitz cell. The calculated results are shown in Fig.6, suggesting that the atomic-scale electric dipoles exist across the interface. The dipole between a Ni atom (2) and a Si atom (4) is the intrinsic dipole and that between a Si atom (1) and a B atom (3) is induced by the B atom. In case2, this dipole overcomes the intrinsic dipole, leading to the large SBH modulation. In Fig.5(b), we plotted the total energy of the interface structure with a B atom at the substitutional site for a Si atom as the difference from that with a B atom substituted for a bulk Si atom. The value at the site farthest from the interface on the NiSi side is indicated as the formation energy difference between the two bulk layers (=-0.05eV). As can be found in Fig.5b, the total energy of the interface structure has the lowest value ($\Delta E$=-0.72eV) in case2. Here, a dipole energy can be estimated as -0.56eV, by employing the equation

$$\Delta E_1 = \frac{q^2}{4\pi\varepsilon a}$$

, where $q$, $a$ and $\varepsilon$ are the effective charge of the B atom (=-0.88), the distance between the B atom and its image charge (=0.166 nm) and dielectric constant, respectively. That is, the dipole generation mainly contributes to the decrease in the interface energy.

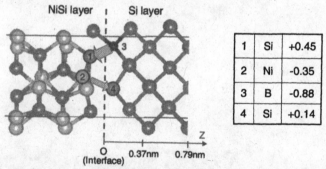

| 1 | Si | +0.45 |
| 2 | Ni | -0.35 |
| 3 | B | -0.88 |
| 4 | Si | +0.14 |

**Figure 6.** Electric dipoles at the interface. They are illustrated by the hatched arrows in the left figure. The calculated effective charges on the atoms from 1 to 4 are shown in the table on the righthand side. The gray and white circles mean the same as those in Fig.2. The black circle stands for a B atom.

As can be clearly seen in Fig.5b, the sites in the bulk NiSi layer are metastable for a B atom. Therefore, in the IBS process, B atoms mostly remain presumably in the bulk NiSi layer after silicidation. On the other hand, in the IAS process, many B atoms exist in the interstice of the NiSi layer after ion implantation and, after the subsequent RTA, move to the energetically most stable sites on the Si side, as shown in Fig.7. This model suggests that ion implantation after silicidation leads to the larger B concentration at the interface than that before silicidation.

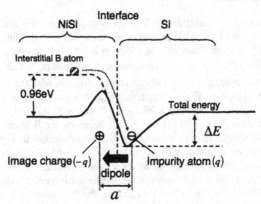

**Figure 7.** Model describing the B migration in the IAS process. The total energy of the interface structure was the same as that in Fig.5b. The calculated formation energy with a B atom in the interstice of the NiSi layer is 0.96eV larger than that in the substitutional site for a Si atom.

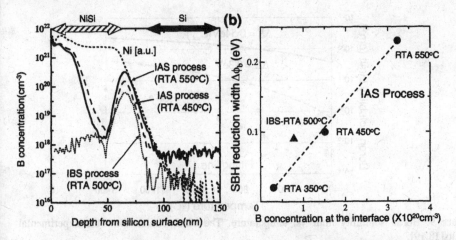

**Figure 8.** Experimetal results in B-doped case (a) B profiles obtained by SIMS for the junction formed through the IAS process. The B profile seems to be slightly broad due to the interface roughness. (b) The dependence of reduction width of SBH $\Delta\phi_b$ on the peak B concentration at the interface.

Actually, we observed the B profiles around the interface by SIMS. The results show the B concentration at the interface in the case of IAS process is larger than that in the case of IBS process in spite of the lower RTA temperature. Furthermore, as shown in Fig.8a, the concentration increases with the RTA temperature in the case of the IAS process and, at 550 °C, it becomes one order of magnitude larger than that in the case of the IBS process. The measured SBH reduction $\Delta\phi_b$ is plotted as a function of B concentration at the interface in Fig.8b, which is evaluated from the SIMS results in Fig.8a. The SBH is modulated by more than 0.2eV (70% of maximum theoretical prediction) after RTA at 550°C, although B atoms seem to be mainly distributed on the NiSi side in Fig.8a. These experimental results suggest that B atoms replace Si atoms on the Si side of the interface, generating dipoles to comfort the SBH.

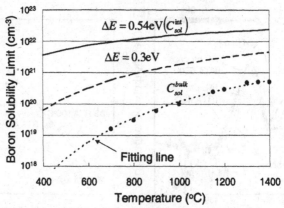

**Figure 9.** Boron solubility limit vs. temperature. The black circles show the experimental results[18,19].

However, the observed B concentration at 550°C exceeds the solubility limit in the bulk Si layer ( $C_{sol}^{bulk}$ ). This value is nearly equal to the B concentration of the (2×2) unit cell containing single B atom(5.1x10$^{20}$cm$^{-3}$). Then, for further understanding on these experimental results, we performed the calculations with the (2×2) unit cell, obtaining $\Delta\phi_b$ of 0.25eV and $\Delta E$ of -0.54eV. Here, we should note that a B atom is able to replace a Si atom in the vicinity of the interface with a larger energy gain ( $\Delta E$ ) than in the bulk Si layer. Therefore, the solubility limit for B at the interface ( $C_{sol}^{int}$ ) can be expressed by

$$ C_{sol}^{int} = C_{sol}^{bulk} \exp\left(-\frac{\Delta E}{kT}\right) $$

[20,21]. The calculated $C_{sol}^{int}$ is plotted by the solid line in Fig.9. Therefore, B atoms are able to be condensed around the interface, as observed in the experiment.

Here, the contact resistance of the Schottky junction is expressed by

$$ \rho_C \propto \exp\left\{\frac{2\sqrt{\varepsilon m^*}}{\hbar}\left(\frac{\phi_B}{\sqrt{N_B}}\right)\right\} $$

, where $\phi_B$ and $N_B$ denote the intrinsic SBH and the B concentration at the interface, respectively. This equation assumes that the contact resistance is inversely proportional to the transmission probability through the Schottky barrier, which is shown in Fig.1. Then, the contact resistance can be plotted as a function of the SBH in Fig.10. In the case of the DCS junction, the band edges are shifted in a few Si layers from the interface, as shown in Fig.11. In our experiment, the measured SBH for a hole is 0.15eV. Therefore, the contact resistance is reduced to about 1/10 of the current value, as shown in Fig.10. These results suggest that NiSi can be used as an electrode material of the p-MOSFETs of hp22nm.

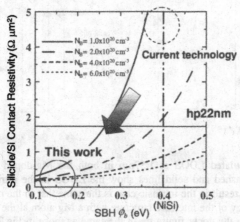

**Figure 10.** The calculated contact resistance as a function of the SBH of the NiSi/Si junction formed in the contact region. The horizontal solid line shows the contact resistance required in the ITRS roadmap.

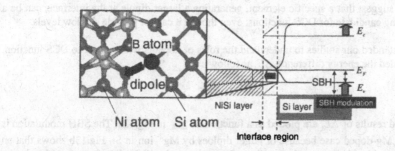

**Figure 11.** Schematic view of a DCS junction. The bold black arrow stands for a dipole, which is generated around a doped B atom at the interface. The circles mean the same as those in Fig.6. The calculated band edges are illustrated in this figure.

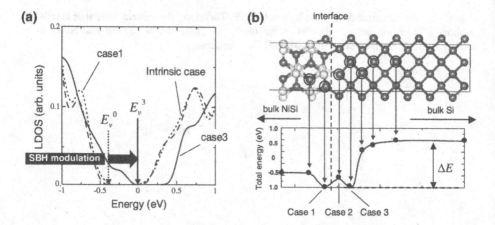

**Figure 12.** (a) The alculated LDOS at 1.59nm in both the Mg-doped case for the NiSi/Si interface. The dotted, dashed and solid lines show the LDOS in the intrinsic case, case1 and case3, respectively. The result in the intrinsic case is the same that in the case that z=1.59nm in Fig.5. (b) The total energy of the interface structure with a Mg atom at the substitutional site for a Si atom. The circles in the upper figure mean the same as those in Fig.2. Black circles in the lower figure show the calculated total energies of the interface structure with a Mg atom at the substitutional site for a Si atom.

Furthermore, we have applied this idea to group II and VI elements because of potentially larger dipole generation at the interfaces in DCS junctions[22]. We note in Fig.12b that substitutions of Mg atoms for Si atoms in both cases 1 and 3 are energetically most stable. The calculated LDOS in Fig.12a indicates a large SBH modulation of 0.38eV for a hole in case3. These results suggest that a specific element, generating a larger dipole at the interface, can be a more promising candidate for DCS junctions, even though it cannot provide shallow levels.

We extended our studies to understand the roles of dopants with respect to DCS junction. First, we divided the energy difference $\Delta E$ as follows:

$$\Delta E = \Delta E_1 + \Delta E_2 .$$

The calculated results of $\Delta E_1$ are plotted as a function of $\Delta \phi_b$ in Fig.13a. The SBH modulation is largest in the Mg-doped case because of larger diploes by $Mg^{2-}$ ion in Si. Fig.13b shows that an impurity atom with a larger covalent radius relaxes the strain of the interface, resulting in the preference on the Si side of the interface with higher energy gain. The additional energy $\Delta E_2$ is associated with the strain of the interface structure. For an atom with a larger value of $\Delta E_2$, the solubility limit at the interface may be expected to be larger than that in the bulk Si layer. A dopant with both $\Delta E_1$ and $\Delta E_2$ larger than those of ordinary dopants is preferable for DCS junctions (Fig.14).

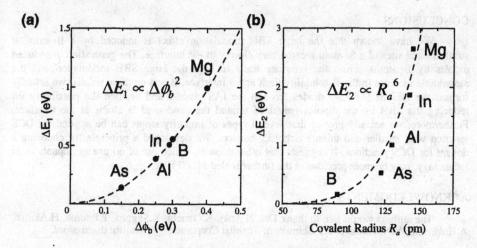

**Figure 13.** (a) Dependence of $\Delta E_1$ on the modulation width of SBH for impurity atoms (b) Dependence of $\Delta E_2$ on the covalent radius of an impurity atom.

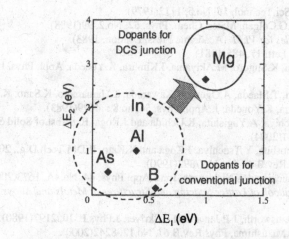

**Figure 14.** The values of $\Delta E_1$ and $\Delta E_2$ for impurity atoms.

## CONCLUSIONS

We have shown that the large SBH modulation effect is induced by a B atom at substitutional sites of a Si atom located very closely to the interface. The generation of induced dipoles by B atoms across the interface leads to both the large SBH modulation and the remarkable increase in the B solubility limit at the interface. Based on these results, we actually formed the NiSi/Si Schottky diodes through the IAS process and verified the possibility for reducing the SBH by the dipoles generated around the condensed B atoms at the interface. Furthermore, we actually proved that several types of impurity atoms can be applied for DCS junction with smaller and thinner Schottky barriers. We developed a principle for choosing a dopant for DCS junctions. It suggests that other dopants, which are of no use as dopants in Si technology, may be more precious in the further scaled MOSFETs.

## ACKNOWLEDGMENTS

The authors would like to thank Drs. N.Aoki, K.Ohuchi, K.Suguro, T.Iinuma, H.Akutsu, A.Hokazono, Y.Nakazaki and T.Shimizu of Toshiba Corporation for fruitful discussions.

## REFERENCES

1. S.-D.Kim and J.C.S.Woo, Ext.Abst.of International Workshop on Junction Technology, 1 (2002).
2. G.Ottaviani, J.Vac.Sci.Technol. **16**, No.5, 1112(1979).
3. J.P.Gambino and E.G.Colgan, Mater. Chem. Phys. **52**, No.2, 99(1998).
4. S.P.Murarka, *Silicides for VLSI* (Academic Press, London, 1983).
5. R.L.Thornton, Elec.Lett. **17**, 485(1981).
6. I.Ohdomari, K.N.Tu, K.Suguro, M.Akiyama, I.Kimura, K.Yoneda, Appl. Phys. Lett., **38**, no.12, 1015(1981).
7. I.Ohdomari, M.Hori, T.Maeda, A.Ogura, H.Kawarada, T.Hamamoto, K.Sano, K.N.Tu, M.Wittmer, I.Kimura, K.Yoneda, J. Appl. Phys. **54**, no.8 : 4679(1983).
8. A.Kinoshita, T.Tsuchiya, A.Yagishita, K.Uchida and J.Koga, Ext.Abst.of Solid State Devices and Materials, A-5-1(2004).
9. T.Yamauchi, A.Kinoshita, Y.Tsuchiya, J.Koga and K.Kato, IEDM Tech.Dig., 2006, pp.385.
10. D.Vanderbilt, Phys.Rev.B **41**, No.11, 7892(1990).
11. M.Tsuchiaki, K.Ohuchi and A.Nishiyama, Jpn.J.Appl.Phys. **44**, No.4A, 1673(2005).
12. W.B.Pearson, *Handbook of Lattice Spacing and Structures of Metals and Alloys* (Pergamon, New York, 1958).
13. R.M.Boulet, A.E.Dunsworth, J-P Jan and H.L.Skriver, J.Phys.F. **10**, 2197(1980).
14. T.Yamauchi and K.Mizushima, Phys.Rev.B **61**, No.12, 8242(2000).
15. J.Tersoff, Phys.Rev.Lett. **52**, No.6, 465(1984).
16. J.L.Freeouf, Solid State Commun. **33**, 1059, 1980.
17. E.Bucher, S.Schulz, M.Ch.Lux-Steiner and P.Munz, Appl.Phys.A **40**, 71 (1986).
18. G.L.Vick and K.M.Whittle, J.Eelectrochem.Soc. **116**, 1142 (1969).
19. S.M.Sze, *Semiconductor Devices physics and technology* (John Wiley \& Sons Inc., New York, 1985).
20. Chris G.Van de Walle, D.B.Laks, G.F.Neumark and S.T.Pantelides, Phys.Rev.B **47**, No.15, 9425(1993).
21. T.Kawasaki, H.Katayama-Yoshida, Physica B **302**, 163(2001).
22. T.Yamauchi, A.Kinoshita, Y.Tsuchiya, J.Koga and K.Kato, IEDM Tech.Dig., 2007, pp.963.

# AUTHOR INDEX

315

# SUBJECT INDEX

Printed in the United States
By Bookmasters